THE PRINCIPLES OF NEWTONIAN AND QUANTUM MECHANICS

The Need for Planck's Constant, *h*

Second Edition

THE PRINCIPLES OF NEWTONIAN AND QUANTUM MECHANICS

The Need for Planck's Constant, *h*

Second Edition

M A de Gosson
University of Vienna, Austria

Foreword by

Basil Hiley

World Scientific

NEW JERSEY • LONDON • SINGAPORE • BEIJING • SHANGHAI • HONG KONG • TAIPEI • CHENNAI • TOKYO

Published by

World Scientific Publishing Co. Pte. Ltd.
5 Toh Tuck Link, Singapore 596224
USA office: 27 Warren Street, Suite 401-402, Hackensack, NJ 07601
UK office: 57 Shelton Street, Covent Garden, London WC2H 9HE

Library of Congress Cataloging-in-Publication Data
Names: Gosson, Maurice de, author.
Title: The principles of Newtonian and quantum mechanics : the need for Planck's constant, *h* / by Maurice A. de Gosson (University of Vienna, Austria).
Description: 2nd edition. | Hackensack, NJ : World Scientific, [2017] | Includes bibliographical references.
Identifiers: LCCN 2016036280 | ISBN 9789813200968 (hardcover ; alk. paper) | ISBN 9813200960 (hardcover ; alk. paper)
Subjects: LCSH: Lagrangian functions. | Maslov index. | Geometric quantization.
Classification: LCC QC20.7.C3 G67 2017 | DDC 530.15/564--dc23
LC record available at https://lccn.loc.gov/2016036280

British Library Cataloguing-in-Publication Data
A catalogue record for this book is available from the British Library.

Printed in Singapore

To my Charlyne, with all my love

Contents

Preface to the First Edition xv

Preface to the Second Edition xxi

Foreword by Basil Hiley xxiii

1 From Kepler to Schrödinger... and Beyond 1
- 1.1 Classical Mechanics 2
  - 1.1.1 Newton's Laws and Mach's Principle 3
  - 1.1.2 Mass, Force, and Momentum 5
- 1.2 Symplectic Mechanics 6
  - 1.2.1 Hamilton's Equations 6
  - 1.2.2 Gauge Transformations 8
  - 1.2.3 Hamiltonian Fields and Flows 8
  - 1.2.4 The "Symplectization of Science" 9
- 1.3 Action and Hamilton-Jacobi's Theory 12
  - 1.3.1 Action 12
  - 1.3.2 Hamilton-Jacobi's Equation 13
- 1.4 Quantum Mechanics 14
  - 1.4.1 Matter Waves 15
  - 1.4.2 "If There Is a Wave, There Must Be a Wave Equation!" 17
  - 1.4.3 Schrödinger's Quantization Rule and Geometric Quantization 18
- 1.5 The Statistical Interpretation of Ψ 20
  - 1.5.1 Heisenberg's Inequalities 20
- 1.6 Quantum Mechanics in Phase Space 23
  - 1.6.1 Schrödinger's "firefly" Argument 23
  - 1.6.2 The Symplectic Camel 24

1.7 Feynman's "Path Integral" 26
1.7.1 The "Sum Over All Paths" 26
1.7.2 The Metaplectic Group 27
1.8 Bohmian Mechanics 28
1.8.1 Quantum Motion: The Bell-DGZ Theory 29
1.8.2 Bohm's Theory 30
1.9 Interpretations 32
1.9.1 Epistemology or Ontology? 32
1.9.2 The Copenhagen Interpretation 33
1.9.3 The Bohmian Interpretation 35
1.9.4 The Platonic Point of View 36

2 Newtonian Mechanics **37**
2.1 Maxwell's Principle and the Lagrange Form 37
2.1.1 The Hamilton Vector Field 38
2.1.2 Force Fields 38
2.1.3 Statement of Maxwell's Principle 40
2.1.4 Magnetic Monopoles and the Dirac String 42
2.1.5 The Lagrange Form 44
2.1.6 N-Particle Systems 47
2.2 Hamilton's Equations 50
2.2.1 The Poincaré-Cartan Form and Hamilton's Equations 51
2.2.2 Hamiltonians for N-Particle Systems 54
2.2.3 The Transformation Law for Hamilton Vector Fields . 56
2.2.4 The Suspended Hamiltonian Vector Field 57
2.3 Galilean Covariance 59
2.3.1 Inertial Frames 59
2.3.2 The Galilean Group $Gal(3)$ 60
2.3.3 Galilean Covariance of Hamilton's Equations 62
2.4 Constants of the Motion and Integrable Systems 66
2.4.1 The Poisson Bracket 66
2.4.2 Constants of the Motion and Liouville's Equation . . 67
2.4.3 Constants of the Motion in Involution 69
2.5 Liouville's Equation and Statistical Mechanics 72
2.5.1 Liouville's Condition 73
2.5.2 Marginal Probabilities 74
2.5.3 Distributional Densities: An Example 76

3 The Symplectic Group **79**

3.1 Symplectic Matrices and Sp(n) 79

3.2 Symplectic Invariance of Hamiltonian Flows 82

3.2.1 Notations and Terminology 83

3.2.2 Proof of the Symplectic Invariance of Hamiltonian Flows 84

3.2.3 Another Proof of the Symplectic Invariance of Flows* 85

3.3 The Properties of Sp(n) 86

3.3.1 The Subgroups $U(n)$ and $O(n)$ of Sp(n) 87

3.3.2 The Lie Algebra $\mathfrak{sp}(n)$ 88

3.3.3 Sp(n) as a Lie Group 89

3.4 Quadratic Hamiltonians 91

3.4.1 The Linear Symmetric Triatomic Molecule 92

3.4.2 Electron in a Uniform Magnetic Field 93

3.4.3 Small Oscillations Near Equilibrium 95

3.5 The Inhomogeneous Symplectic Group 98

3.5.1 Galilean Transformations and ISp(n) 99

3.6 An Illuminating Analogy 100

3.6.1 The Optical Hamiltonian 100

3.6.2 Paraxial Optics 102

3.7 Gromov's Non-Squeezing Theorem 105

3.7.1 Liouville's Theorem Revisited 107

3.7.2 Gromov's Theorem 109

3.7.3 The Uncertainty Principle in Classical Mechanics . . 112

3.8 Symplectic Capacity and Periodic Orbits 114

3.8.1 The Capacity of an Ellipsoid 117

3.8.2 Symplectic Area and Volume 119

3.9 Capacity and Periodic Orbits 120

3.9.1 Periodic Hamiltonian Orbits 121

3.9.2 Action of Periodic Orbits and Capacity 123

3.10 Phase Space Quantization and Quantum Blobs 124

3.10.1 Stationary States of Schrödinger's Equation 125

3.10.2 Quantum Blobs and the Minimum Capacity Principle 127

3.10.3 Quantization of the N-Dimensional Harmonic Oscillator 128

3.10.4 Quantum Blobs and the Uncertainty Principle 133

4 Action and Phase **137**
4.1 Introduction . . . 137
4.2 The Fundamental Property of the Poincaré-Cartan Form . . 138
4.2.1 Helmholtz's Theorem: The Case $n = 1$ . . . 139
4.2.2 Helmholtz's Theorem: The General Case . . . 140
4.3 Free Symplectomorphisms and Generating Functions . . . 142
4.3.1 Generating Functions . . . 144
4.3.2 Optical Analogy: The Eikonal . . . 147
4.4 Generating Functions and Action . . . 148
4.4.1 The Generating Function Determined by H . . . 148
4.4.2 Action vs. Generating Function . . . 152
4.4.3 Gauge Transformations and Generating Functions . . 153
4.4.4 Solving Hamilton's Equations with W . . . 154
4.4.5 The Cauchy Problem for Hamilton-Jacobi's Equation 156
4.5 Short-Time Approximations to the Action . . . 158
4.5.1 The Case of a Scalar Potential . . . 159
4.5.2 One Particle in a Gauge $(\mathbf{A}, U)$ . . . 162
4.5.3 Many Particle Systems in a Gauge (A, U) . . . 165
4.6 Lagrangian Manifolds . . . 167
4.6.1 Definitions and Basic Properties . . . 167
4.6.2 Lagrangian Manifolds in Mechanics . . . 170
4.7 The Phase of a Lagrangian Manifold . . . 172
4.7.1 The Phase of an Exact Lagrangian Manifold . . . 172
4.7.2 The Universal Covering of a Manifold* . . . 175
4.7.3 The Phase: General Case . . . 176
4.7.4 Phase and Hamiltonian Motion . . . 177
4.8 Keller-Maslov Quantization . . . 179
4.8.1 The Maslov Index for Loops . . . 179
4.8.2 Quantization of Lagrangian Manifolds . . . 184
4.8.3 Illustration: The Plane Rotator . . . 186

5 Semi-Classical Mechanics **191**
5.1 Bohmian Motion and Half-Densities . . . 192
5.1.1 Wave-Forms on Exact Lagrangian Manifolds . . . 192
5.1.2 Semi-Classical Mechanics . . . 194
5.1.3 Wave-Forms: Introductory Example . . . 196
5.2 The Leray Index and the Signature Function* . . . 198
5.2.1 Cohomological Notations . . . 198
5.2.2 The Leray Index: $n = 1$ . . . 199
5.2.3 The Leray Index: General Case . . . 201

5.2.4 Properties of the Leray Index 207
5.2.5 More on the Signature Function 210
5.2.6 The Reduced Leray Index 212
5.3 De Rham Forms 213
5.3.1 Volumes and their Absolute Values 214
5.3.2 Construction of de Rham Forms on Manifolds 216
5.3.3 De Rham Forms on Lagrangian Manifolds 220
5.4 Wave-Forms on a Lagrangian Manifold 225
5.4.1 Definition of Wave Forms 225
5.4.2 The Classical Motion of Wave-Forms 228
5.4.3 The Shadow of a Wave-Form 230

6 Metaplectic Group, Maslov Index, and Quantization 235
6.1 Introduction 235
6.1.1 Could Schrödinger have Done it Rigorously? 235
6.1.2 Schrödinger's Idea 236
6.1.3 Sp(n)'s "Big Brother" Mp(n) 237
6.2 Free Symplectic Matrices and their Generating Functions . . 239
6.2.1 Free Symplectic Matrices 240
6.2.2 The Case of Affine Symplectomorphisms 242
6.2.3 The Generators of Sp(n) 243
6.3 The Metaplectic Group Mp(n) 246
6.3.1 Quadratic Fourier Transforms 246
6.3.2 The Operators $M_{L,m}$ and V_P 248
6.3.3 The Metaplectic Group and Wave Optics 251
6.4 The Projections Π and Π^ε 253
6.4.1 Construction of the Projection Π 254
6.4.2 The Covering Groups Mp$^\varepsilon$(n) 257
6.5 The Maslov Index on Mp(n) 259
6.5.1 Maslov Index: a "Simple" Example 260
6.5.2 Definition of the Maslov Index on Mp(n) 262
6.6 The Cohomological Meaning of the Maslov Index* 264
6.6.1 Group Cocycles on Sp(n) 265
6.6.2 The Fundamental Property of $m(\cdot)$ 267
6.7 The Inhomogeneous Metaplectic Group 270
6.7.1 The Heisenberg Group 271
6.7.2 The Group IMp 273
6.7.3 The Heisenberg-Weyl Operators 275
6.8 The Groups Symp(n) and Ham(n)* 277
6.8.1 A Topological Property of Symp(n) 277

6.8.2 The Group Ham(n) of Hamiltonian Symplectomorphisms . . . 278
6.9 Canonical Quantization . . . 281
6.9.1 The Ordering Problem for Monomials . . . 284
6.9.2 The Case of General Observables . . . 287
6.9.3 The Groenewold-van Hove Theorem . . . 292

7 Schrödinger's Equation and the Metatron 295
7.1 Schrödinger's Equation for the Free Particle . . . 296
7.1.1 The Free Particle's Phase . . . 296
7.1.2 The Free Particle Propagator . . . 297
7.1.3 An Explicit Expression for G . . . 299
7.1.4 The Metaplectic Representation of the Free Flow . . . 301
7.1.5 More Quadratic Hamiltonians . . . 302
7.2 Van Vleck's Determinant . . . 305
7.2.1 Trajectory Densities . . . 305
7.2.2 The Continuity Equation for van Vleck's Density . . . 308
7.2.3 The Continuity Equation . . . 310
7.3 The Short-Time Propagator . . . 312
7.3.1 Properties of the Short-Time Propagator . . . 313
7.4 The Case of Quadratic Hamiltonians . . . 316
7.4.1 Exact Green Function . . . 316
7.4.2 Exact Solutions of Schrödinger's Equation . . . 317
7.5 Solving Schrödinger's Equation: General Case . . . 319
7.5.1 The Short-Time Propagator and Causality . . . 319
7.5.2 Statement of the Main Theorem . . . 321
7.5.3 The Formula of Stationary Phase . . . 322
7.5.4 Two Lemmas – and the Proof . . . 324
7.6 Metatrons and the Implicate Order . . . 329
7.6.1 Unfolding and Implicate Order . . . 330
7.6.2 Prediction and Retrodiction . . . 331
7.6.3 The Lie-Trotter Formula for Flows . . . 333
7.6.4 The "Unfolded" Metatron . . . 336
7.6.5 Quantum Zeno Effect and the Mott Paradox . . . 340
7.6.6 The Generalized Metaplectic Representation . . . 341
7.7 Phase Space and Schrödinger's Equation . . . 343
7.7.1 Mixed Representations in Quantum Mechanics . . . 347
7.7.2 Complementarity and the Implicate Order . . . 351

A Symplectic Linear Algebra **355**

B The Lie-Trotter Formula for Flows **359**
B.1 The Lie-Trotter Formula for Vector Fields 360

C The Heisenberg Groups **363**

D The Bundle of s-Densities **367**

E The Lagrangian Grassmannian **371**
E.1 Lag(n) as a Set of Matrices 371
E.1.1 The Universal Coverings $U_\infty(n, \mathbb{C})$ and $W_\infty(n, \mathbb{C})$ . . . 374

Bibliography **377**

Index **391**

Preface to the First Edition

The aim of this book is to expose the mathematical machinery underlying Newtonian mechanics and two of its refinements, semi-classical and non-relativistic quantum mechanics. A recurring theme is that these three Sciences are all obtained from a single mathematical object, the Hamiltonian flow, viewed as an abstract group. To study that group, we need symplectic geometry and analysis, with an emphasis on two fundamental topics: *Symplectic rigidity* (popularly known as the "principle of the symplectic camel"). This principle, whose discovery goes back to the work of M. Gromov in the middle of the 1980's, says that no matter how much we try to deform a phase-space ball with radius r by Hamiltonian flows, the area of the projection of that ball on a position-momentum plane will never become inferior to πr^2. This is a surprising result, which shows that there is, contrarily to every belief, a "classical uncertainty principle". While that principle does not contradict Liouville's theorem on the conservation of phase space volume, it indicates that the behavior of Hamiltonian flows is much less "chaotic" than was believed. Mathematically, the principle of the symplectic camel shows that there is a symplectic invariant (called Gromov's width or symplectic capacity), which is much "finer" than ordinary volume. Symplectic rigidity will allow us to define a semi-classical quantization scheme by a purely topological argument, and will allow us to give a very simple definition of the Maslov index without invoking the WKB method.

The *metaplectic representation* of the symplectic group. That representation allows one to associate in a canonical way to every symplectic matrix exactly two unitary operators (only differing by their signs) acting on the square integrable functions on configuration space. The group $Mp(n)$ of all these operators is called the metaplectic group, and enjoys very special properties; the most important from the point of view of physics since it allows the explicit resolution of all Schrödinger's equations associated to quadratic Hamiltonians. We will in fact partially extend this metaplectic representation in order to include even non-quadratic Hamiltonians, leading to a precise and mathematically justifiable form of Feynman's path integral.

An important issue that is addressed in this book is that of quantum mechanics in phase space. While it is true that the primary perception we, human beings, have of our world privileges positions, and their evolution with time, this does not mean that we have to use only, mathematics in configuration space. As Basil Hiley puts it "*...since thoughts are not located in space-time, mathematics is not necessarily about material things in space-time*". Hiley is right: it is precisely the liberating power — I am tempted to say the grace — of mathematics that allows us to break the chains that tie us to one particular view of our environment. It is unavoidable that some physicists will feel uncomfortable with the fact that I am highlighting one unconventional approach to quantum mechanics, namely the approach initiated by David Bohm in 1952, and later further developed by Basil Hiley and Bohm himself. To them I want to say that since this is not a book on the epistemology or ontology of quantum mechanics (or, of physics, in general), I had no *états d'âme* when I used the Bohmian approach: it is just that this way of seeing quantum mechanics is the easiest way to relate classical and quantum mechanics. It allows us to speak about "particles" even in the quantum regime which is definitely an economy of language... and of thought! The Bohmian approach has moreover immediately been well-accepted in mathematical circles: *magna est veritas et praevalebit...*

While writing this book, I constantly had in mind two categories of readers: my colleagues — mathematicians, and my dear friends — physicists. The first will, hopefully, learn some physics here (but presumably, not the way it is taught in usual physics books). The physicists will get some insight in the beautiful unity of the mathematical structure, symplectic geometry, which is the most natural for expressing both classical and quantum mechanics. They will also get a taste of some sophisticated new mathematics (the symplectic camel, discussed above, and the Leray index, which is the "mother" of all Maslov indices). This book is therefore, in a sense, a tentative to reconcile what Poincaré called, in his book *Science and Hypothesis*, the "two neighboring powers": Mathematics and Physics. While Mathematics and Physics formed during centuries a single branch of the "tree of knowledge" (both were parts of "natural philosophy"), physicists and mathematicians started going different ways during the last century (one of the most recent culprits being the Bourbaki school). For instance, David Hilbert is reported to have said that "Physics is too difficult to leave to physicists", while Albert Einstein characterized Hilbert's physics (in a letter to Hermann Weyl) as "infantile". To be fair, we must add that Einstein's theory was really based on physical principles, while Hilbert's travail in physics was an exercise in pure mathematics (we all know that even today many mathematical texts, which claim to be of physical interest, are too often just pure mathematics dressed up in a phony physical language).

A few words about the technical knowledge required for an optimal understanding of the text. The mathematical tools that are needed are introduced in due time, and are rather elementary (undergraduate linear algebra and calculus suffice, together with some knowledge of the rudiments of the theory of differential forms). This makes the book easily accessible to a rather large and diversified scientific audience, especially since I tried as much as possible to write a "self-contained" text (a few technical Appendices have been added for the reader's convenience). A word to my colleagues - mathematicians: this book can be read without any particular prior knowledge of physics, but it is perhaps somewhat unrealistic to claim that it is an introduction "from scratch" to the subject. Since I have tried to be intelligible by both mathematicians and physicists, I have made every effort to use rigorous, but simple mathematics. I have, however, made every effort to avoid Bourbachian *rigor mortis*.

This book is structured as follows:

Chapter 1 is devoted to a review of the basic principles of Newtonian and quantum mechanics, with a particular emphasis on its Bohmian formulation, and the "quantum motion" of particles, which is in a sense simpler than the classical motion (there are no "caustics" in quantum mechanics: the latter only appear at the semi-classical level, when one imposes classical motion to the wave functions).

Chapter 2 presents modern Newtonian mechanics from the symplectic point of view, with a particular emphasis on the Poincaré-Cartan form. The latter arises in a natural way if one makes a certain physical hypothesis, which we call, following Souriau, the "Maxwell principle", on the form of the fundamental force fields governing the evolution of classical particles. The Maxwell principle allows showing, using the properties of the Poincaré-Cartan invariant, that Newton's second law is equivalent to Hamilton's equations of motion for these force fields.

In Chapter 3, we study thoroughly the symplectic group. The symplectic group being the backbone of the mathematical structure underlying Newtonian mechanics in its Hamiltonian formulation, it deserves as such a thorough study in its own right. We then propose a semi-classical quantization scheme based on the principle of symplectic rigidity. That scheme leads in a very natural way to the Keller-Maslov condition for quantization of Lagrangian manifolds, and is the easiest way to motivate the introduction of the Maslov index in semi-classical mechanics.

In Chapter 4, we study the so fundamental notion of action, which is most easily apprehended by using the Poincaré-Cartan invariant introduced in Chapter 2. An important related notion is that of generating function (also called Hamilton's "two point characteristic functions"). We then introduce the notion of Lagrangian manifold, and show how it leads to an intrinsic definition of the phase of classical completely integrable systems, and of all quantum systems.

Chapter 5 is devoted to a geometrical theory of semi-classical mechanics in phase space, and will probably be of interest to theoretical physicists, quantum chemists and mathematicians. This Chapter is mathematically the most advanced, and can be skipped in a first reading. We begin by showing how the Bohmian approach to quantum mechanics allows one to interpret the wave function as a half-density in phase space. In the general case, wave forms are (up to a phase factor) the square roots of de Rham forms defined on the graph of a Lagrangian manifold. The general definition of a wave form requires the properties of Leray's cohomological index (introduced by Jean Leray in 1978); it is a generalization of the Maslov index, which it contains as a "byproduct". We finally define the "shadows" of our wave forms on configuration space: these shadows are just the usual semi-classical wave functions familiar from Maslov theory.

Chapter 6 is devoted to a rather comprehensive study of the metaplectic group $Mp(n)$. We show that to every element of $Mp(n)$ we can associate an integer modulo 4, its Maslov index, which is closely related to the Leray index. This allows us to eliminate in a simple and elegant way the phase ambiguities, which have been plaguing the theory of the metaplectic group from the beginning. We then define, and give a self-contained treatment, of the inhomogeneous metaplectic group $IMp(n)$, which extends the metaplectic representation to affine symplectic transformations. We also discuss, in a rather sketchy form, the difficult question of the extension of the metaplectic group to arbitrary (non-linear) symplectic transformations, and Groenewold-van Hove's famous theorem.

The central theme of Chapter 7 is that although quantum mechanic cannot be derived from Newtonian mechanics, it nevertheless emerges from it via the theory of the metaplectic group, provided that one makes a physical assumption justifying the need for Planck's constant h. This "metaplectic quantization" procedure is not new; it has been known for decades in mathematical circles for quadratic Hamiltonians. In the general case, there is, however an obstruction for carrying out this quantization, because of Groenewold-van Hove's theorem. This theorem does however not mean that we cannot extend the metaplectic group to non-quadratic Hamiltonians. This is done by using the Lie-Trotter formula for classical flows, and leads to a general metaplectic representation, from which Feynman's path integral "pops out" in a much more precise form than in the usual treatments.

The titles of a few Sections and Subsections are followed by a star * which indicates that the involved mathematics is of a perhaps more sophisticated nature than in the rest of the book. These (sub)sections can be skipped in a first reading.

Maurice de Gosson
March 2001

Preface to the Second Edition

Fifteen years have gone since the first edition of this book was published. Inevitably, things have evolved and changed; new results have been discovered and new ways of interpreting things have appeared. In this second edition of the "*Principles of Newtonian and Quantum Mechanics*" we take into account recent progresses in quantum mechanics in its Bohmian formulation which have led to a better understanding of the quantum Zeno effect and its understanding in terms of the Implicate Order. We have considerably expanded several chapters by adding new progresses. For instance, in Chapter 3 we have redefined the notion of quantum cells related to the principle of the symplectic camel, which by now are known in the literature as "quantum blobs"; it appears that these are closely linked, via the properties of the quantum covariance matrix, to the uncertainty principle in its strong Robertson–Schrödinger form. In Chapter 6 we discuss with some care the notion of quantization; in particular we show that the almost forgotten Born and Jordan initial ideas lead to a particularly interesting scheme, which, we suggest, could compete with the usual Weyl quantization familiar from advanced quantum mechanics. Since Born-Jordan quantization has made a remarkable come-back these last two years this addition was definitely necessary (we did not consider explicitly any particular quantization scheme in the first edition of this book). In Chapter 7 we have added a section about the quantum Zeno effect; the study of this effect needs little more than the machinery already developed in this Chapter (short-time propagators); we also discuss its application to the so-called Mott paradox (these results have been obtained recently in joint work with Basil Hiley and Eliahu Cohen).

Typos and errors that appeared in the first edition of this book have been corrected, and the text has been improved and edited in many places. The Bibliography and the Index have been expanded and updated.

It is my pleasure to thank Basil Hiley for numerous illuminating conversations, and Charlyne de Gosson for suggestions and a careful reading of the revised manuscript.

Acknowledgements

This second edition has been partially financed by a grant of the Austrian Research Agency FWF (grant No. P27773).

Maurice de Gosson
Vienna, July 2016

Foreword by Basil Hiley

One of perennial problems in the continued specialisation of academic disciplines is that an important but unexpected result in one area can go completely unnoticed in another. This gap is particularly great between theoretical physics and the more rigorous mathematical approaches to the basic formalism employed by physicists. The physicists show little patience with what to them seems to be an obsession with the minute detail of a mathematical structure that appears to have no immediate physical consequences. To mathematicians there is puzzle that sometimes borders on dismay at some of the "vague" structures that physicists use successfully. In consequence, each group can be totally unaware of the important progress made by the other. This is not helped by the development of specialized technical languages, which can prevent the "outsider" seeing immediately the relevance of these advances. At times, it becomes essential to set down these advances in a way that brings the two groups together. This book fits into this category as it sets out to explain how recent advances in quantization procedures for Lagrangian manifolds has relevance to the physicist's approach to quantum theory.

Maurice de Gosson has considerable mathematical expertise in the field of Lagrangian quantization, which involves a detail study of symplectic structures, the metaplectic covering of these structures and Maslov indices, all topics that do not fall within the usual remit of a quantum physicist. It is a mathematicians attempt to show the precise relationship between classical and quantum mechanics. This relationship has troubled physicists for a long time, but in spite of this, the techniques presented in this book are not very familiar to them. They are generally content with the plausible, but somewhat vague notion of the correspondence principle. However any detailed analysis of the precise meaning of this principle has always been beset with problems. Recently decoherence has become a fashionable explanation for the emergence of the classical world even though it, too, has its difficulties. This book provides an alternative and more mathematically rigorous approach of the relationship between the classical and quantum formalisms.

Unsurprisingly the discussion of classical mechanics takes us into a detailed study of the symplectic group. A notable feature of this discussion is centred on Gromov's "non-squeezing" theorem, which although classical, contains the seeds of the uncertainty principle. The common conception of Liouville's theorem is that under a symplectic transformation a volume in phase space can be made as thin as one likes provided the volume remains constant. Thus, it would be possible to pass the proverbial camel through the eye of a needle no matter how small the eye! This is in fact not true for the "symplectic camel". For a given process in phase space, ti is not, repeat not, possible to shrink a cross-section defined by conjugate co-ordinates like x and p_x to zero. In other words, we have a minimum cross-sectional area within a given volume that cannot by shrunk further. It is as if the uncertainty has left a "footprint" in classical mechanics.

Perhaps the most important topic discussed in the book is the rope of the metaplectic group and the Maslov index. Apart from the use of this group in optics to account for phenomenon like the Gouy phase, the metaplectic group is almost a complete stranger to the physics community, yet it is the key to the relationship between classical and quantum mechanics. Indeed, it is argued here that Schrödinger's original derivation of his famous equation could be regarded as the discovery of the metaplectic representation of the symplectic group.

To understand how this comes about we must be aware of two facts. First we must realise that the metaplectic group double covers the symplectic group. This is exactly analogous to the double cover of the orthogonal group by the spin group. In this sense, it can be regarded as the "spin group" for the symplectic group. Secondly, we must discuss classical mechanics in terms of the Hamiltonian flow, f_t,which is simply the family of symplectic matrices generated by the Hamiltonian. In contrast, the time evolution in quantum mechanics is described by the Hamiltonian through the group of unitary operators U_t. What this book shows is that the lift of f_t onto the covering space is just U_t! This is a remarkable result which gives a new way to explore the relationship between classical and quantum mechanics.

Historically it was believed that this procedure only applied to Hamiltonians that were at most quadratic in position and momentum. This limitation is seen through the classic Groenewold–van Hove "no-go" theorem. However, this lift can be generalised to all Hamiltonians by using an iteration process on small time lifts. This approach has similarities with the Feynman path integral method and it is based on the Lie-Trotter formula for flows. It has the advantage over Feynman approach in that it is not a "sum over (hypothetical) paths", but is a mathematically rigorous consequence of the metaplectic representation. This opens up the possibilities of new mathematical questions concerning the existence of generalised metaplectic representations, a topic that has yet to be addressed in detail.

All of this opens up a new mathematical route into quantum theory offering a much clearer relation between the classical and quantum formalisms. As the approach is mathematical, there is no need to get embroiled in the interminable debate about interpretations of the formalism. Indeed, because of this focus on the mathematics without any philosophical baggage, it is possible to see exactly how the Bohm approach fits into this general framework, showing the legitimacy of this approach from a mathematical point of view. Indeed, we are offered further insights into this particular approach, which I find particularly exciting for obvious reasons. I hope others will be stimulated into further explorations of the general structure that Maurice de Gosson unfolds in this volume. I am sure this structure will reveal further profound insights into this fascinating subject.

Basil Hiley, Birkbeck College, London

Chapter 1

From Kepler to Schrödinger... and Beyond

Summary. *The mathematical structure underlying Newtonian mechanics is symplectic geometry, which contains a classical form of Heisenberg's uncertainty principle. Quantum mechanics is based on de Broglie's theory of matter waves, whose evolution is governed by Schrödinger's equation. The latter emerges mathematically from classical mechanics using the metaplectic representation of the symplectic group.*

The purpose of this introductory Chapter is to present the basics of both classical and quantum physics "in a nutshell". Much of the material will be further discussed and developed in the forthcoming Chapters.

The three first sections of this Chapter are devoted to a review of the essentials of Newtonian mechanics, in its Hamiltonian formulation. This will allow us to introduce the reader to one of the recurrent themes of this book, which is the "symplectization" of mechanics. The remainder of the Chapter is devoted to a review of quantum mechanics, with an emphasis on its Bohmian formulation. We also briefly discuss two topics which will be developed in this book: the metaplectic representation of the symplectic group, and the non-squeezing result of Gromov, which leads to a topological form of Heisenberg's inequalities.

It is indeed a discouraging (and perilous!) task to try give a bibliography for the topics reviewed in this Chapter, because of the immensity of the available literature. I have therefore decided to only list a few selected references; no doubt that some readers will felicitate me for my good taste, and that the majority probably will curse me for my omissions –and my ignorance!

The reader will note that I have added some historical data. However, this book is not an obituary: only the dates of birth of the mentioned scholars are indicated. These scientists, who have shown us the way, are eternal because they live for us today, and will live for us in time to come, in their great findings, their papers and books.

1.1 Classical Mechanics

> *I will triumph over mankind by the honest confession that I have stolen the golden vases of the Egyptians to build up a tabernacle for my God far away from the confines of Egypt. If you forgive me, I rejoice; if you are angry, I can bear it; the dice is cast, the book is written either for my contemporaries, or for posterity. I care not which; I can wait a hundred years for a Reader when God has waited six thousand years for a witness* (Johannes Kepler).

Johannes Kepler (*b.*1571) had to wait for less than hundred years for recognition: in 1687, Sir Isaac Newton (*b.*1643) published *Philosophiae Naturalis Principia Mathematica.* Newton's work had of course forerunners, as has every work in Science, and he acknowledged this in his famous sentence:

> "*If I have been able to see further, it was because I stood on the shoulders of Giants.*"

These Giants were Kepler, on one side, and Nicolas Copernicus (*b.*1473) and Galileo Galilei (*b.*1564) on the other side. While Galilei studied motions on Earth (reputedly by dropping objects from the Leaning Tower of Pisa), Kepler used the earlier extremely accurate –naked eyed!– observations of his master, the astronomer Tycho Brahe (*b.*1546), to derive his celebrated laws on planetary motion. It is almost certain that Kepler's work actually had a great influence on Newton's theory; what actually prevented Kepler from discovering the mathematical laws of gravitation was his ignorance of the operation of differentiation, which was invented by Newton himself, and probably simultaneously, by Gottfried Wilhelm Leibniz (*b.*1646). It is however noteworthy that Kepler knew how to "integrate", as is witnessed in his work *Astronomia Nova* (1609): one can say (with hindsight!) that the calculations Kepler did to establish his Area Law involved a numerical technique that is reminiscent of integration (see Schempp [136] for an interesting account of the "Keplerian strategy").

1.1.1 Newton's Laws and Mach's Principle

Newton's *Principia* (a paradigm of the exact Sciences, often considered as being the best scientific work ever written) contained the results of Newton's investigations and thoughts about Celestial Mechanics, and culminated in the statement of the laws of gravitation. Newton has often been dubbed the "first physicist"; the *Principia* were in fact the act of birth of Classical Mechanics. As Newton himself put it:

> *"The laws which we have explained abundantly serve to account for all the motions of the celestial bodies, and of our sea."*

We begin by recalling Newton's laws, almost as Newton himself stated them:

> **Newton's First law**: *a body remains in rest –or in uniform motion– as long as no external forces act to change that state.*

This is popularly known as "Newton's law of inertia". A reference frame where it holds is called an *inertial frame*. Newton's First Law may seem "obvious" to us today, but it was really a novelty at Newton's time where one still believed that motion ceased with the cause of motion! Newton's First Law moreover contains in germ a deep question about the identification between "inertial" and "gravitational" mass.

> **Newton's Second law**: *the change in momentum of a body is proportional to the force that acts on the body, and takes place in the direction of that external force.*

This is perhaps the most famous of Newton's laws. It was rephrased by Kirchhoff in the well-known (and somewhat unfortunate!) form "Force equals mass times acceleration".

> **Newton's Third law**: *if a given body acts on a second body with a force, then the latter will act on the first with a force equal in magnitude, but opposite in direction.*

This is of course the familiar law of "action and reaction": when you exert a push on a rigid wall, it "pushes you back" with the same strength.

Newton's Fourth law: *time flows equally, without relation to anything external and there is an absolute time.*

Newton's Fifth law: *absolute space, without relation to anything external, remains always similar and immovable.*

These last two laws are about absolute time and absolute space. They were never widely accepted by physicists, because they pose severe epistemological problems, especially because of the sentence "without relation to anything external." In fact, one does not see how something which exists without relation to anything "external" could be experimentally verified (or falsified, for that!). In fact, Newton's fourth and fifth laws are *ad hoc* postulates. It is interesting to note that Newton himself wrote in his *Principia*:

> *"It is indeed a matter of great difficulty to discover and effectually to distinguish the true from the apparent motion of particular bodies; for the parts of that immovable space in which bodies actually move, do not come under observation of our senses"*

This quotation is taken from Knudsen and Hjorth's book [96], where it is recommended (maybe with insight...) that we think about it for the rest of our lives!

Ernst Mach (*b*.1838) tried to find remedies to these shortcomings of Newton's fourth and fifth laws in his work *The Science of Mechanics*, published in 1883. Mach insisted that only relative motions were physically meaningful, and that Newton's concept of absolute space should therefore be abandoned. He tried to construct a new mechanics by considering that all forces were related to interactions with the entire mass distribution in the Universe (this is known as "Mach's principle"). Following Mach, our galaxy participates in the determination of the inertia of a massive particle: the overall mass distribution of the Universe is thus supposed to determine local mass. This belief is certainly more difficult to refute than it could appear at first sight (see the discussion of Mach's principle in [96]). Let us mention *en passant* that the Irish bishop and philosopher George Berkeley (*b*.1684) had proposed similar views (he argued that all motion was relative to the distant stars). In his *Outline of a general theory of relativity* (1913) Einstein claimed that he had formulated his theory in line with "*Mach's bold idea that inertia has its origin in an interaction of the mass point observed with all other points*" meaning that the inertia of a given body derives from its interaction with all masses in the Universe. To conclude, we remark that the non-locality of quantum mechanics (which Einstein disliked, because it impled the existence of "spooky actions at a distance") shows that Mach was after all right (but in a, by him, certainly unexpected way!).

Although Newton's discoveries were directly motivated by the study of planetary motion, the realm of mechanics quickly expanded well beyond particle or celestial mechanics. It was developed (among many others) by Leonhard Euler (*b*.1707), Joseph Louis Lagrange (*b*.1736), William Rowan Hamilton (*b*.1805) and, later, Jules Henri Poincaré (*b*.1854) and Albert Einstein (*b*.1879). Poincaré, who introduced the notion of manifold in mechanics also made substantial contributions to Celestial Mechanics, and introduced the use of divergent series in perturbation calculations. It seems today certain that Poincaré can be viewed as having discovered special relativity, but he did not, however, fully exploit his discoveries and realize their physical importance, thus leaving all the merit to Einstein. (Auffray's book [7] contains a careful analysis of Poincaré's and Einstein' ideas about Relativity. Also see Fölsing's extremely well written Einstein biography [47].)

1.1.2 Mass, Force, and Momentum

The concept of "force" and "mass" are notoriously difficult to define without using unscientific periphrases like "a force is a push or a pull", or "mass is a measure of stuff". Kirchhoff's statement of Newton's second law as

"Force = Mass × Acceleration"

makes things no better because it is a *circular* definition: it defines "force" and "mass" in terms of each other! The conceptual problems arising when one tries to avoid such circular arguments is discussed with depth and humor – yes, humor! – in Chapter VI of Poincaré's book *Science and Hypothesis* (Dover Publications, 1952). Here is one excerpt from this book (pages 97–98):

> *What is mass? Newton replies: "The product of the volume and the density." "It were better to say," answer Thomson and Tait, that density is the quotient of the mass by the volume." What is force? "It is," replies Lagrange, "that which moves or tends to move a body." "It is," according to Kirchoff, "the product of the mass and the acceleration." Then why not say that mass is the quotient of the force by the acceleration? These difficulties are insurmountable.*

We will occult these conceptual difficulties by using the following légerdemain: we *postulate* that there are two basic quantities describing the motion of a particle, namely: (1) the *position vector* $\mathbf{r} = (x, y, z)$ and (2) the *momentum vector* $\mathbf{p} = (p_x, p_y, p_z)$. While the notion of position is straightforward (its definition only requires the datum of a frame of reference and of a measuring device), that of momentum is slightly subtler. It can however be motivated by physical observation: the momentum vector $\mathbf{p}$ is a quantity which is conserved during free motion and under some interactions (for instance elastic collisions). Empirical evidence also shows that $\mathbf{p}$ is proportional to the velocity $\mathbf{v} = (v_x, v_y, v_z)$, that is $\mathbf{p} = m\mathbf{v}$, where the proportionality constant m is an intrinsic characteristic of the particle, called *mass*. The force $\mathbf{F}$ acting on the particle at time t is then defined as being the rate of change of momentum:

$$\mathbf{F} = \dot{\mathbf{p}} = \frac{d\mathbf{p}}{dt}$$

and Newton's second law can then be stated as the system of first order differential equations

$$\dot{\mathbf{r}} = \mathbf{v} \quad , \; \dot{\mathbf{p}} = \mathbf{F}. \tag{1.1}$$

1.2 Symplectic Mechanics

1.2.1 Hamilton's Equations

Most physical systems can be studied by using two specific theories originating from Newtonian mechanics, and having overlapping –but not identical– domains of validity. The first of these theories is "Lagrangian mechanics", which essentially uses variational principles (e.g., the "least action principle"); it will not be discussed at all in this book; we refer to Souriau [148] (especially page 140) for an analysis of some of the drawbacks of the Lagrangian approach. The second theory, "Hamiltonian mechanics", is based on Hamilton's equations of motion

$$\dot{\mathbf{r}} = \nabla_{\mathbf{p}} H(\mathbf{r}, \mathbf{p}, t) \quad , \quad \dot{\mathbf{p}} = -\nabla_{\mathbf{r}} H(\mathbf{r}, \mathbf{p}, t) \tag{1.2}$$

where the *Hamiltonian*[1] *function*

$$H = \frac{1}{2m}(\mathbf{p} - \mathbf{A}(\mathbf{r}, t))^2 + U(\mathbf{r}, t) \tag{1.3}$$

[1]The letter H was proposed by Lagrange to honor C. Huygens (*b.*1629), *not* Hamilton!

is associated to the "vector" and "scalar" potentials **A** and U. For **A**=0, Hamilton's equations are simply

$$\dot{\mathbf{r}} = \frac{\mathbf{p}}{m} \quad , \quad \dot{\mathbf{p}} = -\nabla_{\mathbf{r}} U(\mathbf{r}, \mathbf{p}, t)$$

and are immediately seen to be equivalent to Newton's second law for a particle moving in a scalar potential. The most familiar example where one has a non-zero vector potential is of course the case of a particle in an electromagnetic field; U is then the Coulomb potential $-e^2/|\mathbf{r}|$ whereas **A** is related to the magnetic field **B** by the familiar formula $\mathbf{B} = \nabla_{\mathbf{r}} \times \mathbf{A}$ in a convenient choice of units. There are however other interesting situations with $\mathbf{A} \neq \mathbf{0}$, one example being the Hamiltonian of the Coriolis force in a geocentric frame. We will see in Chapter 2 that Hamilton's equations are *equivalent* to Newton's Second Law even when a vector potential is present, provided that the latter satisfies a certain condition called by Souriau [148] the "Maxwell principle" in honor of the inventor of electromagnetic theory, James Clerk Maxwell (*b*.1831). One of the appeals of the Maxwell principle is that it automatically incorporates Galilean invariance in the Hamiltonian formalism; it does not however allow the study of physical systems where friction is present, and can thus be considered as defining "non-dissipative mechanics".

Hamiltonian mechanics could actually already be found in disguise in the work of Lagrange in Celestial Mechanics. Lagrange discovered namely that the equations expressing the perturbation of elliptical planetary motion due to interactions could be written down as a simple system of partial differential equations (known today as Hamilton's equations, but Hamilton was only six years old at that time!). It is however undoubtedly Hamilton who realized, some twenty four years later the theoretical importance of Lagrange's discovery, and exploited it fully.

Hamilton's equations form a system of differential equations, and we may thus apply the ordinary theory of existence and uniqueness of solutions to them. We will always make the simplifying assumption that every solution exists for all times, and is unique. This is for instance always the case when the Hamiltonian is of the type

$$H = \frac{1}{2m}\mathbf{p}^2 + U(\mathbf{r})$$

where the potential U satisfies a lower bound of the type

$$U(\mathbf{r}) \geq A - B\mathbf{r}^2$$

where $B \geq 0$, and this condition is actually satisfied in many cases. (See [1].)

In practice, Hamilton's equations are notoriously difficult to solve exactly, outside a few exceptional cases. Two of these lucky exceptions are: (1) the time-independent Hamiltonians with quadratic potentials (they lead to Hamilton equations which are *linear*, and can thus be explicitly solved); (2) the Kepler problem in spherical polar coordinates and, more generally, all "integrable" Hamiltonian systems (they can be solved by successive quadratures).

1.2.2 Gauge Transformations

The pair of potentials $(\mathbf{A}, U)$ appearing in the Hamiltonian given by Eq. (1.3) is called a *gauge*. Two gauges $(\mathbf{A}, U)$ and $(\mathbf{A}', U')$ are equivalent if they lead to the same motion in configuration space. This is always the case when there exists a function $\chi = \chi(\mathbf{r}, t)$ such that the gauges $(\mathbf{A}', U')$ and $(\mathbf{A}, U)$ are related by

$$\mathbf{A}' = \mathbf{A} + \nabla_{\mathbf{r}}\chi \quad , \quad U' = U - \frac{\partial \chi}{\partial t};$$

the mapping $(\mathbf{A}, U) \longmapsto (\mathbf{A}', U')$ is called a *gauge transformation.* The Hamiltonian function in the new gauge (A', U') is denoted by H'. It is related to H by the formula

$$H'(\mathbf{r}, \mathbf{p}, t) = H(\mathbf{r}, \mathbf{p} - \nabla_{\mathbf{r}}\chi, t) - \frac{\partial \chi}{\partial t}. \tag{1.4}$$

The notion of gauge was already implicit in Maxwell's work on electromagnetism; it was later clarified and developed by Hermann Weyl (*b*.1885). The effect of gauge transforms on the fundamental quantities of mechanics (momentum, action, etc.) will be studied later in this book.

1.2.3 Hamiltonian Fields and Flows

Using the letter z to denote the phase space variables $(\mathbf{r}, \mathbf{p})$, Hamilton's equations (1.2) can be written in the compact form

$$\dot{z} = X_H(z, t) \tag{1.5}$$

where X_H is the vector field "Hamiltonian vector field" defined by

$$X_H = (\nabla_{\mathbf{p}}H, -\nabla_{\mathbf{r}}H). \tag{1.6}$$

If H is time-independent, then Eq. (1.5) is an autonomous system of differential equations, whose associated flow is denoted by (f_t): f_t is the mapping that takes an "initial" point $z_0 = (\mathbf{r}_0, \mathbf{p}_0)$ to the point $z_t = (\mathbf{r}_t, \mathbf{p}_t)$ after time t, along the trajectory of X_H through z_0. It is customary to call the trajectory $t \mapsto f_t(z_0)$ the *orbit* of z_0. The mappings f_t obviously satisfy the one-parameter group property:

$$f_t \circ f_{t'} = f_{t+t'} \ , \ (f_t)^{-1} = f_{-t} \ , \ f_0 = I_d. \tag{1.7}$$

When H depends explicitly on time t, Hamilton's equations (1.5) no longer form an autonomous system, so that the mappings f_t no longer satisfy the group property (1.7). One then has advantage in modifying the notion of flow in the following way: given "initial" and "final" times t' and t, we denote by $f_{t,t'}$ the mapping that takes a point $z' = (\mathbf{r}', \mathbf{p}')$ at time t' to the point $z = (\mathbf{r}, \mathbf{p})$ at time t along the trajectory determined by Hamilton's equations. The family $(f_{t,t'})$ of phase-space transformations thus defined satisfies the *Chapman-Kolmogorov law*

$$f_{t,t'} \circ f_{t',t''} = f_{t,t''} \quad , \quad (f_{t,t'})^{-1} = f_{t',t} \quad , \quad f_{t,t} = I_d \tag{1.8}$$

which expresses *causality* in classical mechanics. When the initial time t' is 0, it is customary to write f_t instead of $f_{t,0}$ and call (f_t) the "time-dependent flow", but one must then be careful to remember that in general $f_t \circ f_{t'} \neq f_{t+t'}$.

1.2.4 The "Symplectization of Science"

The underlying mathematical structure of Hamiltonian Mechanics is *symplectic geometry*. (The use of the adjective "symplectic" in mathematics seems to go back to Hermann Weyl, who coined the word by replacing the Latin roots in "*com*-plex" by their Greek equivalents "*sym*-plectic".)

While symplectic methods seem to have been known for quite a while (symplectic geometry was already implicit in Lagrange's work), it has undergone an explosive evolution since the early 1970's, and has now invaded almost all areas of mathematics and physics. For further reading, I recommend Gotay and Isenberg's *Gazette des Mathématiciens* paper [75] (it is written in English!) which gives a very nice discussion of what the authors call the "symplectization of Science".

Symplectic geometry is the study of *symplectic forms*, that is, of antisymmetric bilinear non-degenerate forms on a (finite, or infinite-dimensional) vector space. More explicitly, suppose that E is a vector space (which we assume real). A mapping

$$\Omega : E \times E \longrightarrow \mathbb{R}$$

is a symplectic form if, for all vectors z, z', z'' and scalars $\alpha, \alpha', \alpha''$ we have

$$\begin{cases} \Omega(\alpha z + \alpha' z', z'') = \alpha\, \Omega(z, z'') + \alpha'\, \Omega(z', z'') \\ \Omega(z, \alpha' z' + \alpha'' z'') = \alpha'\, \Omega(z, z') + \alpha''\, \Omega(z, z'') \end{cases}$$

(bilinearity),

$$\Omega(z, z') = -\,\Omega(z', z)$$

(antisymmetry), and

$$\Omega(z, z') = 0,\ \forall z \in E \implies z' = 0$$

(non-degeneracy).

The real number $\Omega(z, z')$ is called the *symplectic product* (or *the skew-product*) of the vectors z and z'. When E is finite-dimensional, the non-degeneracy condition implies that E must have *even* dimension.

The most basic example of a symplectic form on the phase space $\mathbb{R}^3_{\mathbf{r}} \times \mathbb{R}^3_{\mathbf{p}}$ is the following:

$$\Omega(\mathbf{z}, \mathbf{z}') = \mathbf{p} \cdot \mathbf{r}' - \mathbf{p}' \cdot \mathbf{r} \tag{1.9}$$

for $\mathbf{z}' = (\mathbf{r}, \mathbf{p})$, $\mathbf{z}' = (\mathbf{r}', \mathbf{p}')$ (the dots $\cdot$ denote the usual scalar product; they will often be omitted in the sequel). Formula (1.9) defines the so-called "standard symplectic form" on phase space. We notice that the standard symplectic form can be identified with the differential 2-form

$$d\mathbf{p} \wedge d\mathbf{r} = dp_x \wedge dx + dp_y \wedge dy + dp_z \wedge dz$$

which we will denote also by Ω. In fact, by definition of the wedge product, we have

$$dp_x \wedge dx(\mathbf{r}, \mathbf{p}, \mathbf{r}', \mathbf{p}') = p_x x' - p'_x x$$

and similar equalities for $dp_y \wedge dy$, $dp_z \wedge dz$; summing up these equalities we get $\mathbf{p} \cdot \mathbf{r}' - \mathbf{p}' \cdot \mathbf{r}$. Introducing the matrix

$$J = \begin{pmatrix} 0_{3\times 3} & I_{3\times 3} \\ -I_{3\times 3} & 0_{3\times 3} \end{pmatrix},$$

the symplectic form Ω can be written in short as

$$\Omega(\mathbf{z}, \mathbf{z}') = \mathbf{z}'^T J \mathbf{z} \tag{1.10}$$

($\mathbf{z}$ and $\mathbf{z}'$ being viewed as column vectors). A matrix s which preserves the symplectic form, that is, such that

$$\Omega(s\mathbf{z}, s\mathbf{z}') = \Omega(\mathbf{z}, \mathbf{z}')$$

for all $\mathbf{z}$ and $\mathbf{z}'$ is said to be symplectic. The condition above can be restated in terms of the matrix J as

$$sJs^T = s^TJs = J. \tag{1.11}$$

Moreover, the matrix J can be used to relate the Hamilton vector field X_H to the gradient $\nabla_{\mathbf{z}} = \nabla_{\mathbf{r},\mathbf{p}}$: we have

$$X_H = J\nabla_{\mathbf{z}}H$$

($J\nabla_z$ is called the "symplectic gradient operator") so Hamilton's equations (1.2) can be written:

$$\dot{\mathbf{z}} = J\nabla_z H(\mathbf{z}, t). \tag{1.12}$$

There is a fundamental relation between the symplectic form Ω, the Hamilton vector field X_H and H itself. That relation is that we have

$$\Omega(X_H(\mathbf{z}, t), \mathbf{z}') = \mathbf{z}' \cdot \nabla_{\mathbf{z}}H(\mathbf{z}, t) \tag{1.13}$$

for all $\mathbf{z}$, $\mathbf{z}'$. This relation is fundamental because it can be written very simply in the language of intrinsic differentiable geometry as

$$i_{X_H}\Omega + dH = 0, \tag{1.14}$$

where $i_{X_H}\Omega$ is the contraction of the symplectic form $\Omega = d\mathbf{p} \wedge d\mathbf{r}$ with the Hamiltonian vector field X_H, i.e.:

$$i_{X_H}\Omega(\mathbf{z})(\mathbf{z}') = \Omega(X_H(\mathbf{z}), \mathbf{z}'). \tag{1.15}$$

This "abstract" form of Eq. (1.13) is particularly tractable when one wants to study Hamiltonian mechanics on symplectic manifolds (this becomes necessary, for instance, when the physical system is subjected to constraints). The equation (1.14) also has the advantage of leading to straightforward calculations and proofs of many properties of Hamiltonian flows. It allows, for instance, a very neat proof of the fact that Hamiltonian flows consist of "symplectomorphisms" (or "canonical transformations" as they are often called in physics). (Symplectomorphisms are phase-space mappings whose Jacobian matrices are symplectic.)

1.3 Action and Hamilton-Jacobi's Theory

As we said, it is usually very difficult to produce exact solutions of Hamilton's equations. There is however a method which works in many cases. It is the *Hamilton-Jacobi* method, which relies on the Hamilton-Jacobi equation

$$\frac{\partial \Phi}{\partial t} + H(\mathbf{r}, \nabla_{\mathbf{r}}\Phi, t) = 0 \tag{1.16}$$

and which we discuss below. We will not give any application of that method here (the interested reader will find numerous applications and examples in the literature (see for instance [39, 52, 128])) and we will rather focus on the geometric interpretation of equation (1.16). This will give us the opportunity of saying a few words about the associated notion of *Lagrangian manifold* which plays an essential role in mechanics (both classical, where Lagrangian manifolds intervene in the form of the "invariant tori" (or its variants) associated to integrable systems, and in quantum mechanics, where they are the perfect objects to "quantize".

1.3.1 Action

Let $(f_{t,t'})$ be the flow determined by Hamilton's equations. For a point $z' = (\mathbf{r}', \mathbf{p}')$ of phase space, let Γ be the arc of curve $s \longmapsto f_{s,t'}(z_0)$ when s varies from t' to t: it is thus the piece of trajectory joining z' to $z = f_{t,t'}(z')$. By definition, the line integral

$$\mathcal{A}(\Gamma) = \int_{\Gamma} \mathbf{p}{\cdot}d\mathbf{r} - H dt \tag{1.17}$$

is called the *action* along Γ. Action is a fundamental quantity both in classical and quantum mechanics, and will be thoroughly studied in Chapter 3. Now, a crucial observation is that if time $t - t'$ is sufficiently small, then the phase-space arc Γ will project diffeomorphically onto a curve γ without self-intersections in configuration space $\mathbb{R}^3_{\mathbf{r}}$, and joining $\mathbf{r}'$ at time t' to $\mathbf{r}$ at time t. Conversely, $t - t'$ being a (short) given time, the knowledge of initial and final points $\mathbf{r}'$ and $\mathbf{r}$ uniquely determines the initial and final momenta $\mathbf{p}'$ and $\mathbf{p}$. It follows that the datum of γ *uniquely* determines the arc Γ. This allows us to rewrite definition (1.17) of action as $\mathcal{A}(\Gamma) = \mathcal{A}(\gamma)$ where

$$\mathcal{A}(\gamma) = \int_{\gamma} \mathbf{p}{\cdot}d\mathbf{r} - H dt \tag{1.18}$$

is now an integral calculated along a path in the *state space* $\mathbb{R}^n_{\mathbf{r}} \times \mathbb{R}_t$. If we keep the initial values $\mathbf{r}'$ and t' fixed, we may thus view $\mathcal{A}(\gamma)$ as a function[2] $W = W(\mathbf{r}, \mathbf{r}'; t, t')$ of $\mathbf{r}$ and t. A fundamental property of W is now that

$$dW(\mathbf{r}, t) = \mathbf{p}\cdot d\mathbf{r} - H(\mathbf{r}, \mathbf{p}, t)dt \tag{1.19}$$

where $\mathbf{p}$ is the final momentum, that is the momentum at $\mathbf{r}$, at time t (a word of caution: even if Eq. (1.19) looks "obvious", its proof is *not* trivial!).

1.3.2 Hamilton-Jacobi's Equation

Let us shortly describe the idea underlying Hamilton-Jacobi's method for solving Hamilton's equations of motion; it will be detailed in Chapter 4. Consider the Cauchy problem

$$\begin{cases} \dfrac{\partial \Phi}{\partial t} + H\left(\mathbf{r}, \nabla_{\mathbf{r}}\Phi, t\right) = 0 \\ \Phi(\mathbf{r}, 0) = \Phi_0(\mathbf{r}) \end{cases} \tag{1.20}$$

where Φ_0 is some (arbitrary) function on configuration space. The solution exists, at least for short times t, and is unique. It is given by the formula

$$\Phi(\mathbf{r}, t) = \Phi_0(\mathbf{r}_0) + W(\mathbf{r}, \mathbf{r}_0; t, 0) \tag{1.21}$$

where $W(\mathbf{r}, \mathbf{r}_0; t, 0)$ is the action calculated from the point $\mathbf{r}_0$ *from which* $\mathbf{r}$ *is reached at time* t: the point $\mathbf{r}_0$ is thus not fixed, but depends on $\mathbf{r}$ (and on t).

What good does it do to us in practice to have a solution of the Cauchy problem (1.20)? Well, such a solution allows to determine the particle motion with arbitrary initial position and $\mathbf{r}_0$ initial momentum $\mathbf{p}_0 = \nabla_{\mathbf{r}}\Phi_0(\mathbf{r}_0)$ without solving Hamilton's equations! Here is how. The function $\Phi(\mathbf{r}, t)$ defines a "momentum field" which determines, at each point $\mathbf{r}$ and each time t, the momentum of a particle that may potentially be placed there: that momentum is $\mathbf{p} = \nabla_{\mathbf{r}}\Phi(\mathbf{r}, t)$ and we can then find the motion by integrating the first Hamilton equation

$$\dot{\mathbf{r}} = -\nabla_{\mathbf{p}}H(\mathbf{r}, \nabla_{\mathbf{r}}\Phi(\mathbf{r}, t), t) \tag{1.22}$$

[2]The use of the letter W comes from "Wirkung", the German word for "action."

which is just the same thing as

$$\dot{\mathbf{r}} = \frac{1}{m}\left(\nabla_{\mathbf{r}}\Phi(\mathbf{r},t) - \mathbf{A}(\mathbf{r},t)\right). \tag{1.23}$$

Given a solution Φ of the Cauchy problem (1.20) we can only determine the motion corresponding to "locked" initial values of the momentum, corresponding to the "constraint" $\mathbf{p}_0 = \nabla_{\mathbf{r}}\Phi_0(\mathbf{r}_0)$. However, *in principle*, we can use the method to determine the motion corresponding to an arbitrary initial phase space point $(\mathbf{r}_0, \mathbf{p}_0)$ by choosing *one* function Φ_0 such that $\mathbf{p}_0 = \nabla_{\mathbf{r}}\Phi_0(\mathbf{r}_0)$, then to solve the Hamilton-Jacobi equation with Cauchy datum Φ_0 and, finally, to integrate Eq. (1.22). Of course, the solutions we obtain are a priori only defined for short times, because Φ is not usually defined for large values of t. This is however not a true limitation of the method, because one can then obtain solutions of Hamilton's equations for arbitrary t by repeated use of Chapman-Kolmogorov's law (1.8). *Hamilton-Jacobi theory is thus an equivalent formulation of Hamiltonian mechanics.*

1.4 Quantum Mechanics

> *There are two kinds of truths. To the one kind belong statements so simple that the opposite assertion could not be defended. The other kind, the so-called "deep truths", are statements in which the opposite could also be defended* (N. Bohr)

The history of quantum mechanics can be divided into four main periods. The first began with Max Planck's (*b.*1858) theory of black-body radiation in 1900. Planck was looking for a universal formula for the spectral function of the black-body who could reconciliate two apparently contradictory laws of thermodynamics (the Rayleigh-Jeans, and the Wien laws). This led him to postulate, "in an act of despair", that energy exchanges were discrete, and expressed in terms of a certain constant, h. This first period may be described as the period in which the validity of Planck's constant was demonstrated but its real meaning was not fully understood, until Einstein's trail-blazing work on the theory of light quanta in 1905 (remember that Einstein was awarded the Nobel Prize in 1921 for his work on the photoelectric effect, *not* for relativity theory!). For more historical data, I recommend the interesting article of H. Kragh in *Physics World*, **13**(12) (2000)). A traditional reference for these topics is Jammer [92]; Gribbin [76] and Ponomarev [132] are also useful readings.

The second period began with the quantum theory of atomic structure and spectra proposed by Niels Bohr (*b*.1885) in 1913, and which is now called the "old quantum theory." Bohr's theory yielded formulas for calculating the frequencies of spectral lines, but even if his formulas were accurate in many cases, they did not, however, form a consistent and unified theory. They were rather a sort of "patchwork" affair in which classical mechanics was subjected to extraneous and a priori "quantum conditions" imposed on classical trajectories. It was, to quote Jammer [92] (page 196):

> "*...a lamentable hodgepodge of hypotheses, principles, theorems and computational recipes.*"

The third period, quantum mechanics as a theory with sound mathematical foundations, began in the mid-twenties nearly simultaneously in a variety of forms: the matrix theory of Max Born (*b*.1882) and Werner Heisenberg (*b*.1901), the wave mechanics of Louis de Broglie (*b*.1892) and Erwin Schrödinger (*b*.1887), and the theories of Paul Dirac (*b*.1902) and Pascual Jordan (*b*.1902).

The fourth period –which is still under development at the time this book is being written– began in 1952 when David Bohm (*b*.1917) introduced the notion of quantum potential, which allowed him to reinstate the notion of particle and particle trajectories in quantum mechanics. We will expose Bohm's ideas in a while, but let us first discuss de Broglie's matter wave theory and Schrödinger's equation.

1.4.1 Matter Waves

Louis de Broglie proposed in his 1924 Doctoral thesis that just as photons are associated with electromagnetic waves, material particles are accompanied by "matter waves". De Broglie's idea was very simple, as are most traits of genius. (He had actually already written a paper in 1923 where he suggested that a beam of electrons passing through a sufficiently narrow hole must produce interference effects.) He postulated that to every particle is associated a kind of "internal vibration", whose frequency should be obtained from Einstein's formulas

$$E = h\nu \quad , \quad E = mc^2 \tag{1.24}$$

relating the frequency of a photon to its energy, and the energy of a material particle to its mass $m = m_0/\sqrt{1-(v/c)^2}$. De Broglie equated the right hand sides of these two equations to obtain the formula

$$\nu = \frac{mc^2}{h} \tag{1.25}$$

giving the frequency of the internal vibration in terms of Planck's constant and of the relativistic energy of the particle. This was indeed a very bold step, since the first of the Einstein equations (1.24) is about light quanta, and the second about the energy of matter! (We have been oversimplifying a little de Broglie's argument, who was actually rather subtle, and based on a careful discussion of relativistic invariance.) One year later, de Broglie took one step further and postulated the existence of a *wave* associated with the particle, and whose wavelength was given by the simple formula

$$\lambda = \frac{h}{p}. \tag{1.26}$$

The phase velocity of a de Broglie wave is thus

$$v_\phi = \nu\lambda = \frac{c^2}{v}$$

and is hence superior to that of light. Introducing the wave number $k = 2\pi/\lambda$ and the angular frequency ω, the group velocity of the de Broglie wave is

$$v_g = \frac{d\omega}{dk} = \frac{dE}{dk}$$

and a straightforward calculation gives the value $\mathbf{v}_g = \mathbf{v}$ (and hence $v_\phi v_g = c^2$). The group velocity of a de Broglie wave is thus the velocity of the particle to which it is associated; that wave can thus be viewed as accompanying –or piloting – the particle; this is the starting point of the de Broglie-Bohm pilot-wave theory about we will have more to say below.

Since Planck's constant value is

$$h \approx 6.6260755 \times 10^{-34}\,\mathrm{J\,s}$$

the de Broglie wavelength is extraordinarily small for macroscopic (and even mesoscopic) objects. For instance, if you walk in the street, your de Broglie wavelength will have an order of magnitude of 10^{-35} m, which is undetectable by today's means. However, for an electron ($m \approx 0,9 \times 10^{-30}$ kg) with velocity 10^6 $\mathrm{m\,s}^{-1}$, we have $\lambda \approx 7 \times 10^{-9}$ m, and this wavelength leads to observable diffraction patterns (it is comparable to the wavelength of certain X-rays). In fact, some three years after de Broglie's thesis, the celebrated diffraction experiments of C.J. Davisson and L.H. Germer in 1927 described in all physics textbooks (e.g., Messiah, [117], Ch. 2, §6) showed that de Broglie was right. Davisson and Germer had set out to study the scattering of a collimated electron beam by a crystal of nickel. The patterns they observed were typically those of diffracted waves, the wavelengths of which were found to be, with a very good accuracy, exactly those predicted by de Broglie's theory (this wasn't actually the first experimental confirmation of de Broglie's matter wave postulate, since G.P. Thomson had discovered the diffraction of electrons a few months before. It is amusing to note that while G.P. Thomson was awarded the Nobel prize in 1937 (together with C.J. Davisson) for having shown that electrons "are" waves, his father, J.J. Thomson (*b*.1887) had been awarded the same prize in 1906 for proving that electrons were *particles*!

1.4.2 "If There Is a Wave, There Must Be a Wave Equation!"

Only two years after de Broglie's hypothesis, in 1926, Schrödinger proposed an equation governing the evolution of de Broglie's "matter waves" (reportedly in response to a question by one of his colleagues (reputedly Peter Debye) who had exclaimed *"If there is a wave, then there must be a wave equation!"*). Guided by a certain number of a priori conditions (which we will discuss in a while) Schrödinger postulated that the evolution of the wave function Ψ associated to a single particle moving in a potential U should be governed by the partial differential equation

$$i\hbar\frac{\partial\Psi}{\partial t} = -\frac{\hbar^2}{2m}\nabla^2_{\mathbf{r}}\Psi + U\Psi \tag{1.27}$$

where $\hbar$ is Planck's constant h divided by 2π. The notation "*h-bar*" is due to Dirac; Schrödinger used the capital K to denote $h/2\pi$ in his early work.

The solutions Ψ of Schrödinger's equation describe the time evolution of a matter wave associated with a particle moving in a scalar potential U. If there is a vector potential $\mathbf{A}$, Eq. (1.27) should be replaced by the more general equation

$$i\hbar\frac{\partial\Psi}{\partial t} = \frac{1}{2m}\left(-i\hbar\nabla_{\mathbf{r}} - \mathbf{A}\right)^2\Psi + U\Psi \tag{1.28}$$

whose solutions depend (as do the solutions of Hamilton's equations) on the choice of a gauge. If one replaces the gauge $(\mathbf{A}, U)$ by the equivalent gauge

$(\mathbf{A} + \nabla_x \chi, U - \partial\chi/\partial t)$, then a straightforward calculation shows that the solution Ψ should be replaced by

$$\Psi^{\chi} = e^{\frac{i}{\hbar}\chi}\Psi \tag{1.29}$$

and thus merely corresponds to a change of phase of the wave function (the amplitude is not affected by a change of gauge). This change of phase is at the origin of "geometric" or "topological" phase shifts in quantum mechanics (e.g., the Aharonov-Bohm effect or the occurrence of the "Berry phase", already mentioned in connection with the notion of gauge).

Let us next make a little digression about Schrödinger's quantization rule (and "quantization" in general).

1.4.3 Schrödinger's Quantization Rule and Geometric Quantization

Schrödinger's equation is obtained from a *classical* object (the Hamiltonian function H) by formally replacing the momentum variable $\mathbf{p}$ by the operator $-i\hbar\nabla_{\mathbf{r}}$, which has the effect of transforming the scalar *function*

$$H = \frac{1}{2m}(\mathbf{p} - \mathbf{A})^2 + U$$

into the self-adjoint partial-differential operator

$$\widehat{H} = \frac{1}{2m}\left(-i\hbar\nabla_{\mathbf{r}} - \mathbf{A}\right)^2 + U. \tag{1.30}$$

This "quantization rule" leads to an apparent ambiguity, because if we write the Hamiltonian H in the "expanded" forms

$$H = \frac{1}{2m}\left(\mathbf{p}^2 - 2\mathbf{p}\cdot\mathbf{A} + \mathbf{A}^2\right) + U \tag{1.31}$$

or

$$H = \frac{1}{2m}\left(\mathbf{p}^2 - 2\mathbf{A}\cdot\mathbf{p} + \mathbf{A}^2\right) + U \tag{1.32}$$

before applying the rule $\mathbf{p} \longmapsto -i\hbar\nabla_{\mathbf{r}}$, then we would obtain quite different operators, since in general $\nabla_{\mathbf{r}}\cdot\mathbf{A} \neq \mathbf{A}\cdot\nabla_{\mathbf{r}}$. This apparent paradox is in fact immediately eliminated if one makes the following convention, called in the literature "Schrödinger's quantization rule" (or "normal ordering rule"):

> **Schrödinger's quantization rule:** *Each time products* $\mathbf{p} \cdot \mathbf{A}$ *or* $\mathbf{A} \cdot \mathbf{p}$ *appear in a Hamiltonian, apply Schrödinger's quantization rule to the symmetrized expression* $(\mathbf{p} \cdot \mathbf{A} + \mathbf{A} \cdot \mathbf{p})/2$.

For instance, applying this rule to either (1.31) or (1.32), one has to replace $\mathbf{p}$ by $-i\hbar\nabla_{\mathbf{r}}$ in the symmetrized expression

$$H = \frac{1}{2m}\left(\mathbf{p}^2 - (\mathbf{A} \cdot \mathbf{p} + \mathbf{A} \cdot \mathbf{p}) + \mathbf{A}^2\right) + U$$

and this yields, in either case:

$$\widehat{H} = -\frac{\hbar^2}{2m}\nabla_{\mathbf{r}}^2 + \frac{i\hbar}{2m}(\mathbf{A} \cdot \nabla_{\mathbf{r}}) + \frac{i\hbar}{2m}\operatorname{div}\mathbf{A} + \frac{1}{2m}\mathbf{A}^2 + U \tag{1.33}$$

which is just the operator (1.30) in "expanded form".

The prescriptions above are completely unambiguous (at least as long as we are using Cartesian (or, more generally, symplectic) coordinates; see Messiah [117], Chapter II, §15, for a discussion, of both the "ordering" problem, and of what happens when one goes over to polar coordinates. The Schrödinger quantization rule, as *ad hoc* as it may seem, can in fact be fully justified using modern pseudo-differential operator theory (the Weyl-Leray calculus, or its variants; see for instance [40, 79, 101]). Problems of this nature belong to an area of pure mathematics called *geometric quantization,* and whose expansion started in the early 1970's, due to the contributions of Blattner [15, 16], Kostant [97], Sniatycki [146], Souriau [148]. Geometric quantization is a relatively new and very active branch of pure mathematics that was born in the early 1970's. Loosely speaking, geometric quantization tries to assign a self-adjoint operator ("quantum observable") to a function on phase space ("classical observable"). While it is not difficult to "quantize" observables that are quadratic functions of the phase space coordinates (this is intimately related to the existence of the "metaplectic representation", which will be thoroughly discussed in Chapter 6), the general case rapidly leads to new and unexpected difficulties. For instance, it is well known since a celebrated "no-go" result of Groenewold and van Hove (see [46, 79]) that this "quantization procedure" cannot be pushed beyond quadratic Hamiltonian functions. Moreover, quantization becomes a very difficult business when the phase space is an arbitrary symplectic manifold, because it is then no longer any canonical way to define "position" and "momentum" vectors. Mechanics (both classical, or quantum) on symplectic manifolds is not just a purely academic topic: it is necessary to work on such generalized phase spaces when one studies physical systems subjected to constraints. The theory is far from being complete as time being; there still remains much to do because there are formidable roadblocks on the way to understanding what a general quantization procedure should look like, especially since the solution (if it exists) seems not to be unique. The review paper by Tuynman [153] contains a very well-written introduction to the topic, and many references to further work. Although geometric quantization is a beautiful collection of theories, at the very highest level of mathematical research, I must reluctantly confess that I do not really believe in its usefulness in a deeper understanding of quantum mechanics. This is of course only a personal opinion; there have been so many instances in the history of Science where similar guesses have proven to be totally wrong. A famous example is that of G.H. Hardy, who prognosticated in 1941 that there were two fields of Mathematical Science which would stay without any military application: number theory and Einstein's theory of relativity. As we know, Hardy was unfortunately wrong: number theory is used in cryptography and its military applications, while the special theory of relativity has helped develop nuclear weapons.

1.5 The Statistical Interpretation of Ψ

> *One can view the world with the p eye and one can view it with the q [=* position*] eye, but if one tries to open both eyes together, one gets confused* (Wolfgang Pauli).

The probabilistic interpretation of the wave function is due to Born, and is exposed in every textbook on quantum mechanics (see, for instance, [17, 117, 128]).

1.5.1 Heisenberg's Inequalities

Schrödinger's equation is a first order partial differential equation in the time variable t; it therefore makes sense to consider the Cauchy problem

$$\begin{cases} i\hbar\dfrac{\partial\Psi}{\partial t} = \widehat{H}\Psi \\ \Psi(\mathbf{r},0) = \Psi_0(\mathbf{r}). \end{cases}$$

A fundamental property is that its solution Ψ is square integrable for all values of t if Ψ_0 is, and that the L^2-norm is conserved during time evolution.

In fact, the Hamiltonian operator is self-adjoint, and from this readily follows that we must have

$$\int |\Psi(\mathbf{r},t)|^2 d^3r = \int |\Psi_0(\mathbf{r})|^2 d^3r$$

for all times t. The integrations are performed over configuration space, and we are using the notation $d^3r = dxdydz$. It follows that if we normalize Ψ_0 by requiring that

$$\int |\Psi_0(\mathbf{r})|^2 d^3r = 1 \tag{1.34}$$

then we will also have

$$\int |\Psi(\mathbf{r},t)|^2 d^3r = 1 \tag{1.35}$$

for all times t. The wave function Ψ can thus be viewed as a *probability density*, and this leads to the following statistical interpretation of quantum mechanics:

1) Let $X(\mathbf{r},t)$ be the stochastic variable describing the position of the particle at time t; the corresponding expectation values for the position coordinates are thus

$$\begin{cases} \langle x(t)\rangle = \int x|\Psi(\mathbf{r},t)|^2 d^3r \\ \langle y(t)\rangle = \int y|\Psi(\mathbf{r},t)|^2 d^3r \\ \langle z(t)\rangle = \int z|\Psi(\mathbf{r},t)|^2 d^3r \end{cases}$$

and the probability of finding the particle in a measurable region $\mathcal{R}$ of physical space at time t is given by

$$\Pr(\mathbf{r}\in\mathcal{R},t) = \int_{\mathcal{R}} |\Psi(\mathbf{r},t)|^2 d^3r. \tag{1.36}$$

2) Assuming that the Fourier transform

$$\hat{\Psi}(\mathbf{p},t) = \left(\tfrac{1}{2\pi\hbar}\right)^{3/2} \int e^{-\frac{i}{\hbar}\mathbf{p}\cdot\mathbf{r}}\Psi(\mathbf{r},t)d^3r$$

of the wave function exists, we have, in view of Plancherel's theorem:

$$\int |\hat{\Psi}(\mathbf{p},t)|^2 d^3p = \int |\Psi(\mathbf{r},t)|^2 d^3r$$

(with $d^3p = dp_x dp_y dp_z$) and hence

$$\int |\hat{\Psi}(\mathbf{p},t)|^2 d^3p = 1$$

as soon as the normalization condition (1.34) holds. The function $|\hat{\Psi}(\mathbf{p},t)|^2$ is then viewed as the probability density for the stochastic variable $P(\mathbf{p},t)$ whose values are the possible momenta at time t. The expectation values of the momenta coordinates are

$$\begin{cases} \langle p_x(t)\rangle = \int p_x|\Psi(\mathbf{r},t)|^2 d^3r \\ \langle p_y(t)\rangle = \int p_y|\Psi(\mathbf{r},t)|^2 d^3r \\ \langle p_z(t)\rangle = \int p_z|\Psi(\mathbf{r},t)|^2 d^3r \end{cases}$$

and the probability of finding the momentum vector $\mathbf{p}$ in a region $\mathcal{P}$ of momentum space, at time t, is:

$$\Pr(\mathbf{r}\in\mathcal{P},t) = \int_{\mathcal{P}} |\hat{\Psi}(\mathbf{p},t)|^2 d^3p.$$

It is a classical result (see Messiah, [117], Bohm [17]) that the associated standard deviations

$$\begin{cases} \Delta x(t) = \left(\langle x^2(t)\rangle - \langle x(t)\rangle^2\right)^{1/2} \\ \\ \Delta p_x(t) = \left(\langle p_x^2(t)\rangle - \langle p_x(t)\rangle^2\right)^{1/2} \end{cases}$$

(and similar definitions for the other position and momentum coordinates) satisfy the *Heisenberg inequalities*:

$$\begin{cases} \Delta p_x(t)\Delta x(t) \geq \frac{1}{2}\hbar \\ \Delta p_y(t)\Delta y(t) \geq \frac{1}{2}\hbar \\ \Delta p_z(t)\Delta z(t) \geq \frac{1}{2}\hbar. \end{cases} \tag{1.37}$$

This "uncertainty principle" expresses the impossibility of performing simultaneous measurements of position and momenta with arbitrary precision, and indicates that there is some kind of "measurement barrier" we cannot transgress. This apparent limitation of our knowledge has led to the publication of a huge amount of scientific, philosophical, and metaphysical texts. In particular, it is a widespread opinion among quantum physicists (especially the adepts of the Copenhagen interpretation) that it follows from Heisenberg's inequalities that the notion of phase space does not make sense in quantum-mechanics. We will have more to say about this later.

1.6 Quantum Mechanics in Phase Space

As we said above, Heisenberg's inequalities (which mathematically only reflect the non-locality of the Fourier transform) have led many physicists to believe (and to vigorously advocate) that there can exist no such thing as "quantum mechanics in phase space". Their argument goes as follows: the Heisenberg inequalities "show" that one cannot assign to a physical system both a definite position and a definite momentum, and therefore it is no phase space in quantum mechanics. Period. However, these physicists make a confusion between *observation* and *reality*. There is, of course, nothing wrong a priori with such a positivistic attitude (which amounts to identify "perception" and "existence"), as far as it does not lead to a systematic rejection of any phase-space theory just because it is a phase space theory! It is not because Heisenberg's inequalities indicate that there is a "limitation" in the accuracy of the way we perceive our physical world that we should be forbidden to consider a more clear-cut mathematical world where things have precise positions and momenta. Such an attitude would be as counterproductive as to deny the use of mathematical concepts such as points or lines just because nobody has ever "seen" or will ever "see" a point or a line, except with the "eyes of the mind".

There are actually many more reasons to believe that phase space makes sense in quantum mechanics. For instance, the occurrence of the "geometric phase shifts" (e.g., the Aharonov-Bohm or Berry effects) are typical quantum-mechanical phase space manifestations.

1.6.1 Schrödinger's "firefly" Argument

There is no reason, after all, why the "fuzzy" quantum-mechanical world should necessarily coincide with the clear-cut platonic reality of mathematics. (This was already underlined by Wigner in his famous paper [168] on the "unreasonable effectiveness of mathematics".) Here is an example, due to Schrödinger. Let a free particle be located exactly at a point A at time $t_0 = 0$, and again at a distance d, at a point B, after time t. Then obviously d/t is the velocity with which it has travelled from A to B, thus the one it *had* at A. To this Heisenberg answered, says Schrödinger, by saying that

> *"yes, but this belated information is of no physical significance; it was not forthcoming at the initial moment at A, could not be used for predicting the trajectory; it is only vouchsafed after the trajectory is known..."*

Schrödinger then adds, commenting on Heisenberg's objection:

> "*To this, one would have to say that it is all right, but if one accepts it, one grants to Einstein that quantum mechanical description is incomplete. If it is possible to obtain simultaneous accurate values of location and velocity, albeit belatedly, then a description that does not allow them is deficient. After all the thing has obviously moved from A to B with velocity d/t; it has not been interfered with between A and B. so it must have had this velocity at A. And this is beyond the power of quantum mechanical (or wave mechanical) description.*"

There is however another even more compelling evidence for the existence of phase space in quantum mechanics. This evidence is related to a rather recent discovery in pure mathematics, M. Gromov's non-squeezing theorem, which is a topological and *classical* uncertainty principle, which we will discuss later. This principle, which is an active area of research in symplectic topology, is a newcomer in the physics community; we will discuss its implications in Chapter 3. Here is a brief description of that principle.

1.6.2 The Symplectic Camel

Consider a ball B in phase space, with radius R. The orthogonal projection (or "shadow") of that ball on any plane is always a disk with radius πR^2. For instance, if we project B on, say, the (x, p_x) plane we will get a disk

$$(x - x_0)^2 + (p_x - p_{x_0})^2 \leq R^2$$

and if we project it on the (x, p_y) plane, we will get a disk

$$(x - x_0)^2 + (p_y - p_{y_0})^2 \leq R^2$$

and so on. Suppose now that we start moving the ball B by using a Hamiltonian flow (f_t). The divergence of a Hamiltonian vector field being zero:

$$\operatorname{div} X_H = \nabla_{\mathbf{r}} \cdot \nabla_{\mathbf{p}} H + \nabla_{\mathbf{p}} \cdot (-\nabla_{\mathbf{r}} H) = 0,$$

the mappings f_t are volume-preserving. Since conservation of volume has nothing to do with conservation of shape, one might thus envisage that the ball B will distort, a priori in the most bizarre way (parts of the ball might extend very far away, while other might be squeezed) while keeping the same volume during the whole motion. However, in 1985 Gromov [78] discovered a most surprising mathematical property. He discovered that the shadows of the distorting ball on the "conjugate planes" (x, p_x), (y, p_y), and (z, p_z) will never become smaller than their original value πR^2, while the shadows on the other, non-conjugate planes (say, (x, y), (x, p_y),...) can become arbitrarily small. It does of course not require a huge amount of imagination to recognize that this is a sort of classical variant of Heisenberg's inequalities. This property is known in the mathematical literature as the "non-squeezing property", or as the "principle of the symplectic camel". (The reader who is curious to know the origin of this terminology is invited to read the quotation from *Matthew* **7** at the beginning of Section 3.7 in Chapter 3.) The non-squeezing property implies that the action of Hamiltonian flows on phase space volumes is in a sense much more "rigid" than that of plain volume-preserving diffeomorphisms.

The ignorance of the non-squeezing property has led some physicists to believe that classical mechanics is deeply wrong even at the macroscopic level. This is illustrated by the following statement, quoted from the excellent and deservedly acclaimed popular science book by Penrose [129]:

> "...*Without Liouville's theorem, one might envisage that this undoubted tendency for a region to spread out in phase space could be compensated by a reduction in overall volume...Classical Mechanics is, in this kind of sense, essentially unpredictable... this spreading effect in phase space has another remarkable implication... that classical mechanics cannot actually be true of our world...*"

What Penrose could not know at the time he made this statement, is that in view of the non-squeezing property, flows associated with Hamiltonian vector fields are really much more "tame" than volume-preserving flows.

Gromov's non-squeezing theorem opens new perspectives in both Hamiltonian and quantum mechanics. It raises many interesting questions (most of them still unanswered) because it highlights the fact that the existence and properties of closed orbits is in some mysterious way related to the property of the symplectic camel (see Hofer and Zehnder's book [89]).

We will use Gromov's result in Chapter 3 to efficiently quantize phase space in "quantum blobs", which generalize the usual notion of quantum cell.

1.7 Feynman's "Path Integral"

By definition, the Green function G associated to Schrödinger's equation

$$i\hbar\frac{\partial\Psi}{\partial t} = \frac{-\hbar^2}{2m}\nabla_{\mathbf{r}}^2\Psi + U\Psi \tag{1.38}$$

is the kernel G such that

$$\Psi(\mathbf{r}, t) = \int G(\mathbf{r}, \mathbf{r}', t)\Psi_0(\mathbf{r}')\, d^3r'$$

where $\Psi_0 \in \mathcal{S}(\mathbb{R}^n_{\mathbf{r}})$ is an initial condition for (1.38).

1.7.1 The "Sum Over All Paths"

That Green function can be written in the following suggestive (but mathematically meaningless) form, due to Richard Feynman (b.1918):

$$G(\mathbf{r}, \mathbf{r}', t) = \sum_{\text{paths}} \exp\left(\frac{i}{\hbar}\int_{\mathbf{r}',0}^{\mathbf{r},t} \mathbf{p}\cdot d\mathbf{r} - H dt\right). \tag{1.39}$$

The "sum" is taken over all paths (or "histories") leading from the initial point $\mathbf{r}'$ at time $t = 0$ to the final point $\mathbf{r}$ at time t. This formula should in fact be interpreted as follows: let N be a positive integer, and set $\Delta t = t/N$ and consider the function

$$G_N = \left(\frac{m}{2\pi i\hbar\Delta t}\right)^{3(N+1)/2} \int \exp\left(\tfrac{i}{\hbar}W_N\right) d^3r_{(1)}\cdots d^3r_{(N-1)} \tag{1.40}$$

where $d^3r_{(j)} = dx_{(j)}dy_{(j)}dz_{(j)}$ and

$$W_N = \frac{m}{2\Delta t}\sum_{j=1}^{N}(\mathbf{r}_{(j)} - \mathbf{r}'_{(j)})^2 - U(\mathbf{r}_{(j)}, t_j)\Delta t \tag{1.41}$$

and $t_j = j\Delta t$. The exact Green function is then given by the limit $\lim_{N\to\infty} G_N$ (if it exists). This can be shown using either a Lie-Trotter formula on the operator level, or by a direct calculation. One can in fact prove (see, e.g., Schulman's book [139], page 25) that the function Ψ defined by

$$\Psi(\mathbf{r}, t) = \int \left(\lim_{N\to\infty} G_N(\mathbf{r}, \mathbf{r}', t)\right)\Psi_0(\mathbf{r}')\, d^3r' \tag{1.42}$$

satisfies, for $\Delta t \to 0$, the estimate

$$\Psi(\mathbf{r}, t + \Delta t) = \Psi(\mathbf{r}, t) - i\frac{\Delta t}{\hbar} U \Psi(\mathbf{r}, t) + i\frac{\hbar \Delta t}{2m} \nabla_{\mathbf{r}}^2 \Psi(\mathbf{r}, t) + o(\Delta t)$$

from which follows, letting $\Delta t \to 0$, that

$$i\hbar \frac{\partial \Psi}{\partial t} = -\frac{\hbar^2}{2m} \nabla_{\mathbf{r}}^2 \Psi + U\Psi$$

which is just Schrödinger's equation.

1.7.2 The Metaplectic Group

We will see in Chapter 7 that short-time estimates for the action allow us to interpret Feynman's formula from the point of the metaplectic representation of the symplectic group, without invoking any bizarre "sum over paths" argument as is done in the usual physical literature. We will actually prove a refinement of Feynman's formula, whose convergence is much faster than that of the algorithm described in the previous subsection.

It turns out that even if quantum mechanics cannot be derived from classical mechanics, it *emerges* from Hamiltonian mechanics through a property of Hamiltonian flows. That there is a link between both theories is a priori really not very surprising, because Schrödinger's equation is obtained from a perfectly classical object, namely the Hamiltonian itself! I know that this is an unwelcome statement for most physicists, but quantum mechanics *is* deeply rooted in classical (Hamiltonian) mechanics. Let us explain why in a very sketchy form.

Assume first that the Hamiltonian is a time-independent function which is quadratic in both the position and momentum coordinates (typical examples are the free particle, the harmonic oscillator, or the electron in a uniform magnetic field in the symmetric gauge). The flow determined by Hamilton's equations for H consists of symplectic matrices and the knowledge of that flow totally determines the classical motion of a particle, so that we can as well "forget" the existence of the Hamiltonian function H: the primary mathematical object is now a family (f_t) of symplectic matrices. In quantum mechanics the situation is quite similar, because the wave function at time t is entirely determined by the knowledge of the "evolution operator", that is, of the unitary operators U_t which take the initial wave function $\Psi_0(\mathbf{r})$ to its value $\Psi(\mathbf{r}, t)$ at time t. Once (U_t) is known, we can as well forget the Schrödinger equation, because (U_t) contains all the information we need to propagate wave functions. We now ask whether it is possible, by simple "inspection" to recover (U_t) from (f_t), or vice-versa. The answer is "yes", because there is a simple *canonical* relation between (f_t) and (U_t). That relation is the following. The f_t are symplectic matrices, and thus belong to the *symplectic group* Sp(3). On the other hand, the "evolution operators" U_t belong to a group of unitary operators, the *metaplectic group* $Mp(3)$, which is generated by a class of "generalized Fourier transforms"

$$S_W\Psi(\mathbf{r}) = \left(\tfrac{1}{2\pi i\hbar}\right)^{3/2} \sqrt{\mathrm{Hess}(-W)} \int e^{\frac{i}{\hbar}W(\mathbf{r},\mathbf{r}')}\Psi(\mathbf{r}')\, d^3\mathbf{r}' \tag{1.43}$$

associated to all quadratic forms W for which Hess$(-W)$ (the determinant of the matrix of second derivatives of $-W$) is non-zero. Now, it turns out that $Mp(3)$ is a unitary representation of the *double cover* $\mathrm{Sp}_2(3)$ of Sp(3), and by classical property from the theory of covering groups (the "path lifting property") the one-parameter subgroup (f_t) of Sp(3) can be "lifted" to a *unique* one-parameter subgroup of any of its coverings, and thus in particular to $Mp(3)$. What is that lift? It is just the quantum evolution group (U_t)! This crucial property can be viewed, according to personal preferences, as a magic computational fact, or as an application of Feynman's formula, or as a consequence of a property of the Lie algebra of the symplectic group. Either way, we have here a canonical mathematical procedure for determining the quantum evolution group of a system, knowing its classical evolution group. Thus, up to an isomorphism, the Hamiltonian flow (f_t) and the quantum evolution group (U_t) are *identical*! (But this does of course not mean that they have the same *physical* interpretation.)

The discussion above only applies, strictly speaking, to quadratic Hamiltonians. For general Hamiltonians, in arbitrary dimensions, the f_t no longer are symplectic matrices. However their Jacobians matrices are (this is just another way of saying that the f_t are symplectomorphisms). Now, it is a well-known fact that the mathematical "tricks" leading to the construction of the metaplectic representation cannot be extended to construct a unitary representation of $Ham(n)$ containing the solutions of all Schrödinger equations: this is the celebrated "no-go" result of Groenewold-van Hove (see for instance Guillemin-Sternberg [80], or Folland [46]). We will however see in Chapter 6 that there is a way out of this difficulty.

1.8 Bohmian Mechanics

De Broglie viewed his matter waves as "pilot waves", which somehow governed the motion of particles. He however abandoned his pilot wave theory after strong criticism by Wolfgang Pauli (*b*.1900) at the 1927 Solvay conference, and only returned to it more than two decades later, after his ideas had been rediscovered by Bohm in 1952 (see Bohm [18, 19], Bell [9]).

1.8.1 Quantum Motion: The Bell-DGZ Theory

One of the most elementary ways to access Bohm's theory is the following: consider first a free particle with Hamiltonian $H = \mathbf{p}^2/2m$. The quantum velocity v^Ψ obtained from that particle's wave function Ψ should be both rotation invariant (this is a rather obvious requirement) and homogeneous of degree 0 in Ψ (because the velocity should remain the same if Ψ is replaced by $\lambda\Psi$, in conformity with the usual understanding that these wave functions are physically equivalent). This leads to postulate that the velocity is a function of $\nabla_{\mathbf{r}}\Psi/\Psi$:

$$\mathbf{v}^\Psi = f\left(\frac{\nabla_{\mathbf{r}}\Psi}{\Psi}\right) . \tag{1.44}$$

Now, the velocity is real, so that the simplest choice for (1.44) is either

$$\mathbf{v}^\Psi = K\,\mathrm{Re}\,\frac{\nabla_{\mathbf{r}}\Psi}{\Psi} \quad \text{or} \quad \mathbf{v}^\Psi = K\,\mathrm{Im}\,\frac{\nabla_{\mathbf{r}}\Psi}{\Psi}$$

where K is a real constant. The right choice is in fact the second alternative, because time-reversal must also reverse velocity; but changing t into $-t$ amounts replacing Ψ by its complex conjugate Ψ^* so that we have $\mathbf{v}^{\Psi^*} = -\mathbf{v}^\Psi$, and this is only possible if we choose $\mathbf{v}^\Psi = K\,\mathrm{Im}(\nabla_{\mathbf{r}}\Psi/\Psi)$. There remains to determine the constant K. For this purpose one can invoke Galilean invariance: $\mathbf{v}^\Psi$ should transform like a velocity under boosts $\mathbf{v} \longmapsto \mathbf{v} + \mathbf{v}_0$. Such a boost amounts to perform the transformation $\mathbf{p} \longmapsto \mathbf{p} + \mathbf{p}_0$ in the classical Hamiltonian H, and this has the effect of replacing the gauge $(\mathbf{0}, 0)$ in $H = \mathbf{p}^2/2m$ by the new gauge

$$(\nabla_{\mathbf{r}}\chi, -\partial\chi/\partial t) = (\mathbf{p}_0, 0)$$

so that $\chi = \mathbf{p}_0\mathbf{r}$. It follows from formula (1.29) that the wave function Ψ is then replaced by $\exp(i\mathbf{p}_0\mathbf{r}/\hbar)\Psi$, so that we must have

$$\mathbf{v}^\Psi + \mathbf{v}_0 = K\left(\frac{\mathbf{p}_0}{\hbar} + \mathrm{Im}\,\frac{\nabla_{\mathbf{r}}\Psi}{\Psi}\right)$$

which leads to the condition $K\mathbf{p}_0/\hbar = \mathbf{v}_0$, and hence $K = \hbar/m$. We thus obtain the formula

$$\mathbf{v}^\Psi = \frac{\hbar}{m}\,\mathrm{Im}\,\frac{\nabla_{\mathbf{r}}\Psi}{\Psi}. \tag{1.45}$$

One then posits, as the simplest possibility, that this formula holds even if potentials are present. This leads to the equation

$$\dot{\mathbf{r}}^{\Psi} = \frac{\hbar}{m} \operatorname{Im} \frac{\nabla_{\mathbf{r}} \Psi(\mathbf{r}^{\Psi}, t)}{\Psi(\mathbf{r}^{\Psi}, t)} \tag{1.46}$$

which allows (at least in principle) the determination of particle *trajectories* $t \longmapsto r^{\Psi}(t)$ as soon as an initial point r_0 and a wave function Ψ are prescribed. This equation is very different from the usual equations of motion of dynamics, which is of second order in position, and therefore requires the knowledge of both initial position, and initial velocity.

It turns out that formula (1.46) is totally consistent with Bohm's original theory, as we will see in a moment.

1.8.2 Bohm's Theory

Here is Bohm's original argument (see [18–20]). We again limit ourselves to the case of a single particle with mass m, but everything can be generalized in a straightforward way to many-particle systems. Writing the solutions of Schrödinger's equation

$$i\hbar \frac{\partial \Psi}{\partial t} = -\frac{\hbar^2}{2m} \nabla_{\mathbf{r}}^2 \Psi + U\Psi$$

in polar form

$$\Psi(\mathbf{r}, t) = R(\mathbf{r}, t) e^{\frac{i}{\hbar}\Phi(\mathbf{r}, t)}$$

we see, after some calculations, that the argument R and the phase Φ must satisfy the equations:

$$\begin{cases} R\left(\dfrac{\partial \Phi}{\partial t} + \dfrac{(\nabla_{\mathbf{r}} \Phi)^2}{2m} + U\right) - \dfrac{\hbar^2}{2m} \nabla_{\mathbf{r}}^2 R = 0 \\ \dfrac{\partial R^2}{\partial t} + \operatorname{div}\left(\dfrac{\nabla_{\mathbf{r}} \Phi}{m} R^2\right) = 0. \end{cases} \tag{1.47}$$

Setting $\rho = R^2$ $(\rho > 0)$ and $\mathbf{v} = \nabla_{\mathbf{r}} \Phi / m$, the second equation (1.47) can be rewritten as

$$\frac{\partial \rho}{\partial t} + \operatorname{div}(\rho v) = 0 \tag{1.48}$$

which we immediately recognize as the continuity equation describing the time-evolution of a probability density ρ under the flow arising from the vector field v. The interpretation of the first equation, that is Eq. (1.47), is at first sight less obvious. However, at the points $(\mathbf{r}, t)$ where R does not vanish (that is, outside the "nodes" of the wave function), it can be rewritten as

$$\frac{\partial \Phi}{\partial t} + \frac{(\nabla_{\mathbf{r}}\Phi)^2}{2m} + U - \frac{\hbar^2}{2m}\frac{\nabla_{\mathbf{r}}^2 R}{R} = 0. \tag{1.49}$$

Now, it does not require a huge amount of imagination to see that this equation looks like Hamilton-Jacobi's equation. It *is* in fact Hamilton-Jacobi's equation, not for the Hamiltonian function H, though, but rather for

$$H^{\Psi} = H + Q^{\Psi} \tag{1.50}$$

where we have set

$$Q^{\Psi} = -\frac{\hbar^2}{2m}\frac{\nabla_{\mathbf{r}}^2 R}{R}. \tag{1.51}$$

Bohm called that function Q^{Ψ} (which has the dimension of an energy) the *quantum potential* associated to Ψ. The quantum potential is actually very unlike a usual potential. First, it does not arise from any external source, and secondly it is intrinsically non-local. It is a "self-organizing" potential, in fact a response to the environment in which the quantum process takes place. We can interpret Eq. (1.49) as follows: supposing Ψ (and hence R) *fixed*, the phase Φ is thus the unique solution of the Cauchy problem

$$\begin{cases} \dfrac{\partial \Phi}{\partial t} + H^{\Psi}(\mathbf{r}, \nabla_{\mathbf{r}}\Phi, t) = 0 \\ \Phi(\mathbf{r}, 0) = \Phi_0(\mathbf{r}) \end{cases} \tag{1.52}$$

and nothing prevents us from applying Hamilton-Jacobi's theory to the Hamiltonian H^{Ψ}. The solutions of Hamilton's equations

$$\begin{cases} \dot{\mathbf{r}}^{\Psi} = \nabla_{\mathbf{p}} H^{\Psi}(\mathbf{r}^{\Psi}, \mathbf{p}^{\Psi}, t) \\ \dot{\mathbf{p}}^{\Psi} = -\nabla_{\mathbf{r}} H^{\Psi}(\mathbf{r}^{\Psi}, \mathbf{p}^{\Psi}, t) \end{cases} \tag{1.53}$$

with initial conditions $\mathbf{r}^{\Psi}(0) = \mathbf{r}_0$, $\mathbf{p}^{\Psi}(0) = \nabla_{\mathbf{r}}\Phi_0(\mathbf{r})$ are accordingly obtained by setting

$$\mathbf{p}^{\Psi} = \nabla_{\mathbf{r}}\Phi(\mathbf{r}, t) \tag{1.54}$$

and then integrating the equation

$$\dot{\mathbf{r}}^{\Psi} = \nabla_{\mathbf{p}} H^{\Psi}(\mathbf{r}^{\Psi}, \nabla_{\mathbf{r}}\Phi(\mathbf{r}^{\Psi}, t), t).$$

Expressing $\nabla_{\mathbf{r}}\Phi$ in terms of Ψ, Eq. (1.54) can be written

$$\mathbf{p}^{\Psi} = \hbar \operatorname{Im} \frac{\nabla_{\mathbf{r}}\Psi}{\Psi}$$

which is precisely formula (1.45).

Note that we are not free to choose arbitrary initial conditions $(\mathbf{r}_0, \mathbf{p}_0)$ for the Hamilton equations (1.53), because the momenta are constrained by the condition (1.54), but this is exactly what happens in the usual Hamilton-Jacobi theory. If we want different "Bohmian trajectories", we need another wave function Ψ.

In general the Bohmian trajectories are very different from those predicted by classical mechanics. The discrepancy between the classical trajectories and those predicted by Bohmian mechanics is particularly blatant in text-book experiments such as the diffraction of electrons by a crystal (the already mentioned Davisson-Germer experiment), or the abundantly commented and discussed two-slit experiment. See Holland's treatise [90] for explicit calculations of various Bohmian trajectories; the paper by Philippidis et al. [130] is a classical; it contains beautiful graphical representations of the quantum potential.

1.9 Interpretations

> *The function of an expert is not to be more right than other people, but to be wrong for more sophisticated reasons* (David Butler)

Schrödinger's equation is somewhat unique from an epistemological point of view: it is, as far as I know, the only partial differential equation whose solutions have led to so many epistemological, ontological, and philosophical debates. As H. Montgomery notes[3], it has to be conceded that several interpretations of quantum mechanics now exist, and that their relative merits are controversial.

1.9.1 Epistemology or Ontology?

Let us begin with a (superficial) comparison between the mathematical formalisms of classical and quantum mechanics. In classical mechanics the

[3] H. Montgomery, in *Quantum Concepts, past and present*, IOP Newsletter, Spring 2001, No. 14.

fundamental mathematical object associated to a physical system is Hamilton's function H. However, the datum of H alone does not give us very much information about that system (not even about the energy, which is anyway never unambiguously defined, since it is badly gauge-dependent, and can be conserved in some gauges, and vary with time in other!). What really interests the physicist is the evolution of the system, and this evolution is obtained from H by solving Hamilton's equations of motion. The situation is similar in quantum mechanics: to every Hamiltonian H one associates the linear space $\mathcal{H}$ of all solutions Ψ of the corresponding Schrödinger equation

$$i\hbar\frac{\partial\Psi}{\partial t} = \widehat{H}\,\Psi.$$

However, no more than the function H alone describes the effective motion of the classical system, does the space $\mathcal{H}$ yield a complete description of the quantum system. Exactly as the function H must be complemented by Hamilton's equations if one wants classical mechanics to become effective, the space $\mathcal{H}$ has to be complemented by some rule allowing us to calculate relevant physical quantities.

1.9.2 The Copenhagen Interpretation

The "Copenhagen interpretation" is due to Niels Bohr and his school (it is also called the "standard interpretation of quantum mechanics"). It can be regarded as giving the wave function a role in the behavior of certain macroscopic objects, in particular the measurement instruments: following Bohr, the wave function Ψ *is* the description of the physical system, and it does not make sense to talk about particles as such. According to that interpretation, what one calculates (and really needs in the daily practice of quantum mechanics) are *energy levels* and *transition probabilities*, which are directly obtained from Ψ via Schrödinger's equation, without appealing to other "hidden" variables. The usually accepted system of "axioms" for quantum mechanics are, according to the Copenhagen interpretation:

Axiom 1 *The state of a quantum mechanical system is completely specified by the datum of the wave function Ψ, and $|\Psi|^2$, when normalized, is the probability density for finding the position of the system.*

Axiom 2 *To every observable A in classical mechanics there corresponds a unique Hermitian operator $\hat{A}$, and this operator is obtained from A by Schrödinger's quantization rule.*

Axiom 3 *In any measurement of an observable associated with the operator $\hat{A}$, the only values that will be observed in this measurement are the eigenvalues a_i of that Hermitian operator.*

Axiom 4 *If one can expand a normalized state Ψ in a Fourier series $\sum_j c_j \Psi_j$ where the Ψ_j are a complete set of normalized eigenvectors of $\hat{A}$, then the probability of finding the eigenvalue a_j after a measurement is $|c_j|^2$.*

Axiom 5 *If a system is in a state described by a normalized wave function Ψ, then the average value of the observable A is given by*

$$\langle A \rangle = \int \Psi^* \hat{A} \Psi d^n x.$$

Pushed to its extreme, the Copenhagen interpretation leads to a "shut up and calculate" attitude; this is of course harmless if one is merely concerned with applications, but it leads to severe epistemological problems. As Bohm and Hiley note in their book [20], the Copenhagen interpretation gives an algorithm for computing probabilities of experimental results, but it gives no account of the individual quantum processes. To put it in more philosophical terms, it may be said that quantum mechanics is primarily directed towards *epistemology*. It follows from this that quantum mechanics can say little or nothing about reality itself: it does not give an *ontology* for quantum systems. That is, it seems that quantum mechanics is only concerned with our *knowledge* of reality, and especially how to predict and control the behavior of this reality; it is mainly a *statistical* knowledge. The Copenhagen interpretation was never fully accepted by Einstein. (I recommend the book [127] by A. Pais which contains an exciting historical account, not only about the dispute between Bohr and Einstein, but also about physics in general during that period.) Anyhow, there certainly are difficulties with that interpretation. As R. Omnès [126] notes,

> "... *it is remarkable that so long after the discovery of quantum theory the most complete books devoted to the Copenhagen interpretation are all reprints of original articles or learned commentaries, becoming more and more commentaries upon commentaries as time goes on. In fact, these texts are devoted to an endless discussion of the difficulties of the "measurement problem" and of the difficulties facing interpretation, and philosophy of Science becomes more important than physics itself. Such an attitude has never been seen before, or elsewhere, in physics.*"

1.9.3 The Bohmian Interpretation

> *"If we cannot disprove Bohm, then we must agree to ignore him"* (J.R. Oppenheimer)

An appealing alternative to the purely epistemological Copenhagen interpretation, there is the "ontological" approach initiated by Bohm, and developed in Bohm and Hiley [20] (also see Holland [90], the compilation [9] of John Bell's articles). According to this approach, the wave function Ψ does not provide *per se* a complete description of a quantum system, and views Ψ merely as a mathematical device from which the behavior of more fundamental quantities (for instance, positions) can be extracted. These more fundamental quantities define what S. Goldstein calls the *primitive ontology* of quantum mechanics. Although this interpretation has been –and still is– fiercely opposed by most physicists for various epistemological, philosophical, metaphysical or personal reasons, it can certainly not be dismissed, and steadily gains in popularity, especially in its "shadow phase space" form due to Hiley (see Brown and Hiley [26]). In its original form, "Bohmian mechanics" –or: "quantum theory of motion", as it is also called in [90]– attributed absolute reality to *particles* following definite phase space *trajectories.* In fact, Bohm himself long abandoned that position; in Bohm and Hiley's treatise [17] it is made clear that the particle/trajectory model is too simplistic to be viable. Particularly interesting and fruitful in that respect is the "shadow phase space" approach initiated by Hiley [26] and already mentioned above. It is a reflection of the fact that we cannot construct a global chart for the metaplectic group, when it is viewed as a Lie group, that is, as a manifold equipped with a continuous algebraic structure. It is for that reason we cannot construct simultaneous "position" and "momentum" representations of the quantum mechanical reality. (The usual denial of a quantum mechanical phase space comes from Heisenberg's inequalities, or which amounts to the same, to the non-commutativity of operators, but these are in fact simply manifestations of the manifold structure of $Mp(n)$.) In that sense, the approach of Bohm and Hiley [17, 86, 26] is different in spirit and method from other more traditional approaches where the x-representation is viewed as intrinsic, and take the equation of motion (1.46) as *the* basic equation, the Bohm–Hiley approach does not favor *a priori* any representation. It is in that sense much more in spirit with the usual phase space approach of Hamiltonian mechanics, and allows — and this is an essential advantage — the use of symplectic methods.

1.9.4 The Platonic Point of View

Quantum mechanics, as a physical theory, is plagued by problems of interpretation. Since the physical predictions of Bohmian mechanics are the same as those of traditional quantum mechanics, it seems unlikely that there will be in a foreseeable future an *experiment crucis* showing the existence –or non existence– of Bohm's quantum trajectories. The Bohmian interpretation is often misunderstood and misrepresented in the literature: it does not stand, as is often claimed, diametrically opposite to Bohr's views. In fact, it actually shares some of its conclusions, except Bohr's belief that basic to quantum theory is the impossibility of making a sharp distinction between the observed system and the means of observation.

But, as long as its purely mathematical aspects are concerned, there is no need to enter the fierce debate opposing *pros* and *contra* of the Copenhagen interpretation. Mathematics is about thought, not material reality[4]: it is a *language without "real" semantic*. Bohmian mechanics is therefore an example of "*no case*" among mathematical circles where it has been accepted, without controversy and without provoking any emotional reactions, as just another formulation of quantum mechanics.

[4]As was observed by H. Grassmann in 1884 (see *A New Branch of mathematics: the Ausdehnungslehre of 1884, and other works*, trans. by L.C. Kannenberg, Open Court (1995)).

Chapter 2

Newtonian Mechanics

Summary. *A basic physical postulate, the Maxwell principle, implies that Newton's Second Law can be expressed in terms of the Poincaré-Cartan differential form. This leads to a Galilean covariant Hamiltonian mechanics.*

In this second Chapter we propose a rigorous formalization of Newtonian mechanics, which leads to its Hamiltonian formulation once a physical postulate (the "Maxwell principle") is imposed. While this approach goes historically back to the pioneering work of both Hamilton [82] and Lagrange [99], we will follow (with some minor modifications) Souriau's presentation in [148], which originates in previous work of Gallissot [49].

The reader who wants to access directly Hamiltonian mechanics can skip the first section of this Chapter, and proceed directly to Section 2.2.

2.1 Maxwell's Principle and the Lagrange Form

We begin by expressing Newton's second law in terms of vector fields, and briefly discussing the form of the fundamental force fields of mechanics. We then state the Maxwell principle, and justify it on a few examples. We finally express Maxwell's principle in terms of a differential form which was introduced by Lagrange (with different notations!) in his study of celestial mechanics *Mémoire de la première Classe de l'Institut pour 1808.* Lagrange's original form actually corresponds to the case of a scalar potential (the equations of electromagnetism were written down by Maxwell only in 1876), while we consider here the more general case of an arbitrary gauge. We refer to the treatises of Libermann and Marle [104], von Westenholz [164], (especially Chapter 7, §4), and to Gallissot's article [49] for detailed discussions of the relation between the theory of differential forms and physics.

2.1.1 The Hamilton Vector Field

Newton's second law expressed in the differential form

$$\dot{\mathbf{r}} = \mathbf{v} \quad , \quad \dot{\mathbf{p}} = \mathbf{F} \tag{2.1}$$

defines trajectories $t \mapsto (\mathbf{r}(t), \mathbf{v}(t), t)$, which are curves in the *extended state space* $\mathbb{R}^3_{\mathbf{r}} \times \mathbb{R}^3_{\mathbf{v}} \times \mathbb{R}_t$. These trajectories are the integral curves of the "Newton vector field"

$$X_N(\mathbf{r}, \mathbf{v}, t) = (\mathbf{v}, \mathbf{F}/m, 1) \tag{2.2}$$

since the equations (2.1) can be written in the form

$$\frac{d}{dt}(\mathbf{r}, \mathbf{v}, t) = (\mathbf{v}, \mathbf{F}/m, 1)\,. \tag{2.3}$$

One can also represent the trajectories in the *extended phase space* $\mathbb{R}^3_{\mathbf{r}} \times \mathbb{R}^3_{\mathbf{p}} \times \mathbb{R}_t$ as *curves* $t \mapsto (\mathbf{r}(t), \mathbf{p}(t), t)$ by just multiplying the velocity vector $\mathbf{v}$ by m; these trajectories are the integral curves of the *suspended Hamilton vector field*

$$\tilde{X}_H(\mathbf{r}, \mathbf{p}, t) = (\mathbf{p}/m, \mathbf{F}, 1)\,. \tag{2.4}$$

(Bourbakists would, no doubt, frown with disdain to such a cavalier way of passing from state space to phase space without using the Legendre transform!)

2.1.2 Force Fields

A force field can a priori depend, besides position and time, on many other quantities, e.g., velocity, acceleration, temperature, etc. We will however only consider here force fields $\mathbf{F}$ depending on the state space variables $(\mathbf{r}, \mathbf{v}, t)$, and such that the dependence on $\mathbf{v}$ has the special form

$$\mathbf{F}(\mathbf{r}, \mathbf{v}, t) = \mathbf{E}(\mathbf{r}, t) + (\mathbf{v} \times \mathbf{B}(\mathbf{r}, t)) \tag{2.5}$$

where $\mathbf{E}$ and $\mathbf{B}$ are some new fields only depending on positions and time. The datum of the force field $\mathbf{F}$ determines unambiguously both fields $\mathbf{E}$ and $\mathbf{B}$: we can find $\mathbf{E}$ by measuring $\mathbf{F}$ on a particle at rest, and $\mathbf{B}$ by measuring simultaneously, and at the same point, the forces on an identical particle moving with velocity $\mathbf{v}$. We also notice that while $\mathbf{F}$ and $\mathbf{E}$ are intrinsically defined, $\mathbf{B}$ is a "pseudo vector" depending on the choice of orientation of $\mathbb{R}^3_{\mathbf{r}}$. (See Frankel [48] for a thorough discussion of this notion.)

Our restriction to force fields of the type (2.5) a priori eliminates the consideration of physical systems with friction, because friction in general depends on velocity in a way that cannot be put in the form (2.5). (More precisely, we are exclusively dealing with "non-dissipative systems".) It is, however, an accepted postulate that all physical laws, if analyzed with sufficient precision (taking account of the thermal exchanges due to friction, etc.), can be derived from force fields of the type (2.5) above: the usual formulas for friction appear in this perspective as approximations to more fundamental formulas, were one takes all the possible interactions at the microscopic level into account.

Example 6 ***Charged particle in an electromagnetic field.*** *Consider a particle with charge e placed in an electromagnetic field. The force* $\mathbf{F}$ *exerted on that charge by the field is called the "Lorentz force"; its value is given by formula (2.5) if units are chosen so that* $e/c = 1$. *(That formula was actually written down by Heaviside in 1889.) The Lorentz force depends on the velocity; if there is no electric field present, then a particle at rest will remain at rest (this is restated in Physics by saying that "magnetic forces do not work".)*

One should however not conclude that the existence of force fields of the type described by Eq. (2.5) is a feature of electromagnetism alone. That similar phenomena occur in perhaps more unexpected "everyday" situation is illustrated in the example below, which we will revisit several times in this Chapter:

Example 7 ***The Coriolis force.*** *Let us denote by* $\mathbf{R}$ *the rotation vector at a point O situated on the surface of the Earth:* $\mathbf{R}$ *is a vector which is parallel to the axis of rotation and pointing out of the ground at O in the Northern hemisphere, and into the ground in the Southern hemisphere. Its length is the angular velocity calculated in a heliocentric reference frame. At a point at latitude* ϕ, *this vector is*

$$\mathbf{R} = (0,\ R\cos\phi,\ R\sin\phi)$$

($R = |\mathbf{R}|$*) where the coordinates are calculated in a "lab frame" with origin O on the Earth; the z-axis is the vertical, the y-axis points in the direction of the North, and the x-axis towards the East. Suppose now that an observer at O wants to calculate the total force exerted on a nearby point with mass m and velocity v. He finds that the force is*

$$\mathbf{F} = m\mathbf{g} - \mathbf{F}_C \quad \textit{with} \quad \mathbf{F}_C = 2m\,(\mathbf{R} \times \mathbf{v}) \tag{2.6}$$

and Eq. (2.5) thus holds with

$$\mathbf{E} = mg \quad , \quad \mathbf{B} = 2m\,\mathbf{R}. \tag{2.7}$$

The velocity-dependent term $\mathbf{F}_C$ *is called the "Coriolis force" in honor of G. de Coriolis (b.1792). In the Northern hemisphere the Coriolis force deflects every body moving along the Earth to the right, and every falling body eastward.*

We mention that a perfect illustration of the Coriolis force is the experiment of the Foucault pendulum. That experiment, performed by Léon de Foucault (*b.*1819) at the Pantheon church in Paris, was the first experimental evidence of the Earth's rotation. The Foucault pendulum has been reconstituted at the Pantheon, where we are invited to "come and watch the Earth turn".

2.1.3 Statement of Maxwell's Principle

In both examples above, the fields **E** and **B** satisfy the equations

$$\frac{\partial \mathbf{B}}{\partial t} + \nabla_{\mathbf{r}} \times \mathbf{E} = 0 \quad \text{and} \quad \operatorname{div} \mathbf{B} = 0.$$

When **B** and **E** are magnetic and electric fields, respectively, the first of these equations express Faraday's law of induction, and the second the absence of magnetic monopoles.

These equations motivate the following definition:

Definition 8 *A force field* $\mathbf{F} = \mathbf{E} + (\mathbf{v} \times \mathbf{B})$ *such that*

$$\frac{\partial \mathbf{B}}{\partial t} + \nabla_{\mathbf{r}} \times \mathbf{E} = 0 \quad \text{and} \quad \operatorname{div} \mathbf{B} = 0 \tag{2.8}$$

is said to obey the Maxwell principle. *Any functions* **A** *and* U *satisfying the equations*

$$\mathbf{B} = \nabla_{\mathbf{r}} \times \mathbf{A} \quad , \quad \mathbf{E} = -\nabla_{\mathbf{r}} U - \frac{\partial \mathbf{A}}{\partial t} \tag{2.9}$$

are called, respectively, vector and scalar potentials. The datum of a pair $(\mathbf{A}, U)$ *satisfying Eq. (2.9) is called a* gauge.

When the fields **B** and **E** are everywhere defined in space, the existence of **A** is immediately follows from the equation div **B** = 0; insertion of the expression $\mathbf{B} = \nabla_{\mathbf{r}} \times \mathbf{A}$ in the first equation (2.8) yields

$$\nabla_{\mathbf{r}} \times \left(\frac{\partial \mathbf{A}}{\partial t} + \mathbf{E} \right) = 0$$

so that there must exist a scalar function U such that

$$\frac{\partial \mathbf{A}}{\partial t} + \mathbf{E} = -\nabla_{\mathbf{r}} U. \tag{2.10}$$

The expression of the force field in terms of the gauge is thus:

$$\mathbf{F} = -\nabla_{\mathbf{r}} U - \frac{\partial \mathbf{A}}{\partial t} - (\nabla_{\mathbf{r}} \times \mathbf{A}) \times \mathbf{v}. \tag{2.11}$$

Of course the argument above has nothing to do with the physical interpretation of the fields **E** and **B**; it is in particular not limited to electromagnetic fields. Let us verify this on the Coriolis force:

Example 9 ***Potentials for the Coriolis Force.*** *If* **F** *is the Coriolis force (2.6), we have* $\mathbf{E} = m\mathbf{g}$ *and* $\mathbf{B} = 2m\mathbf{R}$. *A straightforward calculation then shows that*

$$\mathbf{A} = m\,(\mathbf{R} \times \mathbf{r}) \quad , \quad U = -m\mathbf{g} \cdot \mathbf{r} \tag{2.12}$$

are vector and scalar potentials.

The potentials **A** and U are not uniquely determined by the equations (2.9): for any function $\chi = \chi(\mathbf{r}, t)$, these equations are also satisfied by

$$\mathbf{A}' = \mathbf{A} + \nabla_{\mathbf{r}} \chi \ \text{ and } \ U' = U - \frac{\partial \chi}{\partial t}. \tag{2.13}$$

We will call the mappings

$$(\mathbf{A}, U) \longrightarrow (\mathbf{A} + \nabla_{\mathbf{r}} \chi, U - \partial \chi / \partial t) \tag{2.14}$$

gauge transformations. It was believed until rather recently that gauge transformations and vector potentials were unphysical quantities. This was however a misbelief, because there are experiments producing evidence of "geometric phase shifts". In quantum mechanics, the Aharonov-Bohm effect or the occurrence of Berry's "geometric phase shift" are well-known examples (see for instance Berry [11]), but there are also measurable phase shifts in classical mechanics (e.g., the "Hannay angles" (see Hannay [84])).

One should be very careful when one tries to apply the construction of a gauge outlined above when the Lagrange form is only defined on a subset of the state space. A first (educated!) guess would be that the whole construction carries over when the domain of definition is simply connected, but this guess is wrong, as illustrated by the classical example of the magnetic monopole:

2.1.4 Magnetic Monopoles and the Dirac String

We are following here Naber [122], §0.1, and Nakahara [123], Ch.1. Let us begin by considering a point-like electric charge q placed at the origin of an inertial frame. That charge determines the electromagnetic field

$$\mathbf{E} = q\frac{\mathbf{r}}{r^3} \quad , \quad \mathbf{B} = 0$$

where we have set $r = |\mathbf{r}|$. Since $\nabla_{\mathbf{r}} \times \mathbf{E} = 0$, it follows that equations (2.8):

$$\frac{\partial \mathbf{B}}{\partial t} + \nabla_{\mathbf{r}} \times \mathbf{E} = 0 \quad \text{and} \quad \operatorname{div} \mathbf{B} = 0$$

are satisfied, and thus the force field $\mathbf{F} = \mathbf{E} + (\mathbf{v} \times \mathbf{B}) = \mathbf{E}$ satisfies the Maxwell principle. We can moreover easily determine a gauge: just take

$$\mathbf{A} = 0 \quad \text{and} \quad U = -\frac{q}{r}.$$

Although the magnetic analogue of the charged particle has never been (so far) observed, its existence was already discussed by Dirac in [37, 38]. Dirac observed that the condition $\operatorname{div} \mathbf{B} = 0$ introduced an asymmetry in Maxwell's equations. Suppose, in fact, that such a "magnetic monopole" exists, and is located at the origin of an inertial frame. The electromagnetic field it determines is given by

$$\mathbf{E} = 0 \quad , \quad \mathbf{B} = g\frac{\mathbf{r}}{r^3} \tag{2.15}$$

in this frame; here g is the "magnetic charge" determining the strength of the field. Notice that we have

$$\mathbf{B} = 4\pi g \delta(\mathbf{r})$$

where δ is the Dirac function centered at the origin. Both equations (2.8) are again satisfied by the fields (2.15), so the Maxwell principle applies again.

However, there is no vector field $\mathbf{A}$ such that $\mathbf{B} = \nabla \times \mathbf{A}$. To see this, we first note that the magnetic flux through the unit sphere S^2 is

$$\Phi = \iint_{S^2} \mathbf{B} \cdot \mathbf{n} dS = g \iint_{S^2} \frac{\mathbf{r}}{r^3} \cdot \mathbf{n} dS = 4\pi g \tag{2.16}$$

where $\mathbf{n}$ is the (outwards oriented) unit normal vector. Suppose now that there exists $\mathbf{A}$ such $\mathbf{B} = \nabla_{\mathbf{r}} \times \mathbf{A}$; in view of Stoke's theorem we would have, denoting by S^2_+ (resp. S^2_-) the Northern (resp. Southern) hemisphere, and C^+ (resp. C^-) the positively (resp. negatively) oriented equator of the sphere:

$$\begin{aligned}
\Phi &= \iint_{S^2} (\nabla_{\mathbf{r}} \times \mathbf{A}) \cdot \mathbf{n} dS \\
&= \iint_{S^2_+} (\nabla_{\mathbf{r}} \times \mathbf{A}) \cdot \mathbf{n} dS + \iint_{S^2_-} (\nabla_{\mathbf{r}} \times \mathbf{A}) \cdot \mathbf{n} dS \\
&= \oint_{C^+} \mathbf{A} \cdot d\mathbf{r} + \oint_{C^-} \mathbf{A} \cdot d\mathbf{r} \\
&= 0.
\end{aligned}$$

But this contradicts the equality (2.16), and there is thus no $\mathbf{A}$ such that $\mathbf{B} = \nabla_{\mathbf{r}} \times \mathbf{A}$.

The obstruction for finding a vector potential $\mathbf{A}$ in this example comes from the fact that although the domain $\mathcal{D} = \mathbb{R}^3_{\mathbf{r}} \setminus 0$ on which the field is defined is simply connected, the existence of $\mathbf{A}$ requires that not only $\pi_1(\mathcal{D}) = 0$, but also that *second* homotopy group $\pi_2(D)$ vanishes. This is not the case here: we have $\pi_2(\mathcal{D}) \neq 0$ since a sphere in $\mathbb{R}^3_{\mathbf{r}} \setminus 0$ cannot be shrunk to a point. There is, however, a way out of this difficulty; it consists of using a *Dirac string*. A Dirac string is a continuous curve Γ in $\mathbb{R}^3_{\mathbf{r}}$, starting from the origin, which never intersects itself, and that goes to infinity in some (arbitrary) direction. Since

$$\pi_1(\mathbb{R}^3_{\mathbf{r}} \setminus \Gamma) = \pi_2(\mathbb{R}^3_{\mathbf{r}} \setminus \Gamma) = 0$$

(the complement of Γ is simply connected, and no sphere in $\mathbb{R}^3_{\mathbf{r}}$ can enclose points of Γ since Γ proceeds away to infinity), we can always find a vector potential for the magnetic monopole field outside the Dirac string Γ.

Example 10 ***Two choices of Dirac strings.*** *(1)* The negative half-axis. *If we let Γ be the axis $z < 0$ then the vector potential $\mathbf{A}^-$ with components*

$$A^-_x = \frac{-gy}{r(r+z)} \quad , \quad A^-_y = \frac{gx}{r(r+z)} \quad , \quad A^-_z = 0$$

satisfies

$$\nabla_{\mathbf{r}} \times \mathbf{A}^{-} = g\frac{\mathbf{r}}{r^3} + 4\pi g\delta(x)\delta(y)H(-z)$$

so that we have $\nabla_{\mathbf{r}} \times \mathbf{A}^{-} = \mathbf{B}$ *except along the negative half-axis* $z < 0$ *(here* H *is Heaviside's step function). (2)* The positive half-axis. *The vector potential* $\mathbf{A}^{+}$ *with components*

$$A_x^{+} = \frac{gy}{r(r-z)} \quad , \quad A_y^{+} = \frac{-gx}{r(r-z)} \quad , \quad A_z^{+} = 0$$

satisfies

$$\nabla_{\mathbf{r}} \times \mathbf{A}^{+} = g\frac{\mathbf{r}}{r^3} + 4\pi g\delta(x)\delta(y)H(z)$$

and thus $\nabla_{\mathbf{r}} \times \mathbf{A}^{+} = \mathbf{B}$ *except along* $z > 0$. *Notice that using polar coordinates* $(\mathbf{r}, \theta, \phi)$ *the vector potentials* $\mathbf{A}^{\pm}$ *can be expressed as*

$$\mathbf{A}^{\pm} = g\frac{1 \pm \cos\theta}{r\sin\theta}\mathbf{e}_{\phi}$$

where $\mathbf{e}_{\phi} = -\sin\phi\mathbf{e}_x + \cos\phi\mathbf{e}_y$.

2.1.5 The Lagrange Form

We begin by giving a useful relation between differential forms and the "Hodge star operator".

Let $\mathbf{f} = (f_x, f_y, f_z)$ be a vector-valued function (the subscripts x, y, z are not indicating partial derivatives, but are just labeling the components); then

$$d\mathbf{f} \wedge d\mathbf{r} = (\nabla_{\mathbf{r}} \times \mathbf{f}) \cdot (*d\mathbf{r}) \tag{2.17}$$

where the wedge-product $d\mathbf{f} \wedge d\mathbf{r}$ is calculated component-wise:

$$d\mathbf{f} \wedge d\mathbf{r} = df_x \wedge dx + df_y \wedge dy + df_z \wedge dz \tag{2.18}$$

and the star "$*$" is the Hodge operator, defined by

$$*d\mathbf{r} = (dy \wedge dz, dz \wedge dx, dx \wedge dy).$$

Formula (2.17) readily follows from the definition of the exterior product: since

$$\nabla_{\mathbf{r}} \times \mathbf{f} = \left(\frac{\partial f_z}{\partial y} - \frac{\partial f_y}{\partial z}, \frac{\partial f_x}{\partial z} - \frac{\partial f_z}{\partial x}, \frac{\partial f_y}{\partial x} - \frac{\partial f_z}{\partial y}\right)$$

we have

$$(\nabla_{\mathbf{r}} \times \mathbf{f}) \cdot (*d\mathbf{r}) = \left(\frac{\partial f_z}{\partial y} - \frac{\partial f_y}{\partial z}\right)(dy \wedge dz) + \left(\frac{\partial f_x}{\partial z} - \frac{\partial f_z}{\partial x}\right) dz \wedge dx + \left(\frac{\partial f_y}{\partial x} - \frac{\partial f_z}{\partial y}\right) dx \wedge dy.$$

On the other hand,

$$\begin{aligned} df_x \wedge dx &= \left(\frac{\partial f_x}{\partial x}dx + \frac{\partial f_x}{\partial y}dy + \frac{\partial f_x}{\partial z}dz\right) \wedge dx \\ &= \frac{\partial f_x}{\partial y} dy \wedge dx + \frac{\partial f_x}{\partial z} dz \wedge dx \\ &= -\frac{\partial f_x}{\partial y} dx \wedge dy + \frac{\partial f_x}{\partial z} dz \wedge dx \end{aligned}$$

and, similarly:

$$\begin{aligned} df_y \wedge dy &= -\frac{\partial f_y}{\partial z} dy \wedge dz + \frac{\partial f_y}{\partial x} dx \wedge dy \\ df_z \wedge dz &= -\frac{\partial f_z}{\partial x} dz \wedge dx + \frac{\partial f_z}{\partial y} dy \wedge dz. \end{aligned}$$

Formula (2.17) follows.

Let us now return to our business, which is to express Maxwell's principle using the language of differential forms.

Definition 11 *The* Lagrange form *is the differential 2-form*

$$\Omega_{\mathbf{B}} = (md\mathbf{v} - \mathbf{E}dt) \wedge (d\mathbf{r} - \mathbf{v}dt) + \mathbf{B}(*d\mathbf{r}) \tag{2.19}$$

where the wedge products of vector-valued differential forms are, as usual, calculated componentwise (cf. Eq. (2.18)).

We thus have, for any pair of vectors $u = (\mathbf{r}, \mathbf{v}, t)$, $u' = (\mathbf{r}', \mathbf{v}', t')$:

$$\Omega_{\mathbf{B}}(u, u') = (m\mathbf{v} - \mathbf{E}t) \cdot (\mathbf{r}' - \mathbf{v}t') - (m\mathbf{v}' - \mathbf{E}t') \cdot (\mathbf{r} - \mathbf{v}t) + \mathbf{B}(\mathbf{r} \times \mathbf{r}').$$

We have the following very important result which relates both Maxwell's principle and Newton's second law to the Lagrange form:

Theorem 12 *(1) We have the equivalences*

$$\text{[Maxwell's principle]} \Longleftrightarrow d\,\Omega_{\mathbf{B}} = 0 \tag{2.20}$$

$$\text{[Newton's } 2^{nd} \text{ law]} \Longleftrightarrow i_N\,\Omega_{\mathbf{B}} = 0 \tag{2.21}$$

and hence: (2) If the Maxwell principle is satisfied, then the Lie derivative of the Lagrange form in the direction N is zero:

$$\mathcal{L}_N\,\Omega_{\mathbf{B}} = 0\ . \tag{2.22}$$

Proof. **(1)** A straightforward calculation yields

$$d\,\Omega_{\mathbf{B}} = \left(d\mathbf{E} \wedge d\mathbf{r} + \frac{\partial \mathbf{B}}{\partial t} \cdot (*d\mathbf{r})\right) \wedge dt + (\operatorname{div} \mathbf{B})\, dx \wedge dy \wedge dz.$$

Applying formula (2.17) to $f = \mathbf{E}$, this can be rewritten

$$d\,\Omega_{\mathbf{B}} = \left(\nabla_{\mathbf{r}} \times \mathbf{E} + \frac{\partial \mathbf{B}}{\partial t}\right) \cdot (*d\mathbf{r}) \wedge dt + (\operatorname{div} \mathbf{B})\ dx \wedge dy \wedge dz$$

and hence

$$d\,\Omega_{\mathbf{B}} = 0 \Longleftrightarrow \begin{cases} \dfrac{\partial \mathbf{B}}{\partial t} + \nabla_{\mathbf{r}} \times \mathbf{E} = 0 \\ \operatorname{div} \mathbf{B} = \mathbf{0}\ . \end{cases}$$

The equivalence (2.20) follows by definition of the Maxwell principle. Let us next prove the equivalence (2.21). We assume, for notational simplicity, that $\mathbf{B} = 0$. We want thus to show that

$$\text{Newton's } 2^{\text{nd}} \text{ law} \iff i_N \Omega_{\mathbf{0}} = 0$$

where $\Omega_{\mathbf{0}}$ is the differential form defined by the wedge product:

$$\Omega_0 = (md\mathbf{v} - \mathbf{F}dt) \wedge (d\mathbf{r} - \mathbf{v}dt)\,. \tag{2.23}$$

By definition of the contraction operation, we have

$$\left(i_N \Omega_0\right)_u (u') = \Omega_0(\mathbf{v}, \mathbf{F}/m, 1; \mathbf{r}', \mathbf{v}', t')$$

that is

$$\left(i_N \Omega_0\right)_u (u') = (m(\mathbf{F}/m) - \mathbf{F}) \cdot (\mathbf{r}' - \mathbf{v}'t') - (m\mathbf{v}' - \mathbf{F}t') \cdot (\mathbf{v} - \mathbf{v})$$

and this is zero for all vectors $u' = (\mathbf{r}', \mathbf{v}', t')$; the direct implication " $\Longrightarrow$ " follows. The inverse implication " $\Longleftarrow$ " is obvious. **(2)** To prove (2.22) it suffices to use Cartan's homotopy formula

$$\mathcal{L}_N \Omega_{\mathbf{B}} = i_N \, d\Omega_{\mathbf{B}} + d i_N \, \Omega_{\mathbf{B}}$$

relating the Lie derivative, the contraction operator, and the exterior derivative: since $d\Omega_{\mathbf{B}} = 0$ and $i_N \Omega_{\mathbf{B}} = 0$ we have $\mathcal{L}_N \Omega_{\mathbf{B}} = 0$. ∎

Theorem 12 above expresses Newton's second law for systems obeying the Maxwell principle in the very concise form of the two equations (2.20) and (2.21). As we will see, these two "laws" contain, in disguise, both Hamilton's equations and Galilean invariance.

2.1.6 N-Particle Systems

Everything we have said in the previous sections extends without difficulty to the case of systems consisting of many particles. Let us begin by introducing some notations.

Suppose we are dealing with a system consisting of N distinguishable point-like particles with masses $m_1, ..., m_N$. A reference frame being chosen, we label by x_1, x_2, x_3 the coordinates of the first particle, by x_4, x_5, x_6 the coordinates of the second particle, and so on. There are thus $3N$ numbers $x_1, ..., x_{3N}$ describing the positions of the particles at some time t. Setting $n = 3N$, we call the vector $x = (x_1, ..., x_n)$ the *position vector* of the system; x is an element of the *configuration space* $\mathbb{R}^n_x$. Notice that the configuration space only coincides with "physical space" when $n = 3$. We will denote by $\mathbf{r}_1 = (x_1, x_2, x_3)$ the coordinate vector of the first particle, by $\mathbf{r}_2 = (x_4, x_5, x_6)$ the coordinate vector of the second particle, and so on. The velocity and momentum vectors v and p are defined quite similarly: $v = (v_1, ..., v_n)$, where v_1, v_2, v_3 are the components of the velocity vector of the first particle, etc. The total momentum

$$p = m_1 \mathbf{v}_1 + \cdots + m_N \mathbf{v}_n$$

can be written as $p = mv$, where m is the "mass matrix"

$$M = \begin{pmatrix} M_1 & \dots & 0 \\ \vdots & \ddots & \vdots \\ 0 & \cdots & M_n \end{pmatrix} \tag{2.24}$$

whose diagonal blocks M_j are the 3×3 matrices

$$M_j = \begin{pmatrix} m_j & 0 & 0 \\ 0 & m_j & 0 \\ 0 & 0 & m_j \end{pmatrix}$$

(m_j is the mass of the j-th particle). We will often use, in this and the forthcoming Chapters, the notation $1/m$ for the inverse m^{-1} of the mass matrix (2.24).

Replacing Eq. (2.5) by its obvious multi-dimensional generalization

$$\mathbf{F}_j(\mathbf{r}_j, t) = \mathbf{E}_j(\mathbf{r}_j, t) + \mathbf{v}_j \times \mathbf{B}_j(\mathbf{r}_j, t) \tag{2.25}$$

($j = 1, 2, ..., N$), and setting

$$\begin{cases} B(x,t) = (\mathbf{B}_1(\mathbf{r}_1, t), ..., \mathbf{B}_N(\mathbf{r}_1, t)) \\ E(x,t) = (\mathbf{E}_1(\mathbf{r}_1, t), ..., \mathbf{E}_N(\mathbf{r}_1, t)) \end{cases} \tag{2.26}$$

the corresponding Lagrange form is, by definition:

$$\Omega_B = \sum_{j=1}^{N} \Omega_{\mathbf{B}_j}. \tag{2.27}$$

This is a differential 2-form on the generalized $2n+1 = (6N+1)$-dimensional state space (or part of it, if the involved fields are not globally defined).

Example 13 ***Charged particles in an electromagnetic field.*** *Consider N particles with respective charges $q_1, ..., q_N$ in an electromagnetic field $(\mathbf{E}, \mathbf{B})$. The vectors $\mathbf{B}_j$ and $\mathbf{E}_j$ are given by the formulas*

$$\begin{cases} B(x,t) = \left(\frac{q_j}{c}\mathbf{B}(\mathbf{r}_j, t)\right)_{1 \le j \le n} \\ E(x,t) = \left(q_j \mathbf{E}(\mathbf{r}_j, t) + \sum_{k \neq j} q_j q_k \frac{\mathbf{r}_j - \mathbf{r}_k}{|\mathbf{r}_j - \mathbf{r}_k|^3}\right)_{1 \le j \le n} \end{cases}. \tag{2.28}$$

The Maxwell principle of course extends *mutatis mutandis* to N-particle systems. The vector B being defined as above, the generalized Maxwell principle is that Ω_B should be closed:

$$d\Omega_B = 0. \tag{2.29}$$

By calculations quite similar to those in the one-particle case, this is equivalent to the sets of conditions:

$$\begin{cases} \dfrac{\partial \mathbf{B}_j}{\partial v_k} = 0 \quad (1 \le j, k \le n) \\ \dfrac{\partial \mathbf{B}_j}{\partial t} + \nabla_{\mathbf{r}_j} \times \mathbf{E}_j = 0, \quad \operatorname{div} \mathbf{B}_j = 0 \;\; (1 \le j \le n) \\ \left(\dfrac{\partial \mathbf{E}_k}{\partial \mathbf{r}_j}\right)^T = \dfrac{\partial \mathbf{E}_j}{\partial \mathbf{r}_k} \quad (j \neq k). \end{cases} \tag{2.30}$$

Remark 14 *The third set of formulas (2.30) are called "Maxwell's reciprocity law" in electromagnetic theory.*

The notion of gauge extends in an obvious way to the case of multiple-particle systems. Defining vector and scalar potentials $\mathbf{A}_j$ and U by

$$\mathbf{E}_j = -\nabla_j U - \frac{\partial \mathbf{A}_j}{\partial t}, \quad \mathbf{B}_j = \nabla_j \times \mathbf{A}_j \tag{2.31}$$

(∇_j is the gradient operator in $\mathbf{r}_j$), and setting $A = (\mathbf{A}_1, ..., \mathbf{A}_N)$, it is readily checked that these equations are preserved by the generalized gauge transformations

$$A \longrightarrow A + \nabla_x \chi, \quad U \longrightarrow U - \frac{\partial \chi}{\partial t}. \tag{2.32}$$

These properties can be expressed in a more concise way with the language of differential forms. We can identify $B = (\mathbf{B}_1, ..., \mathbf{B}_N)$ with the differential 2-form:

$$B = \sum_{j=1}^{N} \mathbf{B}_j(*d\mathbf{r}_j) \tag{2.33}$$

(that point of view was already implicit in the definition of the Lagrange form). For instance, if $N = 1$, we have:

$$B \equiv B_z\, dx \wedge dy + B_x\, dy \wedge dz + B_y\, dz \wedge dx \tag{2.34}$$

and the formula $\mathbf{B} = \nabla_r \times \mathbf{A}$ is then equivalent to saying that the 2-form $\mathbf{B}$ is exact. More generally, in the case of N particles:

$$\mathbf{B}_j = \nabla_j \times \mathbf{A}_j \Longleftrightarrow B = dA$$

where A is the one-form is obtained by identifying the field with

$$A = \sum_{j=1}^{n} \mathbf{A}_j \, d\mathbf{r}_j.$$

Remark 15 *One should however be careful, when using this identification, to keep in mind that the $\mathbf{B}_j$ are pseudo-vectors. The form (2.34) should actually be interpreted as a "de Rham form" or "twisted form" (see Frankel [48]).*

2.2 Hamilton's Equations

Let us now make a tentative approach to Hamilton's equations. We suppose a particle moves under the action of a force field that can be derived from a potential. By this we mean that there exists some scalar function $U = U(\mathbf{r}, t)$ such that

$$\mathbf{F} = -\nabla_{\mathbf{r}} U. \tag{2.35}$$

Defining the "Hamiltonian"

$$H = \frac{\mathbf{p}^2}{2m} + U \tag{2.36}$$

we can rewrite Newton's second law as the system of differential equations

$$\dot{\mathbf{r}} = \nabla_{\mathbf{p}} H \quad , \quad \dot{\mathbf{p}} = -\nabla_{\mathbf{r}} H. \tag{2.37}$$

These equations are *Hamilton's equations* for a single particle. However, this derivation of Hamilton's equations only works if the force field satisfies the very particular condition (2.35), which excludes systems where the force field is of the general type $\mathbf{F} = \mathbf{E} + (\mathbf{v} \times \mathbf{B})$ (and a fortiori those satisfying the Maxwell principle).

The key for the rigorous derivation of Hamilton's equations (2.37) in the general case lies in the properties of the Lagrange form, which plays the role of a potential in the position-momentum variables for the Lagrange form.

2.2.1 The Poincaré-Cartan Form and Hamilton's Equations

Let us assume for a while that the force field is of the type $\mathbf{F} = -\nabla_{\mathbf{r}} U$ considered above. The Lagrange form is then

$$\Omega_0 = (m\,d\mathbf{v} - \mathbf{F}\,dt) \wedge (d\mathbf{r} - \mathbf{v}\,dt).$$

Let us rewrite Ω_0 in phase-space variables by replacing $m\,d\mathbf{v}$ by $d\mathbf{p}$, $\mathbf{v}$ by $\mathbf{p}/m$, and $\mathbf{F}$ by $-\nabla_{\mathbf{r}} U$; we thus define

$$\tilde{\Omega}_0 = (d\mathbf{p} + \nabla_{\mathbf{r}} U\ dt) \wedge \left(d\mathbf{r} - \frac{\mathbf{p}}{m}\,dt\right). \tag{2.38}$$

(We will systematically use tildes "~" to indicate forms or fields defined on extended phase space.) Expanding the right-hand side, and taking into account the fact that $dt \wedge dt = 0$ and $dt \wedge d\mathbf{r} = -d\mathbf{r} \wedge dt$ Eq. (2.38) becomes

$$\tilde{\Omega}_0 = d\mathbf{p} \wedge d\mathbf{r} - (\mathbf{p}/m)\,d\mathbf{p} \wedge dt + \nabla_{\mathbf{r}} U\ d\mathbf{r} \wedge dt$$

that is, in view of the definition (2.36) of Hamilton's function H:

$$\tilde{\Omega}_0 = d\mathbf{p} \wedge d\mathbf{r} - \nabla_{\mathbf{p}} H\,d\mathbf{p} \wedge dt - \nabla_{\mathbf{r}} H\,d\mathbf{r} \wedge dt. \tag{2.39}$$

Since $dt \wedge dt = 0$ we have

$$\begin{aligned} dH \wedge dt &= \left(\nabla_{\mathbf{p}} H\,d\mathbf{p} + \nabla_{\mathbf{r}} H\,d\mathbf{r} + \frac{\partial H}{\partial t}\,dt\right) \wedge dt \\ &= (\nabla_{\mathbf{p}} H\,d\mathbf{p} + \nabla_{\mathbf{r}} H\,d\mathbf{r}) \wedge dt \end{aligned}$$

so that (2.39) takes the very simple form

$$\tilde{\Omega}_0 = d\mathbf{p} \wedge d\mathbf{r} - dH \wedge dt = d(\mathbf{p} \cdot d\mathbf{r} - H dt). \tag{2.40}$$

This motivates the following definition:

Definition 16 *The differential form on extended phase space defined by*

$$\tilde{\lambda}_H = \mathbf{p}\,d\mathbf{r} - H\,dt \tag{2.41}$$

is called the Poincaré-Cartan form, *and we thus have*

$$\tilde{\Omega}_0 = d\,\tilde{\lambda}_H. \tag{2.42}$$

The constructions above work as well when a vector potential is present, provided that we use the conjugate momentum

$$\mathbf{p} = m\mathbf{v} + \mathbf{A} \tag{2.43}$$

instead of the vector $\mathbf{p} = m\mathbf{v}$. In fact, assuming that the fields $\mathbf{E}$ and $\mathbf{B}$ are defined in some suitable region of state space, we can find potentials $\mathbf{A}$ and U satisfying the conditions

$$\mathbf{B} = \nabla_{\mathbf{r}} \times \mathbf{A} \quad , \quad \mathbf{E} = -\nabla_{\mathbf{r}} U - \frac{\partial \mathbf{A}}{\partial t}. \tag{2.44}$$

Theorem 17 *(1) The Lagrange form is the differential of the Poincaré-Cartan form $\tilde{\lambda}_H = \mathbf{p}\,d\mathbf{r} - H\,dt$ associated to the Hamiltonian function*

$$H = \frac{1}{2m}(\mathbf{p} - \mathbf{A}(\mathbf{r},t))^2 + U(\mathbf{r},t)\ . \tag{2.45}$$

In fact, let $\tilde{\Omega}_H$ be the 2-form $\Omega_{\mathbf{B}}$ expressed in terms of the momentum $\mathbf{p}$, then $\tilde{\Omega}_H = d\lambda_H$. (2) A gauge transformation

$$(\mathbf{A}, U) \longrightarrow \left(\mathbf{A} + \nabla_{\mathbf{r}}\chi, U - \frac{\partial \chi}{\partial t}\right) \tag{2.46}$$

changes the function H into the new Hamiltonian

$$H^{\chi}(\mathbf{r},\mathbf{p},t) = H(\mathbf{r},\mathbf{p}-\nabla_{\mathbf{r}}\chi,t) - \frac{\partial \chi}{\partial t}. \tag{2.47}$$

Proof. (1) Setting $\mathbf{p}' = m\mathbf{v}$, the form $\tilde{\Omega}_H$ is given by

$$\begin{aligned}\tilde{\Omega}_H &= \left(d\mathbf{p}' - \mathbf{E}\,dt\right) \wedge \left(d\mathbf{r} - \frac{\mathbf{p}'}{m}\,dt\right) + \mathbf{B}(*d\mathbf{r}) \\ &= d\mathbf{p}' \wedge d\mathbf{r} - \frac{\mathbf{p}'}{m}d\mathbf{p}' \wedge dt - \mathbf{E}dt \wedge d\mathbf{r} + \mathbf{B}(*d\mathbf{r}).\end{aligned}$$

We have $\mathbf{B}(*d\mathbf{r}) = d\mathbf{A} \wedge d\mathbf{r}$ in view of formula (2.17), and definition (2.44) of the vector potential. Hence, returning to the variable $\mathbf{p}$:

$$\begin{aligned}\tilde{\Omega}_H &= d\mathbf{p}' \wedge d\mathbf{r} - \frac{\mathbf{p}'}{m}d\mathbf{p}' \wedge dt - \left(\mathbf{E} + \frac{\partial \mathbf{A}}{\partial t}\right) dt \wedge d\mathbf{r} + d\mathbf{A} \wedge d\mathbf{r} \\ &= d\mathbf{p}' \wedge d\mathbf{r} - \frac{\mathbf{p}'}{m}\,d\mathbf{p}' \wedge dt - U\ d\mathbf{r} \wedge dt + d\mathbf{A} \wedge d\mathbf{r} \\ &= d\mathbf{p} \wedge d\mathbf{r} - d\left(\frac{1}{2m}(\mathbf{p} - \mathbf{A})^2 + U\right) \wedge dt\end{aligned}$$

that is, $\tilde{\Omega}_H = d\lambda_H$. Part **(2)** of the theorem is obvious, since a gauge transformation (2.46) has the effect of changing $\mathbf{p}$ into $\mathbf{p} + \nabla_{\mathbf{r}}\chi$. ■

Here are two typical examples.

Example 18 ***The Hamiltonian of the Coriolis force.*** *Recall (see (2.6)) that the force acting on a material point close to the surface of the Earth is*

$$\mathbf{F} = m\mathbf{g} - 2m\,(\mathbf{R} \times \mathbf{v})$$

where R is the rotation vector. We have $\mathbf{E} = \mathbf{F} + (\mathbf{v} \times \mathbf{B})$ provided that we choose $\mathbf{E} = m\mathbf{g}$, $\mathbf{B} = 2m\mathbf{R}$. A straightforward calculation shows that

$$\mathbf{A} = m\,(\mathbf{R} \times \mathbf{r}), \quad U = -m\mathbf{g} \cdot \mathbf{r}$$

are potentials, and the associated Hamiltonian is therefore

$$H_C = \frac{1}{2m}\,(\mathbf{p} - m\,(\mathbf{R} \times \mathbf{r}))^2 - m\mathbf{g} \cdot \mathbf{r}. \tag{2.48}$$

For more on the topic of Coriolis force, see for instance Knudsen and Hjorth [96], pages 128–142, Arnold [4], §27, or Goldstein [52], §4-10.

Example 19 ***Charged particle in a uniform magnetic field.*** *A particle with mass m and electric charge e is placed in a uniform magnetic field $\mathbf{B} = (0, B, 0)$. The two following functions*

$$A_1 = (Bz, 0, 0) \quad or \quad A_2 = (0, 0, -Bx)$$

are both vector potentials for $\mathbf{B}$, and lead to the two different Hamiltonian functions:

$$H_1 = \frac{1}{2m}\left[(p_x - eBz)^2 + p_y^2 + p_z^2\right]$$
$$H_2 = \frac{1}{2m}\left[p_x^2 + p_y^2 + (p_z - eBx)^2\right].$$

However, the equations of motion are in both cases

$$\ddot{x} = -\frac{eB\dot{z}}{m}\ ,\ \ddot{y} = 0\ ,\ \ddot{z} = \frac{eB\dot{x}}{m}.$$

We will see in a moment that the choice of gauge has no influence whatsoever on the motion of the particle in "physical" space (this is of course a priori obvious, since Hamilton's equations are equivalent to Newton's second law, which is gauge independent).

2.2.2 Hamiltonians for N-Particle Systems

The construction of the Hamiltonian generalizes in a straightforward way to the case of many-particle systems. By the same argument as in the proof of Proposition 17, one checks that a gauge (A, U) being chosen the Lagrange form Ω_B is the differential of the Poincaré-Cartan form

$$\tilde{\lambda}_H = pdx - Hdt. \tag{2.49}$$

We are using here the notation:

$$pdx = \mathbf{p}_1 d\mathbf{r}_1 + \cdots + \mathbf{p}_n d\mathbf{r}_N.$$

The Hamiltonian function H is then

$$H = \sum_{j=1}^{N} \frac{1}{2m_j}(\mathbf{p}_j - \mathbf{A}_j)^2 + U \tag{2.50}$$

which we can write, using the "mass matrix" defined in Section 2.1.6, as

$$H = \frac{1}{2M}(p - A)^2 + U \tag{2.51}$$

(recall that we are using the convention $1/M = M^{-1}$).

Definition 20 *We will call all Hamiltonian functions of the type (2.50)–(2.51)* Maxwell Hamiltonians.

Hamilton's equations of motion are

$$\dot{x}_j = \frac{\partial H}{\partial p_j} \quad , \quad \dot{p}_j = -\frac{\partial H}{\partial x_j} \quad (1 \leq j \leq n) \tag{2.52}$$

where the conjugate momenta p_j are here defined by

$$p_j = mv_j + A_j. \tag{2.53}$$

Introducing the generalized gradients

$$\nabla_x = \left(\frac{\partial}{\partial x_1}, ..., \frac{\partial}{\partial x_n}\right) \quad , \quad \nabla_p = \left(\frac{\partial}{\partial p_1}, ..., \frac{\partial}{\partial p_n}\right)$$

the Hamilton equations (2.52) can be written as

$$\dot{x} = \nabla_p H \quad , \quad \dot{p} = -\nabla_x H. \tag{2.54}$$

Recall that in Chapter 1, Subsection 1.2.4, we defined the matrix

$$J = \begin{pmatrix} 0_{3\times 3} & I_{3\times 3} \\ -I_{3\times 3} & 0_{3\times 3} \end{pmatrix}$$

to relate Hamiltonian mechanics to symplectic geometry. More generally, we define the $2n \times 2n$ matrix

$$J = \begin{pmatrix} 0 & I \\ -I & 0 \end{pmatrix} \tag{2.55}$$

where 0 and I are, respectively, the $n \times n$ zero and identity matrices. We have $\det J = 1$ and

$$J^2 = -I_{2n} \quad , \quad J^T = J^{-1} = -J.$$

The matrix J allows us to rewrite Hamilton's equations (2.54) in compact form as

$$\dot{z} = J\nabla_z H(z,t). \tag{2.56}$$

As mentioned above, the choice of gauge only influences the phase-space motion of the system, not its motion in configuration space:

Proposition 21 *The motion of a system in a gauge (A, U) is determined by the system of second order differential equations*

$$\ddot{x}_j + \frac{1}{m_j}\left(\frac{\partial A_j}{\partial t}(x,t) + \frac{\partial U}{\partial x_j}(x,t)\right) = 0 \tag{2.57}$$

$(1 \leq j \leq n)$. That system is invariant under every gauge transformation $(A, U) \longmapsto (A', U')$.

Proof. The Hamilton equations for a Maxwell Hamiltonian are, explicitly:

$$\dot{x}_j = \frac{1}{m_j}(p_j - A_j) \text{ and } \dot{p}_j = \frac{1}{m_j}(p_j - A_j)\frac{\partial A_j}{\partial x_j} - \frac{\partial U}{\partial x_j}.$$

Differentiating the first equation with respect to t and then inserting in it the value of $\dot{p}_j$ given by the second equation, we get Eq. (2.57). That equation does not depend on the choice of gauge, for if we replace A_j by $A'_j = A_j + \frac{\partial \chi}{\partial x_j}$ and U by $U' = U - \frac{\partial \chi}{\partial t}$ then

$$\frac{\partial A'_j}{\partial t} + \frac{\partial U'}{\partial x_j} = \frac{\partial A_j}{\partial t} + \frac{\partial U}{\partial x_j}$$

so that the left-hand side of Eq. (2.57) does not change. ∎

2.2.3 The Transformation Law for Hamilton Vector Fields

Let us begin by recalling, from the ordinary theory of dynamical systems, how vector fields transform under diffeomorphisms (= changes of variables). Assume that X is a vector field on $\mathbb{R}^m$, and let $x = x(t)$ be a solution of the differential equation $\dot{x} = X(x)$. If u is a diffeomorphism of $\mathbb{R}^m$, we can define a function $y = y(t)$ by $x(t) = u(y(t))$; differentiation with respect to t then shows that y is a solution of $\dot{y} = Y(y)$ where Y is the vector field

$$Y = u'(y)^{-1} \circ X \circ u(y)$$

($u'(y)$ is the Jacobian matrix of u calculated at y). It follows that the flows (f_t) of X and (g_t) of Y are conjugate:

$$g_t = u^{-1} \circ f_t \circ u.$$

We will denote the transformed vector field Y by u^*X. Let us now specialize to the case where $X = X_H$ is a Hamilton vector field. We have:

Proposition 22 *Suppose that u is a diffeomorphism of phase space $\mathbb{R}^n_x \times \mathbb{R}^n_p$ such that the Jacobian matrix u' has the following property: at every $z = (x,p)$*

$$u'(z)^T J u'(z) = u'(z) J u'(z)^T = J$$

where the matrix J is defined by (2.55). Then $u^*X_H = X_{u^*H}$ where $u^*H = H \circ u$, and the flow of X_{u^*H} is thus $(u^{-1} \circ f_t \circ u)$ if (f_t) is the flow of H.

Proof. Let us prove that $u^*X_H = X_{u^*H}$. Setting $K = H \circ u$ we have, by the chain rule:

$$\nabla_z K = u'(z)^T \nabla_z H \circ u$$

and hence $X_K \equiv J\nabla_z K$ is given by

$$X_K = u'(z)^{-1} J \nabla_z H \circ u = X_{u^*H}.$$

That the flows of X_H and X_{u^*H} are conjugate follows from the discussion preceding the statement of the proposition. ■

2.2.4 The Suspended Hamiltonian Vector Field

The solutions of Hamilton's equations determine curves

$$t \longmapsto (x(t), p(t), t)$$

in the *extended phase space* $\mathbb{R}^n_x \times \mathbb{R}^n_p \times \mathbb{R}_t$. Hamilton's equations are trivially equivalent to

$$\frac{d}{dt}(x, p, t) = \tilde{X}_H(x, p, t) \tag{2.58}$$

where $\tilde{X}_H$ is the so-called *suspended Hamilton vector field*

$$\tilde{X}_H = (\nabla_p H, -\nabla_x H, 1)$$

(cf. the Newton field (2.2)). The point in introducing the redundant variable t in (2.58) is that $\tilde{X}_H$ is a "true" vector field, to which we can apply to it the standard theory of autonomous (= time-independent) systems, whereas the ordinary Hamilton vector field

$$X_H = (\nabla_p H, -\nabla_x H) \tag{2.59}$$

is, in general, a family of vector fields indexed by time t.

The projections of the integral curves $t \longmapsto (x(t), p(t), t)$ of $\tilde{X}_H$ on the ordinary phase space $\mathbb{R}^n_x \times \mathbb{R}^n_p$ are just the integral curves of X_H. If we define, as is customary, the "flow" (f_t) of X_H by

$$\frac{d}{dt} f_t(x, p) = X_H(x, p) \tag{2.60}$$

then we will in general have $f_t \circ f_{t'} \neq f_{t+t'}$, the equality $f_t \circ f_{t'} = f_{t+t'}$ occurring only when H is time-independent. As opposed to this situation, the mappings $\tilde{f}_t$ defined by

$$\frac{d}{dt} \tilde{f}_t(x, p, t) = \tilde{X}_H(x, p, t)$$

automatically satisfy the group property

$$\tilde{f}_t \circ \tilde{f}_{t'} = \tilde{f}_{t+t'} \quad , \quad (\tilde{f}_t^{-1}) = \tilde{f}_{-t}. \tag{2.61}$$

Definition 23 *We will call the family $(\tilde{f}_t)$ the* suspended flow *determined by the Hamiltonian H. The associated* time-dependent flow $(f_{t,t'})$ *is then defined by the formula:*

$$(f_{t,t'}(x', p'), t') = \tilde{f}_{t-t'}(x', p', t'). \tag{2.62}$$

The time-dependent flow has the following straightforward interpretation: $f_{t,t'}$ is the phase-space mapping that takes a point (x', p') in phase space at time t', to the point (x, p) at time t, the motion occurring along the solution curve to Hamilton equations passing through these two points. That is, if we keep t' fixed, the formula

$$(x(t), p(t)) = f_{t,t'}(x', p')$$

defines functions $t \longmapsto x(t)$ and $t \longmapsto p(t)$ satisfying

$$\dot{x} = \nabla_p H(x(t), p(t), t) \quad , \quad \dot{p} = -\nabla_x H(x(t), p(t), t)$$

together with the initial conditions $x(t') = x'$, $p(t') = p'$. Notice that the time-dependent flow defined by (2.62) is related to the suspended flow by the formula:

$$\tilde{f}_t(x', p', t') = (f_{t+t',t'}(x', p'), t + t'). \tag{2.63}$$

The time-dependent flow enjoys a very important property, the *Chapman–Kolmogorov law*, which shows that Hamiltonian motion is causal. That property is actually an immediate consequence of the definition of the flow, but we however state it as a theorem, because of its importance:

Theorem 24 *The mappings $f_{t,t'}$ satisfy the Chapman-Kolmogorov law:*

$$f_{t,t'} \circ f_{t',t''} = f_{t,t''} \quad , \quad (f_{t,t'})^{-1} = f_{t',t} \tag{2.64}$$

for all times t, t' and t'' (for which $f_{t,t'}$, etc., are defined).

Proof. We leave it to the reader to check that Eq. (2.64) immediately follows from the definition (2.62) of $f_{t,t'}$ together with the properties (2.61) of the suspended flow. ■

Before we proceed to examine the behavior of Hamilton's equations under Galilean transformations, we make the following useful remark, which relates the time-dependent flow to the flow determined by X_H:

Remark 25 *Suppose that the Hamiltonian H does not depend on time: $H = H(x, p)$. Then $f_{t,t'} = f_{t-t'}$. In particular, $f_{t,t'}$ only depends on the time difference $t - t'$.*

Remark 26 *In Theorem 12 we showed the equivalence*

$$\text{[Newton's } 2^{nd} \text{ law]} \Longleftrightarrow i_N \Omega_{\mathbf{B}} = 0.$$

We can restate this property in terms of the suspended Hamiltonian vector field as

$$\text{[Newton's } 2^{nd} \text{ law]} \Longleftrightarrow i_{\tilde{X}_H} \tilde{\Omega}_H = 0. \tag{2.65}$$

2.3 Galilean Covariance

The Galilean group plays an essential role in both theoretical and practical considerations. For instance, a precise analysis of the cohomological properties of $Gal(3)$ allows one to justify mathematically the concept of "mass". (This fact was already observed by V. Bargmann [8].) The treatment of Galilean relativity we give here is rather sketchy; the study of what Newton's laws become in accelerated frames is for instance totally ignored. We refer to Knudsen and Hjorth [96] for a discussion of that important topic; also see Mackey [108] (especially pages 189–190) for a review of the conceptual difficulties around the notion of motion. In our discussion of the Maxwell's principle we have been adopting a completely geocentric – one is tempted to say egocentric! – point of view. All our arguments were stated with respect to a "lab frame" whose origin O was implicitly identified by us with the "center of the Universe."

Roughly speaking, the principle of Galilean relativity is the claim that no physical experiment whatsoever can distinguish a reference frame from another moving uniformly with respect to it. Galilei, who already remarked in his book *Dialogue Concerning The Two Chief World Systems* that the laws of the physics were the same on earth as in a ship moving on a quiet sea, was probably the first to explicitly write down this principle. The mathematical object which allows one to make this principle rigorous is the Galilean group $Gal(3)$. Let us begin by recalling the notion of inertial frame.

2.3.1 Inertial Frames

Until now we have been discussing the implications of Newton's second law. But what does Newton's *first* law say? It says that there exist coordinate systems, called *inertial reference frames* in which a body remains in rest, or in uniform motion, as long as no external forces act to change that state. Geocentric frames, for instance, are not inertial frames, because the Earth rotates: the Coriolis force deflects the movement of a free particle. In practice, one considers in physics that the heliocentric frames are, to a very good accuracy, inertial frames. (Heliocentric frames are frames whose origin is the center of the Sun, and with axes passing through three distinct fixed stars.)

We now set out to find rules that allow us to relate observations performed in different frames of reference. This means that we have to find "transformation laws" for the positions and velocities, as well for the various vector fields previously introduced. Once this has been done, we will have to examine the behavior of Newton's laws and of the Lagrange form under these transformations. Specifically we ask the following question:

> *"Suppose that we have found that Newton's three first laws of mechanics hold true in some reference frame (O, x, y, z, t). Under which changes of frame $(O, x, y, z, t) \longrightarrow (O', x', y', z', t')$ do these laws remain true?"*

Equivalently:

> *"What group of transformations of space-time changes an inertial frame into another inertial frame? Does this group act transitively on the set of all inertial frames? That is, given two arbitrary inertial frames, can we find a transformation which takes the first into the second?"*

2.3.2 The Galilean Group $Gal(3)$

Consider the four following types of space-time transformation, called *Galilean transformations* :

(1) *Time translations*:

$$g_{t_0} : (\mathbf{r}, t) \mapsto (\mathbf{r}, t + t_0) \, ; \tag{2.66}$$

(2) *Space translations:*

$$g_{\mathbf{r}_0} : (\mathbf{r}, t) \mapsto (\mathbf{r} + \mathbf{r}_0, t) \, ; \tag{2.67}$$

(3) *Velocity boosts:*

$$g_{\nu_0} : (\mathbf{r}, t) \mapsto (\mathbf{r} + \nu_0 t, t) \, ; \tag{2.68}$$

(4) *Space rotations:*

$$g_R : (\mathbf{r}, t) \mapsto (R\mathbf{r}, t) \quad (R \in SO(3)). \tag{2.69}$$

Definition 27 *The transformations (2.66)–(2.69) are invertible, and thus generate a group $Gal(3)$, called the* Galilean group. *Every element of $Gal(3)$ is thus, by definition, either one of the transformations (2.66)–(2.69), or a product of such transformations. The action of $g \in Gal(3)$ on space-time is given by the formula*

$$g(\mathbf{r}, t) = (R\mathbf{r} + \mathbf{v}_0 t + \mathbf{r}_0, t + t_0) \, . \tag{2.70}$$

We can moreover extend the action of $Gal(3)$ into an action on state-space, by setting

$$\tilde{g}\left(\mathbf{r},\mathbf{v},t\right)=\left(\mathbf{r}',\mathbf{v}',t'\right) \tag{2.71}$$

where $\mathbf{r}',\mathbf{v}',t'$ are defined by:

$$\begin{cases} \mathbf{r}'=R\mathbf{r}+\mathbf{v}_0t+\mathbf{r}_0 \\ \mathbf{v}'=R\mathbf{v}+\mathbf{v}_0 \\ t'=t+t_0. \end{cases} \tag{2.72}$$

It is practical to write this action in matrix form as:

$$\begin{pmatrix} \mathbf{r}' \\ \mathbf{v}' \\ t' \\ 1 \end{pmatrix}=\begin{pmatrix} R & 0_{3\times 3} & \mathbf{v}_0 & \mathbf{r}_0 \\ 0_{3\times 3} & R & 0_{3\times 1} & \mathbf{v}_0 \\ 0 & 0 & 1 & t_0 \\ 0 & 0 & 0 & 1 \end{pmatrix}\begin{pmatrix} \mathbf{r} \\ \mathbf{v} \\ t \\ 1 \end{pmatrix} \tag{2.73}$$

($0_{j\times k}$ is the $j\times k$ zero matrix) or, if one only takes into account the action of Galilean transformations on space-time:

$$\begin{pmatrix} \mathbf{r}' \\ t' \\ 1 \end{pmatrix}=\begin{pmatrix} R & \mathbf{v}_0 & \mathbf{r}_0 \\ 0 & 1 & t_0 \\ 0 & 0 & 1 \end{pmatrix}\begin{pmatrix} \mathbf{r} \\ t \\ 1 \end{pmatrix}. \tag{2.74}$$

The composition of Galilean transformations is given in both cases by ordinary matrix multiplication, and $Gal(3)$ is thus a 10-dimensional matrix Lie group.

The translations (2.67), (2.68) together with the rotations (2.69) generate a subgroup $Euc(3)$ of $Gal(3)$, called the *Euclidean group* (it is the group of orientation preserving space isometries, in fact, the “semi-direct product” of the rotation and translation groups).

We can extend the action of $Gal(3)$ to many-particle systems. If $g\in Gal(3)$, we define $(x',t')=\tilde{g}(x,t)$ by the formula

$$\tilde{g}\left(\mathbf{r}_1,...,\mathbf{r}_N,t\right)=\left(\mathbf{r}_1',...,\mathbf{r}_N',t'\right)$$

where the new position vectors are obtained from the old ones by the formulas

$$\mathbf{r}_j'=R\mathbf{r}_j+\nu_0t+\mathbf{r}_0 \quad , \quad t'=t+t_0$$

for $1\leq j\leq N$.

Remark 28 *As opposed to the case $N = 1$, the action of $Gal(n)$ is not transitive on N-particle states if $N > 1$. This is because the relative velocities of two particles in a physical system cannot be changed by a Galilean transformation.*

Let us now study the behavior of Newton's second law under the action of $Gal(3)$.

2.3.3 Galilean Covariance of Hamilton's Equations

Newton's second law retains its form under Galilean transformations. This property is called the *Galilean covariance* of Newton's second law. In particular, Galilean transformations change any inertial frame into another inertial frame:

Proposition 29 *Newton's second law is covariant under the action defined by Eq. (2.73): if the equations of motion in a frame (O, x, y, z, t) are*

$$d\mathbf{r} = \mathbf{v}\, dt \quad , \quad m\, d\mathbf{v} = \mathbf{F}dt \tag{2.75}$$

then they are

$$d\mathbf{r}' = \mathbf{v}'\, dt' \quad , \quad m\, d\mathbf{v}' = \mathbf{F}'dt' \quad , \quad \mathbf{F}' = R\mathbf{F} \tag{2.76}$$

in the frame (O', x', y', z', t') obtained from (O, x, y, z, t) by Eq. (2.74). In particular, if g is of any of the types (2.66), (2.67) or (2.68), then $\mathbf{F}' = \mathbf{F}$.

Proof. Differentiating both sides of the first equation in (2.72) yields

$$d\mathbf{r}' = Rd\mathbf{r} + \mathbf{v}_0 dt = (R\mathbf{v} + \mathbf{v}_0)\, dt$$

that is $d\mathbf{r}' = \mathbf{v}'dt'$ since $dt = dt'$. Inserting that value in the second equation (2.72), we get

$$md\mathbf{v}' = mRd\mathbf{v} = R\mathbf{F}dt'$$

which proves (2.76). ∎

An apparent difficulty immediately arises when one wants to make the Galilean group act on the extended phase space: one has to incorporate mass in any "reasonable" definition of the transformation of momenta. The most "natural" definition is the following: suppose that we are dealing with a particle whose motion is obtained from Hamilton's equations for the Maxwell Hamiltonian

$$H(\mathbf{r},\mathbf{p},t) = \frac{1}{2m}(\mathbf{p} - \mathbf{A}(\mathbf{r},t))^2 + U(\mathbf{r},t).$$

Then we define the Galilean transformation $\tilde{g}_m$ of extended phase-space induced by

$$\tilde{g} : (\mathbf{r},\mathbf{v},t) \mapsto (R\mathbf{r} + \mathbf{v}_0 t + \mathbf{r}_0, \mathbf{v} + \mathbf{v}_{0,}t)$$

by the formula:

$$\tilde{g}_m : (\mathbf{r},\mathbf{p},t) \mapsto (R\mathbf{r} + \mathbf{v}_0 t + \mathbf{r}_0, R\mathbf{p} + m\mathbf{v}_{0,}t + t_0). \tag{2.77}$$

The "naturality" of that definition comes from the fact that it is consistent with the Galilean covariance of Newton's second law (when there is no vector potential, Eq. (2.77) is immediate, because the momentum $\mathbf{p}$ is then simply "*mass* × *velocity*"). It can also be motivated on a purely mathematical basis. Suppose in fact that f is a change of variable in configuration space $\mathbb{R}^n_x$. Identifying the corresponding phase space $\mathbb{R}^n_x \times \mathbb{R}^n_p$ with the "cotangent bundle" $T^*\mathbb{R}^n_x$, we know from the theory of differential manifolds that the change of coordinates

$$f^* : T^*\mathbb{R}^n_x \longrightarrow T^*\mathbb{R}^n_x$$

induced by f is given by the formula

$$f^*(x,p) = (f(x), (f'(x)^T)^{-1}p).$$

(f^* is sometimes called a "Mathieu transform".) If we take $n = 3$ and $f(\mathbf{r}) = R\mathbf{r}$, then the formula above yields

$$f^*(\mathbf{r},\mathbf{p}) = (R\mathbf{r}, R\mathbf{p})$$

since $R^{-1} = R^T$, which "justifies" mathematically the definition (2.77) of $\tilde{g}_m$.

Proposition 30 *(1) Let $t \mapsto (\mathbf{r}(t),\mathbf{p}(t))$ be an integral curve of the suspended Hamilton vector field associated to H. The image by $\tilde{g}_m$ of that integral curve is an integral curve of the suspended Hamilton vector field associated to the compose $H' = H \circ \tilde{g}_m$, that is*

$$H'(\mathbf{r},\mathbf{p},t) = \frac{1}{2m}\left(\mathbf{p} - \mathbf{A}'(\mathbf{r},\mathbf{p},t)\right)^2 + U'(\mathbf{r},t) \tag{2.78}$$

with gauge defined by

$$\begin{cases} \mathbf{A}'(\mathbf{r},t) = R^{-1}\left[\mathbf{A}(R\mathbf{r}+\mathbf{v}_0 t+\mathbf{r}_0, t+t_0) - m\mathbf{v}_0\right] \\ U'(\mathbf{r},t) = U\left(R\mathbf{r}+\mathbf{v}_0 t+\mathbf{r}_0, t+t_0\right). \end{cases} \tag{2.79}$$

(2) This property in fact characterizes Maxwell Hamiltonians : if the solutions of Hamilton's equations for a given function H transform according to the law

$$u \mapsto \tilde{g}_m \circ u \quad \textit{if} \quad H \mapsto H \circ \tilde{g}_m \tag{2.80}$$

then H is necessarily a Maxwell Hamiltonian.

Proof. (1) That the compose H' is given by Eq. (2.78) immediately follows from the identities

$$\begin{aligned} (R\mathbf{p}+m\boldsymbol{\nu}_0-\mathbf{A})^2 &= (R\mathbf{p}+m\boldsymbol{\nu}_0-\mathbf{A})\cdot(R\mathbf{p}+m\boldsymbol{\nu}_0-\mathbf{A}) \\ &= (R^T R\mathbf{p}+R^T m\boldsymbol{\nu}_0-R^T\mathbf{A})\cdot(R\mathbf{p}+m\boldsymbol{\nu}_0-\mathbf{A}) \\ &= (\mathbf{p}+R^{-1}m\boldsymbol{\nu}_0-R^{-1}\mathbf{A})^2 \end{aligned}$$

(R is a rotation, hence $R^T = R^{-1}$). Setting $z(t) = (\mathbf{r}(t), \mathbf{p}(t))$, and hence $u(t) = (z(t), t)$, Hamilton's equations for H are

$$\frac{du}{dt} = \begin{pmatrix} J & 0_{6\times 1} \\ 0_{1\times 6} & 1 \end{pmatrix} \begin{pmatrix} \nabla_z H(u) \\ 1 \end{pmatrix}.$$

Now, by the chain rule

$$\frac{d}{dt}(\tilde{g}_m \circ u) = \tilde{g}_m' \dot{u}$$

where the Jacobian $\tilde{g}_m'$ is the 7×7 matrix

$$\tilde{g}_m' = \begin{pmatrix} \hat{R} & 0_{6\times 1} \\ 0_{1\times 6} & 1 \end{pmatrix}, \ \hat{R} = \begin{pmatrix} R & 0_{3\times 3} \\ 0_{3\times 3} & R \end{pmatrix},$$

so that, in view of the obvious commutation relation

$$\begin{pmatrix} \hat{R} & 0 \\ 0 & 1 \end{pmatrix} \begin{pmatrix} J & 0 \\ 0 & 1 \end{pmatrix} = \begin{pmatrix} J & 0 \\ 0 & 1 \end{pmatrix} \begin{pmatrix} \hat{R} & 0 \\ 0 & 1 \end{pmatrix}$$

(we have dropped the subscripts for the zero matrices) we have

$$\frac{d}{dt}(\tilde{g}_m \circ u) = \begin{pmatrix} J & 0 \\ 0 & 1 \end{pmatrix} \begin{pmatrix} \hat{R} & 0 \\ 0 & 1 \end{pmatrix} \begin{pmatrix} \nabla_z H(u) \\ 1 \end{pmatrix}.$$

But this is the same as

$$\frac{d}{dt}(\tilde{g}_m \circ u) = \begin{pmatrix} J & 0 \\ 0 & 1 \end{pmatrix} \begin{pmatrix} \nabla_z(H(\tilde{g}_m \circ u)) \\ 1 \end{pmatrix}$$

which proves **(1)**, since this formula is equivalent to Hamilton's equations for $H \circ \tilde{g}_m$. **(2)** Let us consider a Galilean transformation

$$\tilde{g} : (\mathbf{r}, \nu, t) \mapsto (\mathbf{r} + \nu_0 t + \mathbf{r}_0, \boldsymbol{\nu} + \boldsymbol{\nu}_{0,} t) .$$

This has the effect of replacing the momentum $\mathbf{p}$ by $\mathbf{p} + m\mathbf{v}_0$, and this imposes that the function H must satisfy the condition

$$\nabla_{\mathbf{p}} H(\mathbf{r}, \mathbf{p} + \mathbf{p}_0, t) = \nabla_{\mathbf{r}} H(\mathbf{r}, \mathbf{p}, t).$$

Integration of these equations with "unknown" H leads, after a few straightforward calculations, to the existence of potentials $\mathbf{A}$ and U such that

$$H = \frac{1}{2m}(\mathbf{p} - \mathbf{A})^2 + U$$

hence H is a Maxwell Hamiltonian, as claimed. ∎

Remark 31 *If we take a closer look at the proof above, it appears that we have actually shown a sharper result: the only Hamiltonians for which the transformation law (2.80) holds for velocity boosts are precisely Maxwell Hamiltonians.*

Here is a simple illustration of Proposition 30:

Example 32 ***The cart with a spring.*** *Suppose that a point-like particle is attached to a spring, the other end of which is fixed on a massless cart that is being moved with uniform velocity ν_0 along a rectilinear rail. Let x be the position coordinate of the particle, the origin being chosen so that the cart passes through it a time $t = 0$. Taking both the mass of the particle and the spring constant equal to unity, the Hamilton function is*

$$H = \frac{1}{2}\left(p^2 + (x - \nu_0 t)^2\right)$$

and the solutions of Hamilton's equations are

$$\begin{cases} x = x_0 \cos t + p_0 \sin t - \nu_0 t \\ p = x_0 \sin t + p_0 \cos t - \nu_0 . \end{cases}$$

If we instead take for origin the point of the cart to which the spring is attached, the new Hamiltonian is the time-independent function

$$H' = \frac{1}{2}\left(p'^2 + x'^2\right)$$

and the solutions to Hamilton's equations for H' are

$$\begin{cases} x' = x_0' \cos t + p_0' \sin t \\ p' = -x_0' \sin t + p_0' \cos t \end{cases}$$

which are obtained from the (x, p) by the phase space translation $x' = x - v_0 t$, $p' = p - p_0$.

2.4 Constants of the Motion and Integrable Systems

Here is a venerable topic from the theory of Hamiltonian systems (for applications to specific problems, see for instance *[4, 52, 128]*, or any other book dealing with Hamiltonian mechanics).

2.4.1 The Poisson Bracket

Let F and G be two real functions of $z = (x, p)$ and, possibly of time t. By definition, the Poisson bracket of these two functions is

$$\{F, G\} = \nabla_p F \cdot \nabla_x G - \nabla_x F \cdot \nabla_p G \tag{2.81}$$

(in some texts the opposite sign convention is chosen). The Poisson bracket is related to the symplectic form Ω by the simple formula

$$\{F, G\} = \Omega(X_F, X_G), \tag{2.82}$$

where X_F and X_G are the "Hamilton fields" associated with F and G:

$$X_F = (\nabla_p F, -\nabla_x F) \quad , \quad X_G = (\nabla_p G, -\nabla_x G).$$

The Poisson bracket is obviously antisymmetric:

$$\{F, G\} = -\{G, F\}$$

and linear:

$$\begin{cases} \{\lambda F, G\} = \lambda \{F, G\} \quad (\lambda \in \mathbb{R}) \\ \{F, G + H\} = \{F, G\} + \{F, H\}. \end{cases}$$

It satisfies, moreover, the so-called *Jacobi identity*:

$$\{\{F, G\}, H\} + \{\{G, H\}, F\} + \{\{H, F\}, G\} = 0.$$

Poisson brackets are useful when one wants to study the constants of motion of Hamiltonian systems:

2.4.2 Constants of the Motion and Liouville's Equation

Let H be a time-dependent Hamiltonian (not necessarily of Maxwell type), and $(f_{t,t'})$ its time-dependent flow.

Definition 33 *A function $F = F(x, p, t)$ is called a* constant of the motion *for H if it is constant along each extended-phase space trajectory determined by Hamilton's equations, that is, if*

$$F(\tilde{f}_t(x', p', t')) = F(x', p', t') \tag{2.83}$$

for all (x', p', t') and all t.

Setting $z' = (x', p')$, $z = f_{t,t'}(z')$ condition (2.83) is equivalent to the relation

$$\frac{d}{dt} F(z, t) = 0$$

which we can rewrite, using the chain rule, as

$$\frac{\partial F}{\partial t}(z, t) + \nabla_x F(z, t) \cdot \dot{x} + \nabla_p F(z, t) \cdot \dot{p} = 0.$$

Since $\dot{x} = \nabla_p H$ and $\dot{p} = -\nabla_x H$, this can in turn be written, using Poisson brackets, as:

$$\frac{\partial F}{\partial t} + \{F, H\} = 0. \tag{2.84}$$

This equation is widely known in the literature as *Liouville's equation.* Observe that when F is time-independent, then Liouville's equation reduces to

$$\{F, H\} = 0. \tag{2.85}$$

We will come back to these conditions in a while, but let us first study two examples: the angular momentum and the energy.

Example 34 ***Angular momentum and central force fields.*** *Consider a single particle with mass m in three-dimensional configuration space placed in a scalar potential field. By definition, the angular momentum of that particle with respect to the origin is the vector product*

$$\mathcal{L} = \mathbf{r} \times \mathbf{p} = m(\mathbf{r} \times \dot{\mathbf{r}}). \tag{2.86}$$

We have, since $\dot{\mathbf{r}} \times \dot{\mathbf{r}} = 0$:

$$\frac{d}{dt}\mathcal{L} = m(\mathbf{r} \times \ddot{\mathbf{r}})$$

so that $\mathcal{L}$ is a constant of the motion if and only if $\mathbf{r} \times \ddot{\mathbf{r}} = 0$, that is if the acceleration vector is colinear with the position vector. This happens, for instance, when the potential is of the type $U = f(|\mathbf{r}|)$, because in this case

$$m\ddot{\mathbf{r}} = -\nabla_{\mathbf{r}} U = f'(|\mathbf{r}|)\frac{\mathbf{r}}{|\mathbf{r}|}.$$

This result generalizes without difficulty to many-particle systems by defining the total angular momentum as being the sum of the individual angular momenta:

$$\mathcal{L} = \sum_{j=1}^{N} \mathbf{r}_j \times \mathbf{p}_j \ .$$

Let us briefly discuss the notion of energy. By definition, the energy of a system with Maxwell Hamiltonian

$$H = \sum_{j=1}^{n} \frac{1}{2m_j}(p_j - A_j)^2 + U$$

is the value of H along the phase space curve followed by that system. The notion of energy is thus gauge-dependent, and has no absolute meaning whatsoever. Even worse, for a given gauge, the value of the energy is not Galilean invariant, so that in one inertial frame it can be conserved, while it is not in another! Here is typical example of such a situation (it is taken from Goldstein's book [52]):

Example 35 ***The cart with a spring revisited.*** *We consider again the device "cart with a spring" of Example 32 above. The Hamiltonian*

$$H = \frac{p^2}{2} + \frac{k}{2}(x - \nu_0 t)^2.$$

Calculating the solutions of Hamilton equations for H, we get the following value for the energy:

$$E(t) = \frac{1}{2}\left(p(0)^2 + \nu_0^2 + x(0)^2\right) + \nu_0(p(0)\cos t - x(0)\sin t).$$

The energy of the particle-cart system is thus not constant. If we choose instead the time-independent Hamiltonian

$$H' = \frac{p'^2}{2} + \frac{k}{2}x'^2$$

we get this time the constant value

$$E' = \frac{1}{2}\left(p'(0)^2 + x'(0)^2\right).$$

The different values for the energy that we have obtained can be interpreted as follows: in the first case, the observer is sitting along the track at the point O chosen as origin. The energy that observer measures is the variable quantity $E(t)$, because according to Newton's first law energy has to "flow into and out of" the system to keep the cart moving with uniform velocity against the reaction of the oscillating mass. In the second situation, the Hamiltonian corresponds to an observer sitting in the cart and who just observes a plain harmonic oscillator whose energy is constant, and equal to E'.

2.4.3 Constants of the Motion in Involution

We now assume that F, G, H,... are time-independent constants of the motion. We say that F and G are in *involution* if their Poisson bracket is zero:

$$\{F, G\} = 0$$

that is:

$$\nabla_p F \cdot \nabla_x G - \nabla_p G \cdot \nabla_x F = 0.$$

In view of Eq. (2.82) this is equivalent to saying that

$$\Omega(X_F, X_G) = 0.$$

We will say that n constants $F_1, ..., F_n$ of the motion are *independent* if the gradients $\nabla_z F_j$ are linearly independent functions:

$$\sum_{j=1}^{n} \lambda_j \nabla_z F_j = 0 \Longrightarrow \lambda_j = 0 \quad (1 \leq j \leq n).$$

That condition is equivalent to

$$\sum_{j=1}^{n} \lambda_j X_j = 0 \Longrightarrow \lambda_j = 0 \quad (1 \leq j \leq n)$$

where we have set $X_1 = X_{F_1}, ..., X_1 = X_{F_n}$ because $X_j = J\nabla_z F_j$.

Let us define the notion of *Liouville integrable system*:

Definition 36 *A system with Hamiltonian H defined on $\mathbb{R}^n_x \times \mathbb{R}^n_p \times \mathbb{R}_t$ is said to be Liouville integrable (or:completely integrable) if it has n independent constants of the motion $F_1, ..., F_n$ in involution.*

The Hamiltonian H actually can never have more than n independent constants of the motions in involution. For such Hamiltonians, the level sets of the n constants of the motion determine (except for exceptional values) a manifold which is topologically of a very special type. In fact (see, e.g., Arnold [4], §49, or Hofer-Zehnder, [89], Appendix A.2):

Proposition 37 *Let H be a time-independent integrable Hamiltonian (H is thus itself one of the constants of the motion, say $H = F_1$). The set V_f defined by the equations*

$$F_j(x,p) = f_j \quad (1 \leq j \leq n) \tag{2.87}$$

is, when non-empty, a submanifold of $\mathbb{R}^n_x \times \mathbb{R}^n_p$ for almost all $f = (f_1, ..., f_n)$. When it is connected, that manifold V_f can be transformed, using a convenient diffeomorphism, to a product of k circles ($0 \leq k \leq n$) and $n-k$ straight lines.

(One can actually prove that the diffeomorphism in question moreover has the property that its Jacobian matrix is symplectic at every point; this can be rephrased by saying that the diffeomorphism is a *symplectomorphism*: see next Chapter.)

It follows from Proposition 37 that when the manifold $V = V_f$ defined by the equations (2.87) is *compact* and *connected*, then it is basically the torus $T^n = \left(S^1\right)^n$. Let us illustrate this with a school-book example where one effectively has such an "invariant torus":

Example 38 ***The n-dimensional harmonic oscillator***. *Consider the Hamiltonian*

$$H = \frac{1}{2}(p^2 + x^2)$$

where $p = (p_1, ..., p_n)$ and $x = (x_1, ..., x_n)$. The solution of the associated equations of motion are

$$\begin{cases} x_j = x'_j \cos t + p'_j \sin t \\ p_j = -x'_j \sin t + p'_j \cos t \end{cases}$$

$(1 \leq j \leq n)$ where the x'_j and p'_j are the initial position and momentum coordinates. The n functions

$$H_j(x,p) = \frac{1}{2}(p_j^2 + x_j^2)$$

are obviously independent constants of the motion; they are also pairwise in involution, since

$$\{H_j, H_k\} = \sum_{i=1}^{n} \frac{\partial H_j}{\partial p_i} \cdot \frac{\partial H_k}{\partial x_i} - \frac{\partial H_j}{\partial x_i} \cdot \frac{\partial H_k}{\partial p_i}$$

is always equal to zero. Proposition 37 thus applies, and one sees in fact that the sets

$$\frac{1}{2}(p_j^2 + x_j^2) = E_j$$

are circles with radius $\sqrt{2E_j}$ (when all E_j are > 0).

The following elementary example shows that one can as well obtain a product of circles and straight lines (here, a cylinder):

Example 39 ***The "tired" harmonic oscillator**. We consider here the Hamiltonian*

$$H = \frac{1}{2}\left(p_x^2 + p_y^2\right) + \frac{1}{2}x^2.$$

The solutions of Hamilton's equations are here

$$\begin{cases} x = x' \cos t + p'_x \sin t \quad , \quad p_x = -x' \sin t + p'_x \cos t \\ y = p'_y t + y' \quad , \quad p_y = p'_y. \end{cases}$$

The functions

$$H_x = \frac{1}{2}\left(p_x^2 + x^2\right) \quad , \quad H_y = \frac{1}{2}p_y^2$$

are again independent constants of the motion in involution. The set V defined by the equations $H_x = E_x$, $H_y = E_y$ is the product of a circle with radius $\sqrt{E_x}$ and of two half-lines (the energy shells (= surfaces of constant energy) of H are in general not connected).

2.5 Liouville's Equation and Statistical Mechanics

We have studied until now Hamiltonian mechanics for finite systems of distinguishable particles. This means that we have been able, at least in principle, to follow each individual trajectory in phase space. We have thus being implicitly using what Dubin *et al.* [40] call:

> **The Principle of Complete Knowledge:** *All classical observables have a definite value at every point of phase space, and in principle these values may all be known simultaneously, with complete accuracy and without altering the state.*

Suppose now that the number N of particles is a "very large" number, so large that it would be a hopeless task to keep track of each particle individually. What one can do, however, is to make measurements about average properties of certain quantities. This leads to the consideration of "particle densities" and probabilities, which one studies by using statistical methods. It turns out that the same procedure is used for the experimental study of a small number non-interacting particles (or even of a single particle). This is because the only physical knowledge we can have of any "real" system is the result of measurements, which are essentially intervals of numbers: it does not make sense, for instance, to claim that the observation of a particle moving continuously on the x-axis has led to a position measurement which was exactly $\sqrt{2}$: no experiment can ever be performed in such a way that we could embrace infinitely many decimals at once. What one does in these cases, is to examine the properties of "statistical ensembles", i.e., of large numbers of ideally identical systems, and to treat the data thus obtained again by probabilistic and statistical methods. For instance, suppose we want to describe the motion of a single particle, perhaps under the action of some field. To obtain maximum precision, we must perform a great number of measurements of position and velocity, on similar particles, and this under conditions being ideally kept identical. We can then represent the results of our position and velocity measurements as a swarm of points in phase space, to which one can apply statistical methods. (It is a postulate of classical mechanics that this procedure will lead to an accurate description of "Reality".) Furthermore, if the number of observations is very large, we can approximate this swarm of points with a "fluid" in phase space. We can thus speak about the average density of that fluid: it is the average number of points per unit volume in phase space. We thus picture the fluid as a continuous system, i.e., we identify the swarm of points with a fluid having a continuously differentiable density $\rho(x,p,t)$ at the point (x,p) at time t.

2.5.1 Liouville's Condition

Now, a fundamental postulate of classical statistical Mechanics is that along each trajectory from $t \longmapsto z(t)$ followed by a "particle" of the "fluid", the density function ρ satisfies

$$\frac{d}{dt}\rho(z(t),t) = 0 \quad (\text{"}\textit{Liouville's condition}\text{"}). \tag{2.88}$$

The idea of introducing probability densities in phase space is due to J.W. Gibbs (b.1839); see [50]. Gibbs called the Liouville condition the *principle of density in phase*. Following this principle, classical statistical mechanics becomes a theory in which the motion of particles (or systems) is deterministic, but unpredictable individually: the particles move in phase space as if they constituted an incompressible fluid of varying density. Note that conservation of volume does not, however, mean conservation of shape; this was called by Gibbs the *principle of extension in phase*. We will discuss thoroughly these notions in the next Chapter, in connection with Gromov's non-squeezing property.

Liouville's condition can be motivated and justified heuristically in two different ways. We begin with a classical "particle counting" argument. Consider at time $t = 0$ a small volume $\mathcal{D}$ in phase space surrounding some given point-like particle. The boundary $\partial\mathcal{D}$ of $\mathcal{D}$ is formed by some surface of neighboring particles. In the course of time, the measure of the volume will remain constant in view of Liouville's theorem, although the volume $\mathcal{D}$ itself will be moved and distorted. Now, any particle inside $\mathcal{D}$ must remain inside $\mathcal{D}$: if some particle were to cross the boundary of $\mathcal{D}$ it would occupy at some time the same position in phase space as one of the particles defining $\partial\mathcal{D}$. Since the subsequent motion of any particle is entirely and uniquely determined by its location in phase space at a given time, the two particles would then travel together from there on, but this is absurd, and the particle can thus never leave $\mathcal{D}$. Reversing the argument, we also find that no particle can ever enter $\mathcal{D}$, so that the total number of particles within $\mathcal{D}$ must remain constant. Summarizing, both the measure of the volume and the number of particles are constant, and so is thus the density, as claimed. Consequently Liouville's condition (2.88) must hold.

A second possible interpretation of Liouville's condition is of probabilistic nature. Assume that the total mass

$$m \equiv \int \rho(z,t) d^n z$$

is non-zero and finite. Dividing ρ by m, we can thus assume that the normalization condition

$$\int \rho(z,t)\, d^n z = 1 \tag{2.89}$$

holds for all t, and this allows us to view ρ as a probability density. In fact, if we consider, as in the argument above, an "infinitesimal volume" $\mathcal{D}$ with measure ΔV around the point z, then $\rho(z,0)\Delta V$ will be the probability of finding a given particle inside Ω. But the probability of finding that particle in the image $\mathcal{D}_t$ of $\mathcal{D}$ by the flow $z \longmapsto z(t)$ is then

$$\rho(z(t),t)\Delta V_t = \rho(z,0)\Delta V \tag{2.90}$$

and since $\Delta V_t = \Delta V$ because volume is preserved, we must have

$$\rho(z(t),t) = \rho(z,0) \tag{2.91}$$

which is just Liouville's condition.

We see that, either way, Liouville's condition (2.88) appears to be a *conservation law*: no particles in phase space can be created or destroyed in classical statistical mechanics. Noticing that the time evolution of individual particles of the fluid is governed by Hamilton's equations, it follows that Liouville's condition is equivalent to the equation

$$\frac{\partial \rho}{\partial t} + \{H, \rho\} = 0 \tag{2.92}$$

since (2.91) means that ρ is a constant of the motion.

2.5.2 Marginal Probabilities

If we interpret ρ as a probability density, it makes sense to try to define its marginal probability densities in both position and momentum space. We will see that the evolution of these marginal densities is governed by "continuity equations" familiar from fluid mechanics. In what follows, H will be a Maxwell Hamiltonian in n dimensions:

$$H = \sum_{j=1}^{n} \frac{1}{2m_j} (p_j - A_j(x,t))^2 + U(x,t).$$

We assume that the normalization condition

$$\int \rho_0(z)\, d^{2n}z = 1$$

holds, and denote by $\rho = \rho(z,t)$ the solution of Liouville's equation with initial condition ρ_0. In view of Eq. (2.91) that solution satisfies $\rho(z(t),t) = \rho_0(z)$, that is, $\rho(z,t) = \rho_0(z(-t))$ (here $z(t) = (x(t), p(t))$ is the solution of Hamilton's equation

$$\dot{x} = \nabla_p H \quad , \quad \dot{p} = -\nabla_x H$$

with $z(0) = z$). It follows that we have

$$\int \rho(z,t)\, d^{2n}z = 1$$

for all t (this relation expresses the fact that the particle must be "somewhere" in phase space).

Proposition 40 *Assume that ρ_0 is compactly supported or, more generally, belongs to the Schwartz space $\mathcal{S}(\mathbb{R}_z^{2n})$. Then, the marginal probability densities*

$$\rho_X(x,t) = \int \rho(z,t)\, d^n p \quad , \quad \rho_P(x,t) = \int \rho(z,t)\, d^n x \tag{2.93}$$

satisfy the continuity equations:

$$\frac{\partial \rho_X}{\partial t} + \operatorname{div}(\rho_X v_X) = 0 \quad , \quad \frac{\partial \rho_P}{\partial t} + \operatorname{div}(\rho_P v_P) \tag{2.94}$$

where the velocity fields v_X and v_P are defined by:

$$\begin{cases} v_X \rho_X(x,t) = \int \nabla_p H(z,t)\rho(z,t)\, d^n p \\ v_P \rho_P(x,t) = -\int \nabla_x H(z,t)\rho(z,t)\, d^n x. \end{cases} \tag{2.95}$$

Proof. Differentiating the expression of ρ_X in (2.93) we have, taking Liouville's equation into account:

$$\frac{\partial \rho_X}{\partial t} = -\int (\nabla_p H \cdot \nabla_x \rho)\, d^n p + \int (\nabla_x H \cdot \nabla_p \rho)\, d^n p. \tag{2.96}$$

The first integral in the right-hand side of this equation consists of the sum from $j=1$ to $j=n$ of terms

$$\int \frac{\partial H}{\partial p_j} \frac{\partial \rho}{\partial x_j}\, d^n p_j = \frac{1}{m_j} \int (p_j - A_j) \frac{\partial \rho}{\partial x_j}\, d^n p_j.$$

Integrating by parts, the second integral is given by

$$\int \frac{\partial H}{\partial p_j}\frac{\partial \rho}{\partial p_j} d^n p = \left[\frac{\partial H}{\partial x_j}\rho\right]_{p=-\infty}^{p=+\infty} - \int \frac{\partial^2 H}{\partial x_j \partial p_j}\rho \, d^n p$$

that is, since ρ vanishes at infinity:

$$\int \frac{\partial H}{\partial p_j}\frac{\partial \rho}{\partial p_j} d^n p = \frac{1}{m_j}\int \frac{\partial A_j}{\partial x_j}\rho d^n p.$$

Inserting these expressions in (2.96), we get

$$\frac{\partial \rho_X}{\partial t} = \sum_{j=1}^{n}\left[-\frac{1}{m_j}\int \frac{\partial}{\partial x_j}(p_j - A_j)\rho\right] d^n p$$

which is the continuity equation for ρ_X. That the momentum marginal density ρ_γ satisfies the second continuity equation in (2.94) is proven exactly in the same way; the details of the calculations are left to the reader. ■

2.5.3 Distributional Densities: An Example

The results above can be extended without too many technical difficulties to "distribution-valued" densities in phase space. Consider, as an illustration, the case where we have a phase space density "concentrated" on an n-dimensional submanifold V of $\mathbb{R}^n_x \times \mathbb{R}^n_p$ given by n equations

$$p_j = \frac{\partial \Phi_0}{\partial x_j}(x) \quad , \quad 1 \leq j \leq n$$

where Φ_0 is some smooth function of the position variables $x_1, ..., x_n$. (We will see in Chapter 4 that V is the archetypical example of a "Lagrangian manifold".) We assume that the initial phase space density is of the type

$$\rho_0(x,p) = f(x)\delta(p - \nabla_x \Phi_0(x)) \tag{2.97}$$

where f is a smooth function (which is assumed to decay rapidly at infinity), and δ is Dirac's distribution. We demand that ρ_0 be normalized, in the sense that

$$\int f(x)\delta(p - \nabla_x \Phi_0(x)) d^n x d^n p = 1.$$

This condition is equivalent to

$$\int f(x)\left[\delta(p-\nabla_x\Phi_0(x))d^np\right]d^nx = 1$$

and hence to

$$\int f(x)d^nx = 1.$$

(If the reader does not find that our calculations are rigorous enough, he can fill in the gaps by replacing the integrals by distributional brackets $\langle\cdot,\cdot\rangle$.) Solving Liouville's equation with the initial condition ρ_0, we find that

$$\rho(x,p,t) = f(x(-t))\delta(p(-t)-\nabla_x\Phi_0(x(-t)))$$

which we can write in the form

$$\rho(x,p,t) = \rho(x,t)\delta(p-\nabla_x\Phi(x,t))$$

where $\rho(x,t) = f(x(-t))$ and $\Phi(x,t)$ is the solution of Hamilton-Jacobi's equation

$$\frac{\partial\Phi}{\partial t} + H(x,\nabla_x\Phi,t) = 0,\ \ \Phi(x,0) = \Phi_0(x).$$

(That $\delta(p(-t)-\nabla_x\Phi_0(x(-t))) = \delta(p-\nabla_x\Phi(x,t))$ follows from the Hamilton-Jacobi theory.) Let us now apply Proposition 40 to ρ; the marginal density ρ_X is a solution of the first equation (2.94), the velocity field being determined by

$$v_{X,j}(x,t)\rho_X(x,t) = \frac{1}{m_j}\int p_jf(x(-t))\delta(p-\nabla_x\Phi(x,t))d^np.$$

Since we have, by the properties of the Dirac distribution,

$$p_j\delta(p-\nabla_x\Phi(x,t)) = \frac{\partial\Phi}{\partial x_j}(x,t)\delta(p-\nabla_x\Phi(x,t))$$

it follows that

$$\begin{aligned} v_{X,j}(x,t)\rho_X(x,t) &= \frac{1}{m_j}\frac{\partial\Phi}{\partial x_j}(x,t)f(x(-t)) \\ &= \frac{1}{m_j}\frac{\partial\Phi}{\partial x_j}(x,t)\rho(x,t). \end{aligned}$$

On the other hand, a direct calculation yields

$$\rho_X(x,t) = \int f(x(-t))\delta(p-\nabla_x\Phi(x,t))d^np = f(x(-t))$$

so we finally have

$$v_{X,j}(x,t) = \frac{1}{m_j}\frac{\partial \Phi}{\partial x_j}(x,t) \tag{2.98}$$

and the equation satisfied by the marginal density ρ_X satisfies again the continuity equation (2.94):

$$\frac{\partial \rho_X}{\partial t} + \operatorname{div}(\rho_X v_X) = 0$$

the velocity field being given by formula (2.98).

Chapter 3

The Symplectic Group

Summary. *Symplectic matrices form a connected Lie group. Hamiltonian flows consist of symplectomorphisms. Gromov's theorem shows that Hamiltonian flows preserve symplectic capacities. This leads to a topological version of Heisenberg's uncertainty principle in classical mechanics. It also leads to a topological quantization of phase space in "quantum blobs", and to the Maslov quantization of Lagrangian manifolds.*

This Chapter is devoted to a thorough study of the properties of the symplectic group $\mathrm{Sp}(n)$. The importance of that group not only comes from the fact that it is the symmetry group of Hamiltonian mechanics, but also from one of its topological properties, discovered in 1985. That property – the Gromov non-squeezing theorem, *alias* the principle of the symplectic camel – says that the action of symplectic transformations on phase space has a "rigidity" that fundamentally distinguishes them from arbitrary volume-preserving mappings. (In particular, an arbitrary volume-preserving diffeomorphism cannot be approximated by a sequence of symplectic mappings.) The principle of the symplectic camel actually leads to a topological version of Heisenberg's uncertainty principle in *classical mechanics*!

3.1 Symplectic Matrices and $\mathrm{Sp}(n)$

We will work with real $2n \times 2n$ matrices written in "block form"

$$s = \begin{pmatrix} A & B \\ C & D \end{pmatrix} \tag{3.1}$$

where each of the entries A, B, C, D is an $n \times n$ matrix. The transpose of s is:

$$s^T = \begin{pmatrix} A^T & C^T \\ B^T & D^T \end{pmatrix}. \tag{3.2}$$

Recall (formula (2.55)) that the matrix J is defined by:

$$J = \begin{pmatrix} 0 & I \\ -I & 0 \end{pmatrix}$$

where $0 = 0_{n \times n}$ and $I = I_{n \times n}$, and that we have $\det J = 1$.

Definition 41 *We will say that the matrix s is* symplectic *if*

$$s^T J s = s J s^T = J. \tag{3.3}$$

Remark 42 *The conditions $sJs^T = J$ and $s^T J s = J$ are in fact equivalent as can be seen by the following simple argument: the inverse of a symplectic matrix is also symplectic (see below), hence $sJs^T = J$ is equivalent to $s^{-1}J(s^{-1})^T = J$; replacing s with s^T (which is also symplectic: see below) this in turn is equivalent to $(s^{-1})^T J s^{-1} = J$ that is to $s^T J s$.*

The matrix J is itself obviously symplectic, and so is the identity matrix. Note that it immediately follows from Eq. (3.3) that a symplectic matrix has determinant ± 1, and is hence invertible. We will see later on that symplectic matrices actually always have determinant $+1$:

$$s \textit{ symplectic} \implies \det s = +1.$$

A straightforward calculation, using (3.3) and the invertibility of s, shows that a matrix (3.1) is symplectic if and only if any of the three sets of equivalent conditions below holds:

$$\begin{cases} A^T C,\ D^T B \text{ symmetric, } A^T D - C^T B = I \\ AB^T,\ CD^T \text{ symmetric, } AD^T - BC^T = I \\ DC^T,\ AB^T \text{ symmetric, } DA^T - CB^T = I. \end{cases} \tag{3.4}$$

It follows, in particular, that the inverse of a symplectic matrix (3.1) is given by

$$s^{-1} = \begin{pmatrix} D^T & -B^T \\ -C^T & A^T \end{pmatrix}. \tag{3.5}$$

Note that in the case $n = 1$ where A, B, C, D are real numbers, the conditions (3.4) reduce to $AD - BC = 1$, i.e. $\det(s) = 1$ (the proof in the general case is much less straightforward).

The conditions (3.3) can be expressed in terms of the standard symplectic form on $\mathbb{R}^n_x \times \mathbb{R}^n_p$:

Definition 43 *The* standard symplectic form *on phase space* $\mathbb{R}^{2n} \equiv \mathbb{R}^n_x \times \mathbb{R}^n_p$ *is the antisymmetric bilinear form defined by*

$$\Omega(z, z') = -z^T J z' = z'^T J z \tag{3.6}$$

(z and z' being written as column vectors). If $z = (x, p)$, $z' = (x', p')$ we thus have

$$\Omega(x, p; x', p') = p \cdot x' - p' \cdot x$$

where the dot $\cdot$ denotes the standard scalar product of vectors in $\mathbb{R}^2$.

The number $\Omega(z, z')$ is called the *symplectic product* (or *skew-product*) of the vectors z and z'. It is sometimes denoted by $z \wedge z'$. The symplectic product has the following immediate interpretation in the language of differential forms: $\Omega(z, z')$ is the value on (z, z') of the 2-form

$$dp \wedge dx = dp_1 \wedge dx_1 + \cdots + dp_n \wedge dx_n.$$

We will denote that 2-form by Ω.

Observe that when $n = 1$ the number $\Omega(z, z')$ is just minus the determinant of the vectors z and z': $\Omega(z, z') = -\det(z, z')$. In the general case $\Omega(z, z')$ can be expressed, using determinants, as

$$\Omega(x, p; x', p') = -\sum_{j=1}^{n} \begin{vmatrix} x_j & x'_j \\ p_j & p'_j \end{vmatrix}.$$

It is thus the algebraic sum of the areas of the projections of the vectors z, z' on the conjugate coordinates axes.

Remark 44 *In many texts the opposite sign convention is used; formula (3.6) should then be replaced by $\Omega(z, z') = -zJz'^T$. The reason for our choice of sign is that it immediately identifies Ω with the exterior derivative of the action 1-form $pdx = p_1 dx_1 + \cdots + p_n dx_n$:*

$$\Omega = d(pdx) = dp \wedge dx.$$

Since the condition $s^T Js = J$ is equivalent to $(sz)^T J(sz') = z^T Jz'$ for all z, z', a $2n \times 2n$ matrix s is symplectic if and only if we have

$$\Omega(sz, sz') = \Omega(z, z') \tag{3.7}$$

for all vectors z, z' in $\mathbb{R}_x^n \times \mathbb{R}_p^n$. In terms of the differential form $\Omega = dp \wedge dx$ this can be written $s^*\Omega = \Omega$, where the upper star * denotes the "pull-back" of differential forms by mappings.

Exactly as Euclidean geometry is the study of orthogonal transformations, that is, of transformations which preserve the scalar product, symplectic geometry is the study of the transformations preserving the symplectic product. In spite of this similarity, both geometries are fundamentally different. One proves for instance, using a famous theorem due to Darboux (see [4, 89, 116]), that all symplectic manifolds are locally identical to the standard symplectic space, and hence "flat". This is of course in strong contrast with what happens with Riemannian manifolds, which are not all locally identical (they can have different curvatures). See Gotay and Isenberg's paper [75] for a discussion of the deep differences between Riemannian and symplectic geometry.

The inverse of a symplectic matrix s is also symplectic, since the condition $s^T Js = J$ is equivalent to $s^{-1}J(s^{-1})^T = J$ (and the inverse of s is given by Eq. (3.5). If two matrices s and s' are symplectic, then so is their product ss': by repeated use of (3.3) we have

$$(ss')J(ss')^T = s(s'Js'^T)s^T = sJs^T = J.$$

The identity matrix being trivially symplectic, it follows that the set of all symplectic matrices is a *multiplicative group.*

Definition 45 *The group of all symplectic* $2n \times 2n$ *matrices is denoted by* $\mathrm{Sp}(n)$, *and is called the* symplectic group.

We will identify without *états d'âme* symplectic matrices and the linear transformations of phase space they represent (this amounts to choose once for all the canonical basis of $\mathbb{R}_x^n \times \mathbb{R}_p^n$ to represent matrices).

3.2 Symplectic Invariance of Hamiltonian Flows

Let H be a general Maxwell Hamiltonian

$$H = \sum_{j=1}^{n} \frac{1}{2m_j}(p_j - A_j)^2 + U$$

where the potentials $A = (A_1(x,t), ..., A_n(x,t))$ and $U = U(x,t)$ are allowed to be time dependent. Recall that such a Hamiltonian is written, with the usual abuse of notation:

$$H = \frac{1}{2m}(p - A)^2 + U$$

where m is the mass matrix. If H is not a quadratic polynomial in the position and momentum variables, then the associated flow $(f_{t,t'})$ is not linear, and does therefore not consist of symplectic matrices. However, the Jacobian matrix of each of the $f_{t,t'}$, calculated at *every* point $z = (x,p)$, is symplectic. Before we prove this fundamental property, let us introduce some notations that will be used without further comment in the rest of this book.

3.2.1 Notations and Terminology

We will usually denote "initial" points in phase space by (x', p') and "final" points by (x, p). This is of course totally consistent with the notation $(f_{t,t'})$ for the time-dependent flow, since we have:

$$(x,p) = f_{t,t'}(x', p')$$

where the points (x', p') and (x, p) are thus the position of the system described by H at times t' and t, respectively. Allowing (x', p') to be variable, we can view the "final" point (x, p) as being a function the "initial" point (x', p'). Setting $z' = (x', p')$ and $z = (x, p)$ we denote by $s_{t,t'}(z')$ the Jacobian matrix of $f_{t,t'}$ calculated at z':

$$s_{t,t'}(z') = f'_{t,t'}(z') \tag{3.8}$$

that is:

$$s_{t,t'}(z') = \frac{\partial z}{\partial z'} = \frac{\partial(x,p)}{\partial(x',p')}.$$

The matrix $s_{t,t'}(z')$ can be written in block-matrix form as

$$s_{t,t'}(z') = \begin{pmatrix} \frac{\partial x}{\partial x'} & \frac{\partial x}{\partial p'} \\ \frac{\partial p}{\partial x'} & \frac{\partial p}{\partial p'} \end{pmatrix}$$

where $\partial x/\partial x'$ is the Jacobian matrix of the mapping $x' \longmapsto x$, and so on. Obviously $s_{t,t}(z)$ is the identity matrix for all values of z and t.

We will see that $s_{t,t'}(z)$ is symplectic for each z, and express this property by saying that the $f_{t,t'}$ are *symplectomorphisms.* (One also says: *canonical transformations*, especially in the physical literature.) More generally, every mapping $f : \mathbb{R}^n_x \times \mathbb{R}^n_p \longrightarrow \mathbb{R}^n_x \times \mathbb{R}^n_p$ whose Jacobian matrix $Df(x,p)$ is symplectic everywhere is called a symplectomorphism.

Here is an example of a symplectomorphism which has as simple interpretation in differential geometry:

Example 46 ***Mathieu transforms.*** *Let $f : \mathbb{R}^n_x \longrightarrow \mathbb{R}^n_x$ be a diffeomorphism of x-space. Then $M(f)$ where*

$$M(f)(x) = \begin{pmatrix} f(x) & 0 \\ 0 & [Df(x)^{-1}]^T \end{pmatrix}$$

is a canonical transformation, called "Mathieu transform" associated to f. It is the diffeomorphism on the cotangent space $T^*\mathbb{R}^n_x \equiv \mathbb{R}^{2n}_{x,p}$ induced by f.

3.2.2 Proof of the Symplectic Invariance of Hamiltonian Flows

We begin by proving a technical lemma, which shows that the mappings $t \longmapsto s_{t,t'}(z')$ satisfy a simple differential equation:

Lemma 47 *For fixed z' and t' set $z(t) = f_{t,t'}(z')$, $s(t) = s_{t,t'}(z')$. The matrix function $s(t)$ satisfies the differential equation*

$$\dot{s}(t) = J\, H''_{x,p}(z(t),t)s(t) \tag{3.9}$$

where

$$H''_{x,p} = \left(\frac{\partial^2 H}{\partial x_i \partial p_j}\right)_{1\leq i,j\leq n}$$

is the Hessian matrix of H.

Proof. For notational simplicity we give the proof in the case $n = 1$. We have, by definition

$$s(t) = \begin{pmatrix} \frac{\partial x}{\partial x'} & \frac{\partial x}{\partial p'} \\ \frac{\partial p}{\partial x'} & \frac{\partial p}{\partial p'} \end{pmatrix}$$

and hence

$$\dot{s} = \begin{pmatrix} \frac{\partial \dot{x}}{\partial x'} & \frac{\partial \dot{x}}{\partial p'} \\ \frac{\partial \dot{p}}{\partial x'} & \frac{\partial \dot{p}}{\partial p'} \end{pmatrix}. \tag{3.10}$$

Using the chain rule, and taking Hamilton's equations into account, we have

$$\frac{\partial \dot{x}}{\partial x'} = \frac{\partial^2 H}{\partial x \partial p}\frac{\partial x}{\partial x'} + \frac{\partial^2 H}{\partial p^2}\frac{\partial p}{\partial x'}$$

and similar equalities for the other entries in the matrix (3.10). It follows that

$$\dot{s} = \begin{pmatrix} \frac{\partial^2 H}{\partial x \partial p} & \frac{\partial^2 H}{\partial p^2} \\ -\frac{\partial^2 H}{\partial x^2} & -\frac{\partial^2 H}{\partial p \partial x} \end{pmatrix} \begin{pmatrix} \frac{\partial x}{\partial x'} & \frac{\partial x}{\partial p'} \\ \frac{\partial p}{\partial x'} & \frac{\partial p}{\partial p'} \end{pmatrix}$$

which is the same thing as Eq. (3.9). ∎

That the matrices $s(t) = s_{t,t'}(z')$ are symplectic readily follows:

Theorem 48 *The Jacobian matrix $s_{t,t'}(z')$ of $f_{t,t'}(z')$ is symplectic at every point $z' = (x', p')$: $s_{t,t'}(z') \in \mathrm{Sp}(n)$.*

Proof. By definition of $\mathrm{Sp}(n)$, we have to show that the matrix $s(t) = s_{t,t'}(z')$ satisfies the condition

$$s(t)^T J\, s(t) = J. \tag{3.11}$$

It suffices in fact to show that $M(t) = s(t)^T J\, s(t)$ is *constant*, because we will then have $M(t) = M(t') = J$, which is Eq. (3.11), since $s(t') = I$. Calculating the derivative of $M(t)$ we get, using the product rule and (3.9):

$$\dot{M} = (\dot{s}\,)^T\, J\, \dot{s} + s^T J\, \dot{s} = s^T H'' s - s^T H'' s = 0$$

as claimed. ∎

3.2.3 Another Proof of the Symplectic Invariance of Flows*

Recall from Remark 26 that the contraction of the differential $\tilde{\Omega}_H = d\tilde{\lambda}_H$ of the Poincaré-Cartan form with the suspended vector field $\tilde{X}_H$ is zero:

$$i_{\tilde{X}_H}\tilde{\Omega}_H = 0. \tag{3.12}$$

Let now $\mathcal{L}_{\tilde{X}_H}\tilde{\Omega}_H$ be the Lie derivative of $\tilde{\Omega}_H$ in the direction of the field $\tilde{X}_H$, that is

$$\mathcal{L}_{\tilde{X}_H}\tilde{\Omega}_H = \lim_{t \to 0} \frac{(\tilde{f}_t)^*\tilde{\Omega}_H - \tilde{\Omega}_H}{t}.$$

In view of Cartan's homotopy formula, we have

$$\mathcal{L}_{\tilde{X}_H}\tilde{\Omega}_H = i_{\tilde{X}_H}d\tilde{\Omega}_H + di_{\tilde{X}_H}\tilde{\Omega}_H,$$

and hence $d\tilde{\Omega}_H = 0$ since $\tilde{\Omega}_H = d\tilde{\lambda}_H$. Taking Eq. (3.12) into account, we have $\mathcal{L}_{\tilde{X}_H}\tilde{\Omega}_H = 0$, which implies that

$$(\tilde{f}_t)^*\tilde{\Omega}_H = \tilde{\Omega}_H + \alpha \tag{3.13}$$

where α is a constant form. Setting $t = 0$, we see that in fact $\alpha = 0$, and hence

$$(f_{t,t'})^*\Omega = \Omega \tag{3.14}$$

which means that the Jacobian matrix $f'_{t,t'}(z_0)$ is symplectic at every point z_0 where it is defined:

$$\Omega(f'_{t,t'}(z_0)z, f'_{t,t'}(z_0)z') = \Omega(z, z')$$

for all z, z'. This is the same as saying that $f_{t,t'}$ is a symplectomorphism.

3.3 The Properties of $\mathrm{Sp}(n)$

We already mentioned that a symplectic matrix always has determinant +1. Let us prove this.

Proposition 49 *Every symplectic matrix has determinant* +1*:*

$$\mathrm{Sp}(n) \subset S\ell(2n, \mathbb{R}).$$

Proof. We are going to prove this by a topological argument. (An alternative proof, using the generators of $\mathrm{Sp}(n)$ will be given later.) The restriction of the determinant function to $\mathrm{Sp}(n)$ is a continuous function, which can only take the two values +1 and −1. It is thus a locally constant function. It turns out that $\mathrm{Sp}(n)$ is a connected Lie group (this will be established below). It follows that the determinant is actually constant $\mathrm{Sp}(n)$, and equal either +1 or −1. Taking $s = I$, we have in fact $\det(s) = \det I = 1$ for all s.

∎

Notice that since a 2×2 matrix with determinant one automatically is symplectic, we have, as already pointed out above $\mathrm{Sp}(1) = S\ell(2, \mathbb{R})$. However, as soon as $n > 1$, $\mathrm{Sp}(n)$ *never* coincides with $S\ell(2n, \mathbb{R})$: in the general case $\mathrm{Sp}(n) \subsetneq S\ell(2n, \mathbb{R})$. (We will use a "dimension count" argument in a moment to calculate the size of the "deviation".)

3.3.1 The Subgroups $U(n)$ and $O(n)$ of Sp(n)

The complex unitary group $U(n,\mathbb{C})$ and its subgroup, the real orthogonal group $O(n,\mathbb{R})$ can be identified with subgroups of Sp(n). This is done as follows: let $R = A + iB$ (A and B real) be an $n \times n$ matrix. The condition $R \in U(n,\mathbb{C})$ means that $RR^* = R^*R = I$, and this is equivalent to the sets of equivalent conditions

$$\left\{ \begin{array}{l} A^TA + B^TB = I \text{ and } AB^T = BA^T \\ AA^T + BB^T = I \text{ and } A^TB = B^TA. \end{array} \right. \tag{3.15}$$

It follows, using any of the formulas (3.4), that

$$r = \begin{pmatrix} A & -B \\ B & A \end{pmatrix} \tag{3.16}$$

is symplectic. If, conversely, a matrix of the type (3.16) is symplectic, then A and B must satisfy the relations (3.15), and $R = A + iB$ must then be unitary.

The set of all symplectic matrices (3.16) is a subgroup of Sp(n); that subgroup is denoted by $U(n)$. The orthogonal subgroup $O(n,\mathbb{R})$ of $U(n,\mathbb{C})$ is also identified with a subgroup of Sp(n). That subgroup, which we denote by $O(n)$, consists of all matrices

$$r = \begin{pmatrix} A & 0 \\ 0 & A \end{pmatrix}, \quad AA^T = A^TA = I. \tag{3.17}$$

The groups $U(n)$ and $O(n)$ will be called the *unitary* and *orthogonal subgroups of* Sp(n), respectively.

Proposition 50 *The elements of $U(n)$ are the only symplectic transformations that are at the same time rotations:*

$$\mathrm{Sp}(n) \cap O(2n,\mathbb{R}) = U(n). \tag{3.18}$$

Proof. Let us first show that $U(n) \subset \mathrm{Sp}(n) \cap O(2n,\mathbb{R})$. The inverse of $r \in U(n)$ is r^T and hence $r \in O(2n,\mathbb{R})$, so that $r \in \mathrm{Sp}(n) \cap O(2n,\mathbb{R})$. Conversely, let r be an orthogonal symplectic matrix. Then $Jr = rJ$, and one checks that r is a block matrix (3.16) whose entries satisfy the conditions (3.15). ∎

3.3.2 The Lie Algebra $\mathfrak{sp}(n)$

We have seen that the flow determined by a (homogeneous) quadratic Hamiltonian is given by $s_t = \exp(tX)$ where X is given by Eq. (3.30). These matrices are not arbitrary $2n \times 2n$ matrices: we see, by simple inspection, that their off-diagonal blocks are both symmetric, and that the transpose of any of the two diagonal blocks is equal to the other, up to their sign. These properties are in fact characteristic of the elements of the Lie algebra $\mathfrak{sp}(n)$ of the symplectic group $\mathrm{Sp}(n)$, which we study now.

The Lie algebra $\mathfrak{sp}(n)$ consists of all matrices X such that $e^{tX} \in \mathrm{Sp}(n)$ for all $t \in \mathbb{R}$. In view of Definition 41 of a symplectic matrix, we thus have

$$X \in \mathfrak{sp}(n) \Longleftrightarrow e^{tX^T} J e^{tX} = J$$

for all real numbers t. Differentiating both sides of this equality, we get

$$e^{tX^T}(X^T J + JX)e^{tX} = 0$$

and hence $X^T J + JX = 0$. Conversely, if this equality holds, then $e^{tX^T} J e^{tX}$ must be a constant matrix; choosing $t = 0$, that matrix must be J. We have thus proven that

$$X \in \mathfrak{sp}(n) \Longleftrightarrow X^T J + JX = 0 \; ; \tag{3.19}$$

since the transpose of J is $-J$, this equivalence can be rewritten as

$$X \in \mathfrak{sp}(n) \Longleftrightarrow JX = (JX)^T. \tag{3.20}$$

Summarizing:

Proposition 51 *The Lie algebra $\mathfrak{sp}(n)$ of the symplectic group $\mathrm{Sp}(n)$ consists of all matrices*

$$X = \begin{pmatrix} \alpha & \beta \\ \gamma & -\alpha^T \end{pmatrix} \quad , \quad \beta = \beta^T \; , \; \gamma = \gamma^T. \tag{3.21}$$

In particular, the Lie algebra $\mathfrak{sp}(1) = sl(2,\mathbb{R})$ consists of all real 2×2 matrices with trace equal to zero.

We have a one-to-one correspondence between quadratic polynomials in x, p, and elements of the Lie algebra $\mathfrak{sp}(n)$. It is easy to make this correspondence explicit: suppose that we write H as a polynomial in x, p, written as

$$H = \frac{1}{2}\beta p^2 + \alpha x \cdot p - \frac{1}{2}\gamma x^2 \tag{3.22}$$

where the matrices β and γ are symmetric; the associated Hamilton system is

$$\dot{x} = \beta x + \alpha p \quad , \quad \dot{p} = \gamma x - \alpha^T p$$

and its solution is $z(t) = \exp(tX)z(0)$ where X is given by Eq. (3.21).

Remark 52 *This does not mean, of course, that the one-parameter subgroups of* Sp(n) *cover* Sp(n) *(which would be true if* Sp(n) *were compact). For instance, one checks that the matrix*

$$s = \begin{pmatrix} -2 & 0 \\ 0 & -\frac{1}{2} \end{pmatrix}$$

never is of the form e^X *and hence, in particular, no Hamiltonian flow* (f_t) *"passes through"* s*. (See Frankel [48], page 407.)*

3.3.3 Sp(n) as a Lie Group

It turns out that the Sp(n) is in fact one of the "classical Lie groups", that is, it is a closed subgroup of $Gl(2n, \mathbb{R})$. It is closed, because it is defined by a condition of the type

$$s \in \mathrm{Sp}(n) \Longleftrightarrow f(s) = 0$$

where f is a continuous function $G\ell(2n, \mathbb{R}) \longrightarrow \mathbb{R}$, here the function $f(s) = s^T Js - J$. More precisely:

Proposition 53 *The symplectic group* Sp(n) *is a connected Lie group with dimension* $n(2n+1)$*; in fact*

$$\mathrm{Sp}(n) \simeq U(n) \times \mathbb{R}^{n(n+1)/2}. \tag{3.23}$$

(The symbol $\simeq$ *means "homeomorphic to".)*

Proof. The homeomorphism (3.23) can be explicitly constructed by using for instance the polar decomposition theorem for elements of $G\ell(2n, \mathbb{R})$ (see, e.g., [79, 80, 118]). One can also note that since $U(n)$ is a maximal compact subgroup of Sp(n) (see the references above), the result then follows from a deep theorem of E. Cartan, which says that an "algebraic" Lie group is homeomorphic to the product of any of its maximal compact subgroup (they are all conjugate, and hence homeomorphic), and a Euclidean space (see any textbook on the theory of Lie groups, for instance [118], §3.5). The connectedness of $\mathrm{Sp}(n)$ immediately follows from (3.23) since both $U(n)$ and $\mathbb{R}^{n(n+1)/2}$ are connected. That $\mathrm{Sp}(n)$ has dimension $n(2n+1)$ follows by dimension count. We can however give two conceptually simpler proofs of this fact. Writing $X \in \mathfrak{sp}(n)$ in the form (3.21) such a matrix can be parametrized by the n^2 arbitrary entries of α plus the

$$n(n+1)/2 + n(n+1)/2 = n(n+1)$$

arbitrary entries of the symmetric matrices β and γ. Hence

$$\dim \mathfrak{sp}(n) = n^2 + n(n+1) = n(2n+1).$$

Since, as manifolds, a connected Lie group and its Lie algebra have the same dimension, $\mathrm{Sp}(n)$ is thus an $n(2n+1)$-dimensional Lie group, as claimed. ∎

Remark 54 *Here is a direct proof of the equality* $\dim \mathrm{Sp}(n) = n(2n+1)$. *Definition 3.3 of symplectic matrices imposes constraints on the* $4n^2$ *entries of such a matrix. Since* J *is antisymmetric, there are exactly* $2n(2n-1)/2$ *independent conditions, and every element of* $\mathrm{Sp}(n)$ *thus depends on*

$$4n^2 - n(2n-1) = n(2n+1)$$

independent parameters. It follows that $\mathrm{Sp}(n)$ *has dimension* $n(2n+1)$ *as a Lie group.*

Since the condition $s \in S\ell(n,\mathbb{R})$ is equivalent to $\det s = +1$, the group $S\ell(n,\mathbb{R})$ has dimension $4n^2 - 1$; hence:

$$\begin{aligned}\operatorname*{codim}_{S\ell(2n,\mathbb{R})} \mathrm{Sp}(n) &= (4n^2 - 1) - n(2n+1) \\ &= (2n+1)(n-1).\end{aligned}$$

The symplectic group $\mathrm{Sp}(n)$ is thus "much smaller" than $S\ell(n,\mathbb{R})$, except for $n = 1$ (in which case both groups are identical). Since the ratio $n(2n+1) : (4n^2 - 1)$ has limit $1/2$ when $n \to \infty$, there is a 50% chance to choose at random a symplectic matrix from a bag containing all matrices with determinant one.

3.4 Quadratic Hamiltonians

Quadratic Maxwell Hamiltonians, that is, Maxwell Hamiltonians which are quadratic polynomials in the position and momentum coordinates are not just trivial or academic variations on the theme of the harmonic oscillator. They are associated to many interesting and sometimes sophisticated physical systems. Here are a few examples, which will be developed in the forthcoming subsections:

(**1**) *the triatomic molecule*;
(**2**) *the electron in a uniform magnetic field*;
(**3**) *the Coriolis force*;
(**4**) *Oscillations near equilibrium.*

A quadratic Maxwell Hamiltonian can always be written (up to an additive constant) in the form

$$H(z) = \frac{1}{2}Mz \cdot z + k \cdot z \tag{3.24}$$

where M is a real symmetric $2n \times 2n$ matrix, and k a fixed vector. For instance,

$$H(z) = \sum_{j=1}^{n} \frac{1}{2m_j}(p_j - A_j \cdot x)^2 + \frac{1}{2}Kx^2 + a \cdot x, \tag{3.25}$$

where $A_j = A_j(t)$ and $a = a(t)$ are vectors and $K = K(T)$ a symmetric matrix, is of the type above. Using the mass matrix m defined in Section 2.1.6, such a Hamiltonian can always be written in the short-hand form

$$H = \frac{1}{2m}(p - Ax)^2 + \frac{1}{2}Kx^2 + a \cdot x \tag{3.26}$$

where A is the $n \times n$ matrix $(A_j)_{1 \leq j \leq n}$. (We are using again here the abuse of notation $1/m = m^{-1}$.)

In the case $n = 1$, the archetypical example is the one-dimensional harmonic oscillator:

$$H = \frac{1}{2m}(p^2 + m\omega^2 x^2).$$

If we denote the coordinate pair (x, p) by the letter z (viewed in calculations as a column vector), a Hamiltonian (3.26) can always be written in the compact form

$$H = \frac{1}{2}Rz^2 + a \cdot x \tag{3.27}$$

where R is the symmetric block-matrix:

$$R = \begin{pmatrix} K + A^T m^{-1} A & -A^T m^{-1} \\ \\ -m^{-1} A & m^{-1} \end{pmatrix}. \tag{3.28}$$

When H is a homogeneous quadratic polynomial

$$H = \frac{1}{2} z^T R z$$

(R a time-independent symmetric matrix), Hamilton's equations $\dot{z} = J\nabla_z H$ become the *linear* system

$$\dot{z} = JRz \tag{3.29}$$

whose solution is

$$z(t) = e^{tJR} z(0).$$

We denote by (s_t) the flow defined by this formula; in view of definition of R we thus have $s_t = e^{tX}$ where the matrix $X = RJ$ is given by

$$X = \begin{pmatrix} -m^{-1} A & m^{-1} \\ \\ -K - A^T m^{-1} A & A^T m^{-1} \end{pmatrix}. \tag{3.30}$$

3.4.1 The Linear Symmetric Triatomic Molecule

(From Goldstein [52], §6–4.) We consider a molecule consisting of three aligned atoms: two atoms of mass m are symmetrically located on each side of an atom of mass m_0. In the equilibrium configuration the distances apart are equal to d. We suppose that the molecule vibrates along its line, which we choose to be the x-axis after having chosen an orientation on it. A good model for approximating the actual complicated interatomic potentials is to view the molecule as a system of two springs of force constant k joining the atom of mass m_0 to the two other atoms. An origin on the x-axis being chosen, we denote by x and z the coordinates of the particles with mass m, and by y that of the central particle; we denote by $\mathbf{p}$ the momentum vector (p_x, p_y, p_z). With these notations the Hamiltonian of the molecule is

$$H = \frac{1}{2} M^{-1} \mathbf{p}^2 + \frac{k}{2} \left[(y - x - d)^2 + (z - y - d)^2 \right] \tag{3.31}$$

where M is the mass matrix:

$$M = \begin{pmatrix} m & 0 & 0 \\ 0 & m_0 & 0 \\ 0 & 0 & m \end{pmatrix}.$$

Making the change of variables $x_1 = x - x_0$, $x_2 = y - y_0$, $x_3 = z - z_0$ where x_0, y_0, z_0 are the coordinates of the three atoms in equilibrium position, and taking into account the relations $y_0 - x_0 = z_0 - y_0 = d$, the Hamiltonian becomes

$$H = \frac{1}{2} M^{-1} \mathbf{p}^2 + \frac{k}{2} \left[(x_1 - x_2)^2 + (x_2 - x_3)^2 \right].$$

This function is of the type (3.27) with $x = (x_1, x_2, x_3)$, $A = 0$ and

$$K = \begin{pmatrix} k & -k & 0 \\ -k & 2k & -k \\ 0 & -k & k \end{pmatrix}.$$

The matrix (3.28) here is the 6×6 matrix

$$R = \begin{pmatrix} K & 0 \\ 0 & M^{-1} \end{pmatrix}.$$

This example can be extended without difficulty to the case of N aligned particles forming an open or closed chain.

3.4.2 Electron in a Uniform Magnetic Field

Consider a hydrogen atom placed in a magnetic field $\mathbf{B}$. Neglecting spin effects, an approximation for the Hamiltonian of the particle is

$$H = \frac{1}{2m} \left(\mathbf{p} - \frac{e}{c} \mathbf{A} \right)^2 - \frac{e^2}{r} \tag{3.32}$$

where the vector potential $\mathbf{A}$ is, as usual, determined by the equation $\mathbf{B} = \nabla_r \times \mathbf{A}$. Suppose now the atom is prepared in a very highly excited but still bound state, near the ionization threshold. The electron can then be viewed, to a good approximation, as free, except for the presence of the magnetic field. This is a situation encountered in alkali metals (it was first investigated by Landau), but recent experiments have been performed on hydrogen atoms. Since r is large, we can neglect the Coulomb potential $-e^2/r$, and thus assume that the Hamiltonian is

$$H = \frac{1}{2m}\left(\mathbf{p} - \frac{e}{c}\mathbf{A}\right)^2.$$

Suppose now that the magnetic field is uniform in space, and that its direction is the z-axis: $\mathbf{B} = (0, 0, B_z)$. The coordinates of $\mathbf{A}$ will then satisfy the equations

$$\frac{\partial A_y}{\partial z} - \frac{\partial A_z}{\partial y} = \frac{\partial A_x}{\partial z} - \frac{\partial A_z}{\partial x} = 0 \quad , \quad \frac{\partial A_y}{\partial x} - \frac{\partial A_x}{\partial y} = B_z$$

which have (among others) the solutions:

$$A_x = -\tfrac{1}{2}B_z y,\ A_y = \tfrac{1}{2}B_z x,\ A_z = 0.$$

With that choice, the vector fields $\mathbf{A}$ and $\mathbf{B}$ are related by the simple formula

$$\mathbf{A} = \tfrac{1}{2}(\mathbf{r} \times \mathbf{B}). \tag{3.33}$$

It is customary in Physics to call this gauge the *symmetric gauge.* In that gauge the Hamiltonian is

$$H_{sym} = \frac{\mathbf{p}^2}{2m} + \frac{e^2 B_z^2}{8mc^2}(x^2 + y^2) - \frac{eB_z}{2mc} L_z \tag{3.34}$$

where the quantity

$$L_z = xp_y - yp_x \tag{3.35}$$

is the angular momentum in the z-direction. The term

$$\omega_L = \frac{eB_z}{2mc} \tag{3.36}$$

is called the *Larmor frequency;* it is one-half of the cyclotron frequency. The second term in the right-hand side of (3.34) is the "diamagnetic term", and the third, the "paramagnetic term". The matrix (3.28) is here

$$R = \begin{pmatrix} \lambda & 0 & 0 & 0 & \mu & 0 \\ 0 & \lambda & 0 & -\mu & 0 & 0 \\ 0 & 0 & 0 & 0 & 0 & 0 \\ 0 & -\mu & 0 & \frac{1}{m} & 0 & 0 \\ \mu & 0 & 0 & 0 & \frac{1}{m} & 0 \\ 0 & 0 & 0 & 0 & 0 & \frac{1}{m} \end{pmatrix} \tag{3.37}$$

the terms λ and μ being given by

$$\lambda = \frac{e^2B_z^2}{8mc^2} \quad , \quad \mu = -\frac{eB_z}{2mc}.$$

If the magnetic field is extremely strong, the paramagnetic term μ can be neglected, and a good approximation to (3.34) is then

$$H_{diam} = \frac{\mathbf{p}^2}{2m} + \frac{e^2B_z^2}{8mc^2}(x^2+y^2) \tag{3.38}$$

in which case (3.37) reduces to the diagonal matrix

$$R = \begin{pmatrix} \lambda & 0 & 0 & 0 & 0 & 0 \\ 0 & \lambda & 0 & 0 & 0 & 0 \\ 0 & 0 & 0 & 0 & 0 & 0 \\ 0 & 0 & 0 & \frac{1}{m} & 0 & 0 \\ 0 & 0 & 0 & 0 & \frac{1}{m} & 0 \\ 0 & 0 & 0 & 0 & 0 & \frac{1}{m} \end{pmatrix}.$$

While the magnetic fields produced in laboratories rarely exceeds few Teslas, evidence for strong fields motivating the use of the Hamiltonian (3.38) have been found by astronomers in white dwarfs, and extremely strong fields are assumed to exist in neutron stars.

3.4.3 Small Oscillations Near Equilibrium

Let $X = X_H$ be a Hamiltonian vector field. One says that z_0 is an equilibrium point of X_H if and only if $\nabla_z H(z_0) = 0$ that is, if and only if z_0 is a critical point of H. In particular, when the Hamiltonian H is of the standard type

$$H(z) = \sum_{j=1}^{n} \frac{p_j^2}{2m_j} + U(x)$$

then $z_0 = (x_0, p_0)$ is an equilibrium point if and only if $p_0 = 0$ and $\nabla_x U(x_0) = 0$, that is if and only if x_0 is a critical point of the potential function. Shifting if necessary the origin on the x-axis, we may always assume that $x_0 = 0$. On the other hand, Hamilton's equations being insensitive to the addition of a constant term to the Hamiltonian function, we may as well assume that $U(0) = 0$. Performing a second order Taylor expansion of U at $x = 0$ we then have

$$U(x,t) = \frac{1}{2}D^2U(0)x \cdot x + O(|x|^2)$$

and if $|x|$ is small one might envisage replacing the study of H by that of its second order approximation

$$H_0(z) = \sum_{j=1}^{n} \frac{p_j^2}{2m_j} + \frac{1}{2}D^2U(0)x \cdot x \tag{3.39}$$

which is essentially the harmonic oscillator Hamiltonian.

The approximation of a general Hamiltonian function near equilibrium by a quadratic one is best justified by the following "blow-up" or "rescaling" procedure. Assume that

$$H(0) = 0 \quad \text{and} \quad \nabla_z H(0) = 0$$

and define, for $\varepsilon > 0$, the new Hamiltonian

$$H_\varepsilon(z) = \frac{1}{\varepsilon^2}H(\varepsilon z). \tag{3.40}$$

The Taylor expansion of H at $z = 0$ is

$$H(z) = \frac{1}{2}D^2H(0)z \cdot z + h(z)$$

hence, by homogeneity

$$H_\varepsilon(z) = \frac{1}{2}D^2H(0)z \cdot z + h_\varepsilon(z) \tag{3.41}$$

where h_ε and its first and second partial derivatives in z vanish at $z = 0$; moreover $h_\varepsilon(z) = O(\varepsilon)$ uniformly near $z = 0$. In the limit $\varepsilon \to 0$ we obtain the quadratic approximation

$$H_0(z) = \frac{1}{2}D^2H(0)z \cdot z$$

which is of the type (3.24) with $k = 0$. This blow-up technique is useful in related problems, *e.g.* the study of the existence of periodic orbits for Hamiltonian systems, and bifurcation theory.

Proposition 55 *If the Hessian matrix $D^2U(0)$ is positive definite, the system with Hamiltonian*

$$H_0(z) = \frac{1}{2}D^2H(0)z \cdot z$$

will perform oscillations around equilibrium; the frequencies $\omega_1, ..., \omega_n$ of these oscillations are solutions of the eigenvalue equation

$$\det(\lambda M - D^2U(0)) = 0. \tag{3.42}$$

Proof. Let f_m be the canonical transformation of $\mathbb{R}^{2n}$ defined by

$$(x_j, p_j) \longmapsto (m_j^{-1/2}X_j, m_j^{1/2}P_j)$$

for $1 \leq j \leq n$; we have

$$(H_0 \circ f_m)(x,p) = \frac{1}{2}|p|^2 + \frac{1}{2}M^{-1/2}D^2U(0)M^{-1/2}x \cdot x$$

where M is the "mass matrix", *i.e.* the diagonal matrix with diagonal entries $m_1, ..., m_n$. Choose now $R \in O(n, \mathbb{R})$ such that

$$\Lambda = R^T(M^{-1/2}D^2U(0)M^{-1/2})R$$

is diagonal. The diagonal entries of Λ are the eigenvalues $\lambda_1, ..., \lambda_n$ of $M^{-1/2}D^2U(0)M^{-1/2}$; they are the solutions of the characteristic equation

$$\det(\lambda I_n - M^{-1/2}D^2U(0)M^{-1/2}) = 0$$

which is equivalent to (3.42). Let g be the canonical transformation $(x,p) \longmapsto (Rx, Rp)$ and set $f = g \circ f_m$; then

$$(H_0 \circ f_m)(x,p) = \frac{1}{2}|p|^2 + \frac{1}{2}\sum_{j=1}^{n} \lambda_j x_j^2$$

which is the Hamiltonian of an n-dimensional harmonic oscillator with frequencies $\lambda_1, ..., \lambda_n$. ■

The proof above can be extended to study the behavior of more general quadratic Hamiltonians of the type

$$H = \frac{1}{2}Ap \cdot p + \frac{1}{2}Bx \cdot x$$

where A is positive definite by diagonalizing simultaneously the quadratic forms $x \longmapsto Bx \cdot x$ and $p \longmapsto Ap \cdot p$.

Here is an elementary example in the case $n = 1$:

Example 56 ***Bead on a wire.*** *Consider a bead with mass m moving along a rigid wire in the x, y plane under the action of the gravitation force, which is directed towards the negative y-axis. We assume that the bead is represented by the equation $y = f(x)$ where f is a twice differentiable function such that $f''(x) > 0$. At a point (x, y) of the wire the potential energy is $U(x) = mgf(x)$ and the kinetic energy is*

$$T(x) = \frac{1}{2}m(1 + f'(x)^2)\dot{x}^2 = \frac{1}{2m}(1 + f'(x)^2)p_x^2.$$

The Hamiltonian is thus here

$$H(x,p_x)=\frac{1}{2m}(1+f'(x)^2)p_x^2+mgf(x).$$

Assume that $f'(0)=0$. *The critical point* $(x,p_x)=(0,0)$ *is non-degenerate, and we may choose as approximate Hamiltonian the quadratic form*

$$H_0(x,p_x)=\frac{1}{2m}(p_x^2+m^2gf''(0)x^2).$$

The frequency of the small oscillations is $\omega=\sqrt{gf''(0)}$.

3.5 The Inhomogeneous Symplectic Group

An affine symplectic transformation is the product of a translation in phase space and of a symplectic transformation. It follows from the obvious relation

$$\tau(z_0)\circ s=s\circ\tau(s^{-1}z_0) \tag{3.43}$$

that the set of all affine symplectic transformations is a group for the usual composition law of automorphisms. In fact, an immediate calculation, using (3.43), shows that we have the relations

$$(\tau(z_0)\circ s)\circ(\tau(z_0')\circ s')=\tau(z_0+sz_0')\circ ss' \tag{3.44}$$

$$(\tau(z_0)\circ s)^{-1}=\tau(-s^{-1}z_0)\circ s^{-1}. \tag{3.45}$$

That group, which we denote by $\mathrm{ISp}(n)$, is called the *inhomogeneous* (or sometimes *affine*) *symplectic group*. Of course, it contains $\mathrm{Sp}(n)$ as a subgroup of $\mathrm{ISp}(n)$. The following result shows that there is a convenient "realization" of $\mathrm{ISp}(n)$ as a group of matrices:

Proposition 57 *The inhomogeneous symplectic group* $\mathrm{ISp}(n)$ *is isomorphic to the group of all* $(2n+1)\times(2n+1)$ *matrices of the type*

$$\langle s,z_0\rangle\equiv\begin{pmatrix} s & z_0\\ 0_{1\times 2n} & 1\end{pmatrix}. \tag{3.46}$$

The inverse of such a matrix is given by:

$$\langle s,z_0\rangle^{-1}\equiv\begin{pmatrix} s^{-1} & -s^{-1}(z_0)\\ 0_{1\times 2n} & 1\end{pmatrix}. \tag{3.47}$$

Proof. The mapping

$$f : \mathrm{ISp}(n) \longrightarrow \mathcal{M}(2n+1, \mathbb{R})$$

of the inhomogeneous symplectic group into the space of all $(2n+1)\times(2n+1)$ real matrices defined by

$$f(\tau(z_0) \circ s) = \langle s, z_0 \rangle \tag{3.48}$$

is clearly bijective. Using Eq. (3.44) one immediately checks that f in fact is a group homomorphism. Formula (3.47) is trivially obtained, either by matrix inversion, or directly from (3.45). ∎

Remark 58 *The inhomogeneous symplectic group is the "semi-direct product" of* Sp(n) *and of the translation group.*

Identifying ISp(n) with the group of all matrices (3.46), we have the following sequence of homeomorphisms:

$$\mathrm{ISp}(n) \simeq \mathrm{Sp}(n) \times \mathbb{R}^{2n} \simeq U(n) \times \mathbb{R}^{n(3n+1)/2}. \tag{3.49}$$

In particular, ISp(n) is connected and contractible to $U(n)$.

3.5.1 Galilean Transformations and ISp(n)

Recall from Subsection 2.3.3 of the last Chapter, that a Galilean transformation

$$g : (R\mathbf{r} + \mathbf{v}_0 t + \mathbf{r}_0, t_0)$$

of space time induces the extended phase-space transformation of the type

$$\tilde{g}_m : (\mathbf{r}, \mathbf{p}, t) \mapsto (R\mathbf{r} + \mathbf{v}_0 t + \mathbf{r}_0, R\mathbf{p} + m\mathbf{v}_{0,} t + t_0) \tag{3.50}$$

(formula (2.77)). We noticed in the proof of Proposition 30 that

$$\begin{pmatrix} \hat{R} & 0 \\ 0 & 1 \end{pmatrix} \begin{pmatrix} J & 0 \\ 0 & 1 \end{pmatrix} = \begin{pmatrix} J & 0 \\ 0 & 1 \end{pmatrix} \begin{pmatrix} \hat{R} & 0 \\ 0 & 1 \end{pmatrix}$$

where

$$\hat{R} = \begin{pmatrix} R & 0 \\ 0 & R \end{pmatrix}$$

was related to the Jacobian of $\tilde{g}_m$ by

$$D(\tilde{g}_m) = \begin{pmatrix} \hat{R} & 0 \\ 0 & 1 \end{pmatrix}.$$

This is because the phase space transformation

$$(\mathbf{r}, \mathbf{p}) \mapsto (R\mathbf{r} + \mathbf{v}_0 t + \mathbf{r}_0, R\mathbf{p} + m\nu_0)$$

induced by g_m is *symplectic*. Thus, the extended Galilean group, consisting of the transformations

$$(\mathbf{r}, \mathbf{p}) \mapsto (R\mathbf{r} + \mathbf{v}_0 t + \mathbf{r}_0, R\mathbf{p} + \mathbf{p}_0) \tag{3.51}$$

form a subgroup of the inhomogeneous symplectic group ISp(3).

3.6 An Illuminating Analogy

In this section we briefly study the optical-mechanical analogy which historically goes back to Hamilton [83]. For more on the topic see Arnold [4] or Guillemin–Sternberg [80], Chapter 1.

Geometrical optics (also called "ray optics") views light as having a corpuscular nature; its propagation can then be defined in terms of rays, which are the trajectories of these corpuscles. The concern of geometrical optics is the location and direction of these rays. A. Fresnel (*b.*1788), using previous work of T. Young (*b.*1773) on interference patterns, showed that light also has a wavelike behavior; the study of the related properties belongs to the area of physical (or wave) optics. Geometrical optics can be viewed as the short-wave limit of physical optics, where interference and other wave phenomena can be neglected. A particular simple theory of geometrical optics is the paraxial linear approximation. A luminous (if we dare say so) introduction to the topic can be found in the first Chapter of Guillemin and Sternberg's book [80]. This book moreover contains an interesting historical account of the evolution of optics.

3.6.1 The Optical Hamiltonian

We consider a three-dimensional optical medium (air, vacuum, glass...) in which the speed of light is a function $\mathbf{v} = \mathbf{v}(x, y, z)$ of position. By definition, the *index of refraction* $n = n(x, y, z)$ of that medium at a point $M(x, y, z)$ is the quotient $c/v(x, y, z)$, where

$$c \approx 2.99792458 \times 10^8 \mathrm{m\,s}^{-1}$$

is the speed of light in vacuum. We thus always have $n \geq 1$. Suppose now that a light ray originates at an initial point M' of the medium and moves in a direction specified by a unit vector $\mathbf{u}'$. The ray arrives at a final point M in a direction specified by a final unit vector $\mathbf{u}$. The "imaging problem" of geometrical optics is the problem of the determination of the relation between the pairs $(M', \mathbf{u}')$ and $(M, \mathbf{u})$. Assuming that the coordinate z can be chosen as a parameter for the light ray going from M to M', the ray can thus be described by two functions of the variable z:

$$x = x(z'') \ , \ y = y(z'') \ , \ \ z' \leq z'' \leq \ z \tag{3.52}$$

which we assume piecewise continuously differentiable. We will refer to the z-axis as the *optical axis*. The planes $z = z'$ and $z = z$ will be called the "object" and "image" planes. By definition, the *optical length* along a ray from M' to M is the integral

$$A(M, M') = \int_{z'}^{z} n(x, y, z'')\sqrt{1 + \dot{x}(z'')^2 + \dot{y}(z'')^2}dz''. \tag{3.53}$$

A basic law of optics is Fermat's principle which says that the optical path minimizes or maximizes the time of propagation of light between two points. Thus, among all functions (3.52), the choices that correspond to possible light ray (for fixed z and z') are those which maximize or minimize the integral (3.53). One can show that Fermat's principle is equivalent to the Euler-Lagrange equations

$$\begin{cases} \left(\frac{\partial L}{\partial \dot{y}}\right) - \frac{\partial L}{\partial y} = 0 \\ \frac{d}{dz}\left(\frac{\partial L}{\partial \dot{x}}\right) - \frac{\partial L}{\partial x} = 0 \end{cases} \tag{3.54}$$

where the function

$$L = n(x, y, z)\sqrt{1 + \dot{x}^2 + \dot{y}^2} \tag{3.55}$$

is called the "optical Lagrangian"; $\dot{x}$ and $\dot{y}$ are viewed here as independent variables. Defining "conjugate momenta" by the formulas

$$\begin{cases} p_x = n(x, y, z)\frac{\dot{x}}{\sqrt{1+\dot{x}^2+\dot{y}^2}} \\ p_y = n(x, y, z)\frac{\dot{y}}{\sqrt{1+\dot{x}^2+\dot{y}^2}} \end{cases} \tag{3.56}$$

it is rather straightforward to check that Euler-Lagrange's equations (3.54) are equivalent to Hamilton's equations:

$$\begin{cases} \dot{x} = \dfrac{\partial H}{\partial p_x} \ , \ \dot{y} = \dfrac{\partial H}{\partial p_y} \\ \\ \dot{p}_x = -\dfrac{\partial H}{\partial x} \ , \ \dot{p}_y = -\dfrac{\partial H}{\partial p_y} \end{cases} \tag{3.57}$$

for the function

$$H = p_x\dot{x} + p_y\dot{y} - L = -\sqrt{n(x,y,z) - p^2} \tag{3.58}$$

where $p^2 = p_x^2 + p_y^2$. That function H, which only depends on the coordinates and the conjugate momenta, is called the *optical Hamiltonian.* The momenta (3.56) and the optical Hamiltonian (3.58) have the following geometric interpretation: if θ is the angle at the point $M(x,y,z)$ of the light ray with the optical axis, then p and θ satisfy the relations

$$p = n\sin\theta \ , \ H = -n\cos\theta.$$

3.6.2 Paraxial Optics

We consider a simple optical system consisting of refracting surfaces separated by regions where the refraction index remains constant. A typical example is a "cascade" of lenses separated by vacuum, or air and the surface of the sea. A light ray entering that device will propagate along a broken line, as it is being refracted by the various surfaces separating the media with constant index. We now make the following three simplifying restrictions: first, we exclusively consider optical systems where all the refracting surfaces are rotationally symmetric about the optical axis. This hypothesis is for instance satisfied by a sequence of ordinary parallel round lenses. Moreover, we suppose that all rays are coplanar, more precisely that they all lie in some plane containing the optical axis. (These two restrictions are by no means essential restrictions, because the general case can be reduced to this one without difficulty.) Finally, and *this* is indeed a serious restriction, we only study light rays that travel at small inclinations around the optical axis. More specifically, we assume that the angles that the rays form with that axis are so small that we can disregard their squares in all our calculations, thus neglecting all terms of order two, or higher, which appear in the expansions of the trigonometric functions of these angles. Such rays are called *paraxial rays* in optics. That assumption allows us to replace the exact version of Snell's law of refraction

$$n \sin i = n' \sin i' \tag{3.59}$$

(which is deduced from Fermat's principle) by its linear approximation

$$ni = n'i'. \tag{3.60}$$

Here n and n' are the respective refraction indices of two adjacent regions; the speed of light in these regions are thus $v = c/n$ and $v' = c/n'$. The angles i and i' are the angles of the light ray with the normal to the surface separating these two regions.

Let us now choose an origin O and a length unit on the optical axis. Both a point of the axis, and its coordinate, will be denoted by t. Thus, the optical axis is the "t-axis". By definition, a "reference line" t is then a plane orthogonal to the optical axis and passing through t. We next introduce coordinates on each reference line. This can be done by specifying a ray by two numbers as it passes through the line t. These numbers are the height q of the point above the optical axis where the ray hits the line t, and the quantity $p = nq$, where n is the index of refraction at that point; q is the angle of the ray with the optical axis. It will become clear in a moment why we are choosing nq, and not q as a variable. Suppose now that we pick one reference line t' at the entrance of the optical system, and another, t, at the exit. We will call these particular reference lines the "input line" and the "output line". We can thus specify the ray by the two coordinates (x', p') when it enters the system by the input line, and by (x, p) when it leaves it by the output line. We now want to know what type of dependence can be expected between (x', p') and (x, p). It is actually not difficult to see that the relation must be linear, because we are using the first order formulation (3.60) of Snell's law. Thus, the new coordinates (x, p) are related to the old coordinates (x', p') by a formula of the type

$$\begin{pmatrix} x \\ p \end{pmatrix} = \begin{pmatrix} A & B \\ C & D \end{pmatrix} \begin{pmatrix} x' \\ p' \end{pmatrix} \tag{3.61}$$

where A, B, C and D are some real numbers depending on both the optical system, and on the reference lines t and t' that are being used. The 2×2 matrix appearing in (3.61) is called the *optical matrix* (or, also, the *ray-transfer matrix*) of the system, relatively to the reference lines t and t'. Of course, the choice of two reference lines, one at the "input", the other at the "output", has nothing imperious. In fact, by viewing the optical system as a juxtaposition of adjacent subsystems ("components", in the optical literature), we are actually free to choose as many intermediate reference lines as we like, and we can describe the light ray when it passes through each of these planes. If there are, for instance, three lines t, t' and t'', and if we denote by

$$\begin{pmatrix} A & B \\ C & D \end{pmatrix}, \begin{pmatrix} A' & B' \\ C' & D' \end{pmatrix} \text{ and } \begin{pmatrix} A'' & B'' \\ C'' & D'' \end{pmatrix} \tag{3.62}$$

the optical matrices relatively to (t', t), (t', t''), and (t'', t), respectively, then these matrices are related by the formula

$$\begin{pmatrix} A & B \\ C & D \end{pmatrix} = \begin{pmatrix} A' & B' \\ C' & D' \end{pmatrix} \begin{pmatrix} A'' & B'' \\ C'' & D'' \end{pmatrix} \tag{3.63}$$

(to the first subsystem corresponds the first matrix on the right). It follows that the most general optical matrix can be reduced to the calculation of products of matrices corresponding to arbitrarily small parts of the optical system under consideration. It thus suffices to determine the optical matrices in the two following elementary cases:

(1) A light ray travels in a straight line between two reference lines t, t' in the same medium with index n. If the index of refraction of that medium is n, then the optical matrix is

$$u_d = \begin{pmatrix} 1 & d \\ 0 & 1 \end{pmatrix} \quad \text{with } d = (t - t')/n \tag{3.64}$$

(the number d is called the "reduced distance");

(2) A light ray is refracted by the surface separating two regions of constant indices n' and n. Assume that the right and left reference planes are "infinitely close" to the surface, so that the free propagation effects in (**1**) can be neglected. The optical matrix is then

$$\nu_p = \begin{pmatrix} 1 & 0 \\ -p & 1 \end{pmatrix} \quad \text{with } \ p = k(n - n'). \tag{3.65}$$

(P is called the "lens power" in optics; k is a constant associated to the curvature of the surface at its intersection with the optical axis.)

Both formulas (3.63) and (3.64) are proven by using elementary plane geometry; to derive (3.64) it is sufficient to consider lenses whose surface cut the plane of the ray and the optical axis following a parabola (see the explicit calculations in [80]).

The matrices u_d and v_P both have determinant one, hence the matrix associated to an arbitrary optical system will also have determinant one, because it can be written as a product of matrices of these two types. It is not difficult to prove that, conversely, every unimodular matrix can be factorized as a product of matrices (3.63), (3.64). It follows that the unimodular group $S\ell(2, \mathbb{R})$ is the natural reservoir for all optical matrices. This truly remarkable result justifies a posteriori the introduction of the variable $p = nq$. Had we instead worked with the variable q, then the "refraction matrix" v_P in (3.64) would have been replaced by

$$\begin{pmatrix} 1 & 0 \\ -P' & \frac{n'}{n} \end{pmatrix}, \quad P' = P/n.$$

The latter has determinant n'/n, which is different from one, except in the uninteresting case $n = n'$ where it reduces to the identity.

We have only been considering paraxial optics for coplanar rays, but the whole discussion above goes through in the non-coplanar case as well; "reference lines" are then replaced by "reference planes" and 2×2 optical matrices by 4×4 matrices which not only have determinant one, but truly are symplectic (see Guillemin-Sternberg [80] for details).

3.7 Gromov's Non-Squeezing Theorem

It is easier for a camel to pass through the eye of a needle than for a wealthy man to access the Kingdom of God (Matthew, **9**)

We begin by sketching the meaning of the non-squeezing theorem by a metaphor. Suppose that we are performing some simple experiments with a spherical balloon containing an incompressible fluid (e.g., water), and a circular cylinder (for instance, a piece of pipe). We assume – and this is essential in our metaphor – that the balloon has a *larger* radius than the cylinder. We want to check the laws of deformation of incompressible objects by deforming that balloon in various ways, so it enters the piece of pipe. We first let the pipe stand vertically on the table, and proceed to deform the balloon between our palms. After a few efforts, we have of course successfully squeezed the balloon totally inside the pipe. We now repeat the experiment, with the pipe this time lying horizontally on the table (or, alternatively, we can glue its base on the wall). To our greatest surprise and dissatisfaction, we must accept, after many unfruitful attempts, that there is this time no way we can make the balloon fit into the horizontal cylinder. Does this sound ridiculous? Well, it *is* ridiculous in ordinary "physical" space (and in n-dimensional configuration space, as well!), but this is exactly what happens in phase space! In fact, Gromov's non-squeezing theorem says that there is no way we can deform an elastic incompressible ball so that its "shadow" on any plane of conjugate coordinates (that is, x, p_x, or y, p_y, or z, p_z) decreases, if we use symplectomorphisms (and, in particular, Hamiltonian flows).

Let us give a heuristic "explanation" of that, a priori, strange phenomenon, which was discovered by M. Gromov in 1985 (Gromov, [78]). Consider the isotropic two-dimensional harmonic oscillator with Hamiltonian

$$H = \frac{1}{2}\left(p_x^2 + p_y^2 + x^2 + y^2\right)$$

(but everything in this argument actually applies in an arbitrary number of dimensions). The solutions of the associated Hamilton equations are the 2π-periodic functions:

$$\begin{cases} x = x' \cos t + p_x' \sin t \quad , \quad y = y' \cos t + p_y' \sin t \\ p_x = -x' \sin t + p_x' \cos t \quad , \quad p_y = -y' \sin t + p_y' \cos t. \end{cases}$$

Suppose now we fix the initial point (x', y', p_x', p_y') on the sphere with radius R in $\mathbb{R}^2_{x,t} \times \mathbb{R}^n_{p_x,p_y}$. Since H is a constant of the motion we will have

$$\frac{1}{2}\left(p_x^2 + p_y^2 + x^2 + y^2\right) = R^2$$

for all times, so that the orbit will stay forever on the sphere. If we fix the initial and final times t' and t so that $t - t' = 2\pi$, we will have a closed orbit, which is a big circle of the sphere; the action of that orbit is $A = \pi R^2$. Suppose next that the initial point (x', y', p_x', p_y') is on a symplectic cylinder, say $Z_1(r) : x^2 + p_x^2 = r^2$. In that case, the trajectory will wind around that cylinder. (Note that if we had chosen instead a cylinder based on the plane x, y, then the orbits would have been straight lines, and hence not periodic.) Suppose now that we deform the sphere, using symplectomorphisms (for instance, a Hamiltonian flow), so that it "fits exactly" inside the cylinder, touching it along a circle. Since the actions are unaffected by symplectic deformations (action is a symplectic invariant), we must have $\pi R^2 = \pi r^2$ (r the radius of the cylinder) so that $R = r$.

Gromov's non-squeezing result is also called the *property of the symplectic camel*. One of its consequences in particular, that "chaos" is severely limited in Hamiltonian mechanics, because phase space volumes cannot distort in arbitrary, uncontrolled ways. In fact, the property of the symplectic camel can be interpreted as a classical, topological, form of Heisenberg's uncertainty principle!

Let us begin the study of this property by briefly revisiting Liouville's theorem, which we already encountered in Chapter 2 (Section 2.5). This will help us in underlining the similarities as well as the differences between volume-preserving and Hamiltonian flows.

3.7.1 Liouville's Theorem Revisited

Let X be a vector field on phase space (or, more generally, on any space $\mathbb{R}^m$). If X is "incompressible", that is, if $\operatorname{div} X = 0$, then its flow (f_t) consists of volume-preserving diffeomorphisms. That is, if we choose a measurable subset $\mathcal{D}$ of $\mathbb{R}^m$ and set $\mathcal{D}_t = f_t(\mathcal{D})$, then we will have, by Liouville's theorem

$$\operatorname{Vol}(\mathcal{D}_t) = \operatorname{Vol}(\mathcal{D}) \tag{3.66}$$

for all t, whether $\operatorname{Vol}(\mathcal{D})$ is finite or infinite. Liouville's theorem applies to Hamiltonian vector fields since we have

$$\operatorname{div} X_H = \nabla_x \cdot \nabla_p H - \nabla_p \cdot \nabla_x H = 0$$

and this fact has led (and still leads) to frequent misunderstandings of the actual behavior of classical systems. This is because the flow (f_t) of a Hamiltonian vector field X_H consists of symplectomorphisms, which is a much stronger property than being just volume-preserving! Let me explain why. To say that f_t is a symplectomorphism means that at each point z the Jacobian matrix $f_t'(z)$ is symplectic, that is:

$$\Omega(f_t'(z)u, f_t'(z)u') = \Omega(u, u')$$

for all u, u'. Identifying the symplectic form Ω with the differential 2-form

$$dp \wedge dx = dp_1 \wedge dx_1 + \cdots + dp_n \wedge dx_n$$

this amounts to say that the symplectic form is preserved by each f_t: $f_t^*\Omega = \Omega$. It follows that every exterior power

$$\Omega^k = \underbrace{\Omega \wedge \cdots \wedge \Omega}_{k \text{ factors}}$$

is also preserved by the f_t since we have

$$f_t^*\Omega^k = f_t^*\Omega \wedge \cdots \wedge f_t^*\Omega = \Omega^k$$

and hence, in particular, $f_t^*\Omega^n = \Omega^n$. Noting that

$$\Omega^n = (-1)^{n(n-1)/2} n!\, dp_1 \wedge \cdots \wedge dp_n \wedge dx_1 \wedge \cdots \wedge dx_n$$

it follows that the f_t also preserves the standard volume form

$$\mu = dp_1 \wedge \cdots \wedge dp_n \wedge dx_1 \wedge \cdots \wedge dx_n$$

on phase space: $f_t^*\mu = \mu$. This property immediately yields an alternative proof of Liouville's theorem for Hamiltonian flows. In fact, setting $\mathcal{D}_t = f(\mathcal{D})$ we have:

$$\operatorname{Vol}(\mathcal{D}_t) = \int_{\mathcal{D}_t} \mu = \int_{\mathcal{D}} f_t^*\mu = \int_{\mathcal{D}} \mu = \operatorname{Vol}(\mathcal{D}).$$

However, one should be aware of the fact that this new proof uses the property $f_t^*\mu = \mu$ which is much weaker than $f_t^*\Omega = \Omega$ as soon as $n > 1$.

The discussion above is related to the following deep result from differential topology, which says that, conversely, if two regions have the same volume, then each can be mapped onto the other by using a volume-preserving diffeomorphism. (If you want to convince yourself that it is a highly nontrivial result, try to prove it first for $n = 1$. That is, try to prove rigorously that two surfaces in the plane with same area can be mapped diffeomorphically onto each other.)

Theorem 59 (Dacorogna-Moser) *Let $\mathcal{D}$ and $\mathcal{D}'$ be two compact and connected subsets of $\mathbb{R}^m$ with smooth boundaries. Equipping $\mathbb{R}^m$ with the standard volume form $dx_1 \wedge \cdots \wedge dx_m$ we assume that there exists an orientation preserving diffeomorphism $f : \mathcal{D} \longrightarrow \mathcal{D}'$ such that* $\operatorname{Vol}(\mathcal{D}) = \operatorname{Vol}(\mathcal{D}')$. *Then there exists a volume-preserving diffeomorphism $g : \mathcal{D} \longrightarrow \mathcal{D}'$.*

That theorem was proved by Moser [119] for manifolds without boundary, and extended by Dacorogna [33]. To really appreciate this theorem, one should realize that it is not a priori obvious that if $\mathcal{D}$ and $\mathcal{D}'$ have the same volume and are diffeomorphic, there must exist a volume-preserving diffeomorphism between them, i.e., a diffeomorphism whose Jacobian determinant is one at each point! It could very well happen, after all that every diffeomorphism "contracts" $\mathcal{D}$ in some regions and "expands" it in some other regions, while keeping the total volume constant.

Remark 60 *The Dacorogna-Moser theorem says that volume is the* only *invariant associated to volume-preserving diffeomorphisms. This is in strong contrast with symplectomorphisms, for which not only volume, but also capacities are invariants. One can in fact prove that (see [89], §2.2) if a diffeomorphism f preserves the capacity of all open sets, then it is either a symplectomorphism: $f^*\Omega = \Omega$ or an antisymplectomorphism: $f^*\Omega = -\Omega$.*

We next go to the Heart of the subject of this section.

3.7.2 Gromov's Theorem

We will use the following notations:

$$B(R) = \left\{z : x^2 + p^2 < R^2\right\}$$

is the open ball centered at the origin and with radius $R > 0$; for $1 \leq j \leq n$ the sets

$$Z_j(r) = \left\{z : x_j^2 + p_j^2 < r^2\right\}$$

are called *symplectic cylinders.* One proves (see Hofer and Zehnder [89]) that $B(R)$ and $Z_j(r)$ are "symplectic submanifolds" of $\mathbb{R}^n_x \times \mathbb{R}^n_p$.

We are going to prove the non-squeezing theorem for linear and affine symplectomorphisms. The proof relies on the following straightforward property of symplectic matrices:

Lemma 61 *Let s be a symplectic matrix:*

$$s = \begin{pmatrix} A & B \\ C & D \end{pmatrix}$$

and $(a,b) = (a_1, ..., a_n, b_1, ..., b_n)$, $(c,d) = (c_1, ..., c_n, d_1, ..., d_n)$ its j-th line and $(n+j)$-th line, respectively. We then have

$$a \cdot d - b \cdot c = 1$$

where the dot $\cdot$ is the usual scalar product in $\mathbb{R}^n$.

Proof. Since s is symplectic, A, B, C and D satisfy the conditions

$$AD^T - BC^T = I.$$

(see the equivalences (3.4)). The equality $a \cdot d - b \cdot c = 1$ follows. ■

Our main result is then:

Theorem 62 (Gromov) *There exists a symplectomorphism f of $\mathbb{R}^n_x \times \mathbb{R}^n_p$ such that $f(B(R)) \subset Z_j(r)$ if and only if $R \leq r$.*

Proof. We will content ourselves with proving a weaker form of the theorem, namely that it is not possible to squeeze $B(R)$ into $Z_j(r)$ if $R \geq r$ using affine symplectomorphisms. The proof we give completes and clarifies that of McDuff and Salamon in [116] (p. 55). For the general case we refer to Gromov's original paper [78], or to Hofer and Zehnder's book [89]. (Viterbo gives in his pioneering paper a completely different proof, using generating functions.) We thus set out to show that

$$\left.\begin{array}{c} f(B(R)) \subset Z_j(r) \\ f \in \mathrm{ISp}(n) \end{array}\right\} \Rightarrow R \leq r. \tag{3.67}$$

Since $f(z) = s(z) + z_0$ for a symplectic matrix

$$s = \begin{pmatrix} A & B \\ C & D \end{pmatrix}$$

and a translation vector $z_0 = (x_0, p_0)$, it is sufficient, by homogeneity in the x, p variables, to assume that $R = 1$. It is moreover no restriction to assume $j = 1$. It is thus sufficient to prove the implication

$$\left.\begin{array}{c} x_1^2 + p_1^2 < r^2 \\ (x,p) \in B(1) \end{array}\right\} \Rightarrow 1 \leq r. \tag{3.68}$$

Denoting by a, b, c, d the first lines of the matrices A, B, C, D we have

$$\begin{pmatrix} x_1 \\ p_1 \end{pmatrix} = \begin{pmatrix} a & b \\ c & d \end{pmatrix} \begin{pmatrix} x \\ p \end{pmatrix} + \begin{pmatrix} x_{0,1} \\ p_{0,1} \end{pmatrix}$$

so that the condition $x_1^2 + p_1^2 < r^2$ for all $(x,p) \in B(1)$ is equivalent to

$$(u \cdot z + x_{0,1})^2 + (\nu \cdot z + p_{0,1})^2 \leq r^2$$

for all $\|z\| \leq 1$, where we have set $u = (a,b), \nu = (c,d)$. In particular, choosing respectively $z = \pm u/\|u\|$ and $z = \pm v/\|v\|$, we must thus have both

$$(\pm\|u\| + x_{0,1})^2 \leq r^2 \text{ and } (\pm\|v\| + x_{0,1})^2 \leq r^2. \tag{3.69}$$

Let us show that these inequalities imply that we must have $1 \leq r^2$; the implication (3.68) will follow. In view of Lemma 61, it follows by Cauchy-Schwarz's inequality that

$$1 \leq |a \cdot d - b \cdot c| \leq \|u\| \|\nu\| \tag{3.70}$$

and hence at least one of the vectors u or v has length superior or equal to one. Suppose for instance that $\|u\| \geq 1$. We must then have

$$1 \leq (\|u\| + x_{0,1})^2 \quad \text{or} \quad 1 \leq (-\|u\| + x_{0,1})^2$$

and hence $1 \leq r^2$, for otherwise we would have

$$(\|u\| + x_{0,1})^2 + (-\|u\| + x_{0,1})^2 = 2(\|u\|^2 + x_{0,1}^2) < 2$$

and hence $\|u\| < 1$, contradicting the assumption $\|u\| \geq 1$. ■

One should be very careful to note that Gromov's result ceases to hold when the x_j, p_j plane is replaced by non-conjugate coordinate planes:

Example 63 ***Non-symplectic cylinders***. *Consider the non-symplectic cylinder*

$$Z_{12}(R) = \left\{(x,p) : x_1^2 + x_2^2 < R^2\right\}.$$

Every symplectic transformation $m_\lambda : (x,p) \longmapsto (\lambda x, p/\lambda)$ sends $B(R)$ into $Z_{12}(r)$ for all R provided that $\lambda \leq R/r$.

On the other hand, it is always possible to squeeze the ball $B(R)$ inside a symplectic cylinder $Z_j(r)$ if one uses general volume-preserving diffeomorphisms:

Example 64 ***Non-symplectic diffeomorphisms***. *The diffeomorphism f of $\mathbb{R}^4$ defined, for $\lambda > 0$, by*

$$f(x,p) = (\lambda x_1, \lambda^{-1} x_2, \lambda p_1, \lambda^{-1} p_2)$$

sends $B(R)$ inside $Z(r)$ if $\lambda \leq r/R$. That diffeomorphism is obviously volume-preserving (it has Jacobian determinant one), but is symplectic only for $\lambda = 1$ (that is, if $R \leq r$, in conformity with Gromov's theorem).

Gromov's theorem is equivalent to the following property:

Theorem 65 *Let Pr_j be the projection of phase space on the symplectic plane $\mathbb{R}_{x_j} \times \mathbb{R}_{p_j}$. Then, for every symplectomorphism f, we have:*

$$\text{Area}(Pr_j f(B(R))) \geq \pi R^2. \tag{3.71}$$

Proof. The projection $Pr_j f(B(R))$ is a compact and connected submanifold of $\mathbb{R}_{x_j} \times \mathbb{R}_{p_j}$ with smooth boundary γ. Set

$$\text{Area}(Pr_j f(B(R)) = \pi r^2$$

so that $Pr_j f(B(R))$ is diffeomorphic to the disk $D_j(r) : x_j^2 + p_j^2 \leq r^2$. In view of Dacorogna and Moser's theorem there exists a volume-preserving diffeomorphism $h : Pr_j f(B(R)) \longrightarrow D_j(R)$. Define now a diffeomorphism g of $\mathbb{R}_x^n \times \mathbb{R}_p^n$ by $g(x', p') = (x, p)$ where

$$\begin{cases} x_k = x'_k \ , \ p_k = p'_k \ \text{ if } \ k \neq j \\ (x_j, p_j) = h(x'_j, p'_j). \end{cases}$$

Since h is area-preserving we have $dp_j \wedge dx_j = dp'_j \wedge dx'_j$ so that g is in fact a symplectomorphism. Let now $T(r)$ be the set of all lines orthogonal to the x_j, p_j plane and passing through the boundary γ of $Pr_j f(B(R))$: $T(r)$ is thus a cylinder in phase space containing $f(B(R))$, and this cylinder is transformed into $Z_j(r)$ by the symplectomorphism g. But then

$$g(f(B(R))) \subset Z_j(r)$$

so that we must have $R \leq r$ in view of Gromov's theorem; the inequality (3.71) follows. ∎

3.7.3 The Uncertainty Principle in Classical Mechanics

Consider now a point in phase space $\mathbb{R}_x^n \times \mathbb{R}_p^n$, and suppose that by making position and momentum measurements we are able to find out that this point lies in a ball B with radius R. Then the "range of uncertainty" in our knowledge of the values of a pair (x_j, p_k) of position and momentum coordinates lies in the projection of that ball on the x_j, p_k plane. Since this projection is a circle with area πR^2, one might thus say that πR^2 is a lower bound for the uncertainty range of joint measurements of x_j and p_k. Suppose now that the system moves under the influence of a Hamiltonian flow (f_t). The ball B will in general be distorted by the flow into a more or less complicated region of phase space, while keeping the same volume. Since conservation of volume does not imply conservation of shape, a first guess is that one can say nothing about the time-evolution of the uncertainty range of (x_j, p_k), which can a priori become arbitrarily small. This guess is however wrong because of Gromov's theorem: as B is getting distorted by the Hamiltonian flow (f_t), the projection $\Pr_j(f_t(B))$ of $f_t(B)$ on each *conjugate* variable plane x_j, p_j will however never shrink, and always have an area superior or equal to πR^2. This is in contrast with the areas of the projections of $f_t(B)$ onto the non-conjugate planes x_j, p_k $(j \neq k)$, which can take arbitrarily small values. One can thus say that the uncertainty range of every pair (x_j, p_j) of conjugate variables can never be decreased by Hamiltonian motion, and this property can of course be viewed as a *classical* topological form of Heisenberg's uncertainty principle.

Let us quantify the argument above. Assume H, for simplicity, quadratic. We denote by X_i' and P_j' the stochastic variables whose values are the results of the measurements, at initial time $t' = 0$, of the i-th position and j-th momentum coordinate, respectively. We assume that these variables are independent, so that their covariance is zero:

$$\mathrm{Cov}(X_i', P_j') = 0. \tag{3.72}$$

Let $\Delta x_i' = \sigma(X_i')$ and $\Delta p_j' = \sigma(P_j')$ be the standard deviations at time t'; and Δx_i, Δp_j those at time t. We ask the following question:

> *What happens to Δx_i and Δp_j during the motion? More precisely what can we predict about Δx_i and Δp_j, knowing $\Delta x_1', ..., \Delta x_n'$, $\Delta p_1', ..., \Delta p_n'$?*

We claim that:

Proposition 66 *Suppose that we have $\Delta p_j' \Delta x_j' \geq \varepsilon$ for $1 \leq j \leq n$. Then we also have $\Delta p_j \Delta x_j \geq \varepsilon$ for $1 \leq j \leq n$ and all times t.*

Proof. (*Cf.* the proof of Gromov's theorem.) Since the Hamiltonian H is quadratic, the flow consists of symplectic matrices; writing

$$\begin{pmatrix} x \\ p \end{pmatrix} = \begin{pmatrix} A & B \\ C & D \end{pmatrix} \begin{pmatrix} x' \\ p' \end{pmatrix}$$

the coordinates x_j, p_j are given by the formulas

$$x_j = a \cdot x' + b \cdot p' \quad , \quad p_j = c \cdot x' + d \cdot p'$$

where (a, b) is the j-th line of the matrix s and (c, d) its $(n + j)$-th line. Writing $a = (a_1, ..., a_n)$, and so on, condition (3.72) implies that

$$\begin{cases} (\Delta x_j)^2 = \sum_{i=1}^n a_i^2 (\Delta x_i')^2 + b_i^2 (\Delta p_i')^2 \\ (\Delta p_j)^2 = \sum_{i=1}^n c_i^2 (\Delta x_i')^2 + d_i^2 (\Delta p_i')^2. \end{cases} \tag{3.73}$$

Setting

$$\alpha = (a_1\Delta x_1, ..., a_n\Delta x_n) \ , \ \beta = (b_1\Delta p_1, ..., b_n\Delta p_n)$$
$$\gamma = (c_1\Delta x_1, ..., c_n\Delta x_n) \ , \ \delta = (d_1\Delta p_1, ..., d_n\Delta p_n)$$

the equalities (3.73) can be written

$$(\Delta x_j)^2 = \alpha^2 + \beta^2 \quad , \quad (\Delta p_j)^2 = \gamma^2 + \delta^2$$

and hence, by Cauchy-Schwartz's inequality:

$$(\Delta p_j)^2(\Delta x_j)^2 \geq (\alpha \cdot \delta - \beta \cdot \gamma)^2.$$

Since we have, by definition of $\alpha, \beta, \delta, \gamma$:

$$\alpha \cdot \delta - \beta \cdot \gamma = \sum_{i=1}^{n}(a_i d_i - b_i c_i)\Delta p_i' \Delta x_i'$$

it follows that

$$\Delta p_j \Delta x_j \geq \left| \sum_{i=1}^{n}(a_i d_i - b_i c_i) \right| \inf_i \left\{ \Delta p_i' \Delta x_i' \right\}$$

and hence

$$\Delta p_j \Delta x_j \geq |a \cdot d - b \cdot c| \, \varepsilon \geq \varepsilon \tag{3.74}$$

in view of Lemma 61. ∎

There is, of course, just a little step to take if one wants to enter quantum mechanics using the result above. This step will be (gladly) taken in a moment.

It turns out that the notion of capacity is, rather unexpectedly, related to the theory of periodic orbits. This is the subject of the next section.

3.8 Symplectic Capacity and Periodic Orbits

Let $\mathcal{D}$ be a subset of phase space $\mathbb{R}^n_x \times \mathbb{R}^n_p$ (we do not require $\mathcal{D}$ to be open). Gromov's non-squeezing theorem motivates the following definition:

Definition 67 *(1) The* symplectic radius *of $\mathcal{D}$ is the radius R of the largest ball that can be symplectically embedded inside $\mathcal{D}$. (2) The* symplectic capacity *or* symplectic area *is then*

$$\operatorname{Cap}(\mathcal{D}) = \pi R^2. \tag{3.75}$$

It is a non-negative real number, or $+\infty$.

We will often say simply "capacity" instead of "symplectic capacity".

While the capacity can be zero, we have $0 < \operatorname{Cap}(\mathcal{D}) < +\infty$ if $\mathcal{D}$ is a non-empty open and bounded set: translating if necessary $\mathcal{D}$, we can namely find r and R such that $B(r) \subset \mathcal{D} \subset B(R)$ and hence

$$\pi r^2 \leq \operatorname{Cap}(\mathcal{D}) \leq \pi R^2.$$

The notion of symplectic capacity highlights the deep differences between volume-preserving and symplectic diffeomorphisms. For instance, properties **(1)** and **(2)** in Proposition 68 below implies that if $\mathcal{D}$ is a subset of $\mathbb{R}_x^n \times \mathbb{R}_p^n$, then we must have:

$$B(R) \subset \mathcal{D} \subset Z_j(R) \Longrightarrow \operatorname{Cap}(\mathcal{D}) = \pi R^2$$

showing that sets of very different shapes and volumes can have the same capacity.

Proposition 68 *The symplectic capacity has the following properties:* ***(1)*** *For all R and j we have*

$$\operatorname{Cap}(B(R)) = \operatorname{Cap}(Z_j(R)) = \pi R^2; \tag{3.76}$$

(2) If f is a symplectomorphism, then

$$f(\mathcal{D}) \subset \mathcal{D}' \implies \operatorname{Cap}(\mathcal{D}) \leq \operatorname{Cap}(\mathcal{D}');$$

(3) For every $\lambda \neq 0$ we have:

$$\operatorname{Cap}(\lambda\mathcal{D}) = \lambda^2 \operatorname{Cap}(\mathcal{D});$$

(4) A symplectomorphism f preserves the symplectic capacity:

$$\operatorname{Cap}(f(\mathcal{D})) = \operatorname{Cap}(\mathcal{D}).$$

Proof. **(1)** The equality $\text{Cap}(Z_j(R)) = \pi R^2$ is equivalent to Gromov's theorem. That $\text{Cap}(B(R)) = \pi R^2$ is obvious: no ball with radius superior to R can be sent into $B(R)$ since symplectomorphisms are volume-preserving. Properties **(2)** and **(3)** are immediate consequences of the definition of the symplectic area. Property **(4)** follows from property **(2)**. ■

Note that immediately follows from Property **(2)**, taking $f = I_d$, that

$$\mathcal{D} \subset \mathcal{D}' \implies \text{Cap}(\mathcal{D}) \leq \text{Cap}(\mathcal{D}').$$

More generally, any function c associating a non-negative number (or $+\infty$) to the subsets of $\mathbb{R}^n_x \times \mathbb{R}^n_p$ is called a *symplectic capacity* if it satisfies the properties **(1)**–**(3)** (and hence **(4)**) in Proposition 68:

Axiom 1: *For all R and j we have:*

$$c(B(R)) = c(Z_j(R)) = \pi R^2;$$

Axiom 2: *If f is a symplectomorphism, then*

$$f(\mathcal{D}) \subset \mathcal{D}' \implies \text{Cap}(\mathcal{D}) \leq \text{Cap}(\mathcal{D}');$$

Axiom 3: *For every $\lambda \neq 0$ we have:*

$$c(\lambda \mathcal{D}) = \lambda^2 c(\mathcal{D}).$$

The notion of symplectic capacity was first introduced by Ekeland and Hofer [42] in connection with Gromov's symplectic area. There are actually infinitely many different capacities on $\mathbb{R}^n_x \times \mathbb{R}^n_p$ (see [89, 116]); however Cap is the smallest (see Hofer and Zehnder [89]):

Proposition 69 *The capacity* Cap *is the smallest symplectic capacity: for every symplectic capacity c we have*

$$\text{Cap}(\mathcal{D}) \leq c(\mathcal{D})$$

for all subsets $\mathcal{D}$ of $\mathbb{R}^n_x \times \mathbb{R}^n_p$.

Another example of symplectic capacity is provided by considering only affine symplectomorphisms (as we did in the partial proof we gave of Gromov's theorem):

Example 70 ***Linear capacity***. *The function* $\mathrm{Cap}_{\mathrm{Sp}}$ *defined by:*

$$\mathrm{Cap}_{\mathrm{Sp}}(\mathcal{D}) = \sup_{f \in \mathrm{ISp}(n)} \left\{ \pi R^2 : f(B(R)) \subset \mathcal{D} \right\} \tag{3.77}$$

is a symplectic capacity, called the linear symplectic capacity.

Since $\mathrm{Cap}(\mathcal{D}) \leq \mathrm{Cap}_{\mathrm{Sp}}(\mathcal{D})$ in view of Proposition 69, one might think that we might have more flexibility in "squeezing" balls into symplectic cylinders by using general symplectomorphisms rather than linear symplectomorphisms. However, the first axiom in the general definition of a capacity just says this is not the case: all capacities agree on balls and symplectic cylinders of same radius R, and are equal to πR^2.

3.8.1 The Capacity of an Ellipsoid

Let Q be a quadratic form on $\mathbb{R}^n_x \times \mathbb{R}^n_p$. We say that the set

$$\mathcal{E} : Q(z) \leq 1$$

is an *ellipsoid* if Q is positive definite. One can show, using symplectic geometry (Hofer and Zehnder assure us in [89] that this was already known to K. Weierstrass (*b*.1815), and give [161] as earliest reference) that there exists $s \in \mathrm{Sp}(n)$ and a unique finite sequence $0 < R_1 \leq \cdots \leq R_n$ of real numbers such that if $z = s(z')$ then

$$Q(s(z')) = \sum_{j=1}^{n} \frac{1}{R_j^2} \left(p_j^2 + x_j^2\right)$$

and hence $s(\mathcal{E})$ is the ellipsoid

$$B(R_1, ..., R_n) : \quad \sum_{j=1}^{n} \frac{1}{R_j^2} \left(p_j^2 + x_j^2\right) \leq 1.$$

The sequence $R = (R_1, ..., R_n)$ is called the *symplectic spectrum* of $\mathcal{E}$. Notice that if $R_1 = ... = R_n = R$, then $B(R_1, ..., R_n)$ is the ball $B(R)$. The volume of $B(R_1, ..., R_n)$ (and hence of $\mathcal{E}$) is thus

$$\mathrm{Vol}\,\mathcal{E} = \frac{\pi^n}{n!} R_1^2 \cdots R_n^2. \tag{3.78}$$

We have the following generalization of the formula (3.76) for the capacity of a ball:

Proposition 71 *Let $\mathcal{E}$ be an ellipsoid with symplectic spectrum $R = (R_1, ..., R_n)$. Then*

$$\operatorname{Cap}\mathcal{E} = \operatorname*{Cap}_{\mathrm{Sp}}\mathcal{E} = \pi R_1^2. \tag{3.79}$$

($\operatorname{Cap}_{\mathrm{Sp}}$ *being the linear capacity defined by Eq. (3.77).)*

Proof. We only prove the equality

$$\operatorname*{Cap}_{\mathrm{Sp}}\mathcal{E} = \pi R_1^2$$

here (the proof of the equality $\operatorname{Cap}\mathcal{E} = \operatorname{Cap}_{\mathrm{Sp}}\mathcal{E}$ is much more delicate; see [89]). We will in fact show that

$$\sup_{B\subset\mathcal{E}} \operatorname*{Cap}_{\mathrm{Sp}}(B) = \operatorname{Cap}\mathcal{E} = \inf_{Z\supset\mathcal{E}} \operatorname*{Cap}_{\mathrm{Sp}}(Z) \tag{3.80}$$

where the supremum (resp. infimum) is taken over all balls $B = B(R)$ (resp. symplectic cylinders $Z = Z_j(r)$) containing $\mathcal{E}$ (resp. contained in $\mathcal{E}$). This will prove formula (3.79): since the intersection of $\mathcal{E}(R)$ with the x_j, p_j is a disk with radius R_j, no ball with radius superior to R_1 can be contained in $\mathcal{E}$, and $\mathcal{E}$ cannot be contained in a symplectic cylinder with radius inferior to R_1. Since symplectic capacities are symplectic invariants, it suffices to prove (3.80) in the case $\mathcal{E} = B(R_1, ..., R_n)$, in which case we have

$$B(R_1) \subset \mathcal{E} \subset Z_1(R_1).$$

Suppose that $Z \supset \mathcal{E}$; then $Z \supset B(R_1)$ and so $\operatorname{Cap}_{\mathrm{Sp}}(Z) \geq \pi R_1^2$. Similarly, if $B \subset E$ then $B \subset Z$ and $\operatorname{Cap}_{\mathrm{Sp}}(B) \leq \pi R_1^2$. Hence

$$\inf_{Z\supset\mathcal{E}} \operatorname*{Cap}_{\mathrm{Sp}}(Z) \leq \pi R_1^2 \leq \sup_{B\subset\mathcal{E}} \operatorname*{Cap}_{\mathrm{Sp}}(B). \tag{3.81}$$

Now suppose that $B(R) \subset \mathcal{E}$. Then $B(R) \subset Z_1(R_1)$ and so $R \leq R_1$. Similarly, if $\mathcal{E} \subset Z_1(r) \subset Z_1(R_1)$ then $r \leq R_1$ and hence

$$\sup_{B\subset\mathcal{E}} \operatorname*{Cap}_{\mathrm{ISp}}(B) \leq \pi R_1^2 \leq \inf_{Z\supset\mathcal{E}} \operatorname*{Cap}_{\mathrm{ISp}}(Z). \tag{3.82}$$

Combining (3.81) and (3.82) yields (3.80). ∎

3.8.2 Symplectic Area and Volume

Suppose first that $n = 1$. Then the symplectic capacity of a measurable set $\mathcal{D}$ is just its area:

$$\operatorname{Cap}(\mathcal{D}) = \left| \int_{\mathcal{D}} dp\, dx \right| .$$

This is, in spite of the apparent simplicity of the statement, not a trivial property; it was first proven by Siburg [145] (also see [89], pages 100–103). Siburg's result does not extend to higher dimensions: if $n > 1$ the function

$$(\operatorname{Vol}(\mathcal{D}))^{1/n} = \left| \int_{\mathcal{D}} d^n p\, d^n x \right|^{1/n}$$

cannot be a symplectic capacity on $\mathbb{R}^n_x \times \mathbb{R}^n_p$, because every symplectic cylinder then has infinite volume:

$$\operatorname{Vol}(Z_j(R) = +\infty \quad \text{if} \quad n > 1.$$

Let us compare the volume and the symplectic capacity of balls in $\mathbb{R}^n_x \times \mathbb{R}^n_p$. By definition,

$$\operatorname{Cap}(B(R)) = \pi R^2$$

while the volume of a $2n$-dimensional ball is

$$\operatorname{Vol}(B(R)) = \frac{\pi^n R^{2n}}{n!}.$$

We thus have the formula:

$$\operatorname{Vol}(B(R)) = \frac{1}{n!} \left[\operatorname{Cap}(B(R))\right]^n \tag{3.83}$$

so that volume and capacity of phase space balls only agree when $n = 1$. We also observe that the capacity of a ball $B(R)$ is independent of the dimension of the phase space, as it is always πR^2.

Notice that if $\mathcal{E}$ is an ellipsoid with spectrum $(R_1, ..., R_n)$, then

$$\operatorname{Vol}(\mathcal{E}) = \frac{\pi^n}{n!} R_1^2 \cdots R_n^2 \geq \frac{\pi^n}{n!} R_1^{2n}$$

and hence

$$\operatorname{Vol}(\mathcal{E}) \geq \frac{1}{n!} \left[\operatorname{Cap}(\mathcal{E})\right]^n . \tag{3.84}$$

3.9 Capacity and Periodic Orbits

Let γ be a smooth curve in phase space. By definition the "action" along this curve is the value of the integral of pdx along γ :

$$\mathcal{A} = \int_{\gamma} pdx. \tag{3.85}$$

Thus, once a parametrization $t \longmapsto \gamma(t)$, $a \leq t \leq b$, is chosen:

$$\mathcal{A} = \int_{a}^{b} p(\gamma(t)\dot{\gamma}(t)dt. \tag{3.86}$$

This terminology is not quite standard, because what one calls "action" in physics is usually the integral

$$\mathcal{A} = \int_{\gamma} pdx - Hdt \tag{3.87}$$

of the Poincaré-Cartan form (one sometimes calls (3.85) the "reduced action").

Periodic orbits play a fundamental role not only in quantum mechanics, but also in classical (especially Celestial) mechanics. They seem to be related in some crucial (but not fully understood) way to problems in various areas of pure mathematics, the most notable being Riemann's hypothesis about the zeroes of the zeta function (see Brack and Bhaduri's treatise [25] for a thorough and up-to-date discussion of this relationship).

Let H be a Hamiltonian function (not necessarily of Maxwell type) on phase space $\mathbb{R}^n_x \times \mathbb{R}^n_p$, and X_H the associated Hamilton vector field. We will assume that H is time-independent. Recall that an "energy shell" is a non-empty level set of the Hamiltonian H. We will always denote an energy shell by the symbol ∂M, whether it is the boundary of a set M or not. Thus:

$$\partial M = \left\{ z \in \mathbb{R}^n_x \times \mathbb{R}^n_p : H(z) = E \right\}.$$

We notice that any smooth hypersurface of phase space is the energy shell of some Hamiltonian function H: it suffices to choose for H any smooth function on $\mathbb{R}^n_x \times \mathbb{R}^n_p$ and keeping some constant value E on ∂M.

3.9.1 Periodic Hamiltonian Orbits

We will call *periodic orbit of* X_H any solution curve $t \longmapsto z(t)$ of Hamilton's equations for H such that there exists $T > 0$ such that $z(t+T) = z(t)$ for all t. Such an orbit may, or may not exist, but if it exists it is carried by an "energy shell" (as are all orbits).

It is a remarkable result, well-known from the regularization theory of collision singularities in Kepler's two-body problem, that Hamiltonian periodic orbits on a hypersurface are independent of the choice of the Hamiltonian having that hypersurface as energy shell. Periodic orbits are thus intrinsically attached to any hypersurface in phase space:

Proposition 72 *Let H and K be two functions on $\mathbb{R}^n_x \times \mathbb{R}^n_p$. Suppose that there exist two constants h and k such that*

$$\partial M = \{z : H(z) = h\} = \{z : K(z) = k\} \tag{3.88}$$

with $\nabla_z H \neq 0$ and $\nabla_z K \neq 0$ on ∂M. Then the Hamiltonian vector fields X_H and X_K have the same periodic orbits on ∂M.

Proof. The intuitive idea underlying the proof is simple: the vector fields $\nabla_z H$ and $\nabla_z K$ being both normal to the constant energy hypersurface ∂M, the Hamiltonian fields X_H and X_K must have the same flow lines, and hence the same periodic orbits. Let us make this "proof" precise. Since $\nabla_z H(z) \neq 0$ and $\nabla_z K(z) \neq 0$ are both normal to ∂M at z, there exists a function $\alpha \neq 0$ such that $X_K = \alpha X_H$ on ∂M. Let now (f_t) and (g_t) be the flows of H and K, respectively, and define a function $t = t(z,s)$, $s \in \mathbb{R}$, as being the solution of the ordinary differential equation

$$\frac{dt}{ds} = \alpha(f_t(z)) \quad , \quad t(z,0) = 0$$

where z is viewed as a parameter. We claim that

$$g_s(z) = f_t(z) \;\; \textit{for} \;\; z \in \partial M. \tag{3.89}$$

In fact, by the chain rule

$$\frac{d}{ds} f_t(z) = \frac{d}{dt} f_t(z) \cdot \frac{dt}{ds} = X_H(f_t(z))\alpha(f_t(z))$$

that is, since $X_K = \alpha X_H$:

$$\frac{d}{ds} f_t(z) = X_K(f_t(z))$$

which shows that the mapping $s \longmapsto f_{t(z,s)}(z)$ is a solution of the differential equation $\dot{z} = X_H(z)$ passing through z at time $s = t(z,0) = 0$. By the uniqueness theorem on solutions of systems of differential equations, this mapping must be identical to the mapping $s \longmapsto g_s(z)$; hence the equality (3.89). Both Hamiltonians H and K thus have the same orbits; the proposition follows. ■

In view of this result, we will in the sequel talk about the "periodic orbits of a set ∂M" without obligatorily singling out some particular Hamiltonian.

The problem of the existence of periodic orbits on a given energy shell ∂M is a very difficult one, and has not yet been solved in the general case at the time this book is being written. There are, however, some partial results. One of the oldest is due to Seifert [143] in 1948: he showed that every *compact* energy shell for a Hamiltonian

$$H = \frac{p^2}{2m} + U$$

contains at least one periodic orbit, provided it is homeomorphic to a convex set.

The following general result is due to Rabinowitz [133] (also see Weinstein [163]):

Proposition 73 *Let M be a compact and convex region in $\mathbb{R}^n_x \times \mathbb{R}^n_p$. If ∂M is C^2, then it contains at least one periodic orbit.*

The compactness of M in these criteria cannot be relaxed in general. Suppose for instance $n = 1$ and take for M the half-plane $p > 0$. Then ∂M is the line $p = 0$, which is an energy shell for the free particle Hamiltonian, but it has no periodic orbits. There are, however, simple examples of non-compact energy shells bearing periodic orbits:

Example 74 ***Periodic orbits on symplectic cylinders.*** *Any symplectic cylinder $Z_j(R)$ $(1 \leq j \leq n)$ is a convex but unbounded set. It is an energy shell of the Hamiltonian*

$$H_j(x,p) = \frac{1}{R^2}(p_j^2 + x_j^2)$$

which has periodic orbits

$$\begin{cases} x_j = x'_j \cos \frac{t}{R} + p'_j \sin \frac{t}{R} \\ p_j = -x'_j \sin \frac{t}{R} + p'_j \cos \frac{t}{R} \\ x_k = p_k = 0 \quad \textit{if} \quad k \neq j \end{cases}$$

lying on $Z_j(R)$ *if* $x_j'^2 + p_j'^2 = R^2$. *Notice that the action of such a periodic orbit equals the capacity* πR^2 *of* $Z_j(R)$.

For a more detailed review of the known conditions for the existence of periodic orbits, see Hofer and Zehnder [89] and the numerous references therein.

3.9.2 Action of Periodic Orbits and Capacity

There is a fundamental relation between the action of periodic orbits and capacities of subsets of phase space. Suppose for instance that M is the ball $B(R) = \{z : |z| \leq R\}$ in $\mathbb{R}_x^n \times \mathbb{R}_p^n$. Then every periodic orbit is an orbit of the isotropic oscillator Hamiltonian

$$H = \frac{1}{2}(p^2 + x^2)$$

whose solutions precisely have action $\pi R^2 = \mathrm{Cap}(B(R))$. In Example 74 above, we constructed a periodic orbit on the symplectic cylinder $Z_j(R)$, and remarked that its action was equal to the capacity of this cylinder.

It turns out that these results are by no means a particularity of the harmonic oscillator. They are, in fact, a general feature of systems with compact and convex energy shells:

Theorem 75 *Let* M *be a compact and convex region in phase space. Then: (1) There exists at least one periodic orbit* γ^* *on* ∂M *whose action is the capacity of* M*:*

$$|A(\gamma^*)| = \mathrm{Cap}(M).$$

(2) For every periodic orbit γ *on* ∂M*, the following inequality holds:*

$$|A(\gamma)| \geq \mathrm{Cap}(M). \tag{3.90}$$

The proof of this theorem is by no means trivial; it relies on the use of a particular choice of capacity, distinct from the symplectic area, and for which equality occurs for some special period orbits. (See Hofer and Zehnder [89].)

Let us check this theorem on ellipsoids. Recall that $B(R_1, ..., R_n)$ ($R_1 \leq \cdots \leq R_n$) denotes the ellipsoid in $\mathbb{R}_x^n \times \mathbb{R}_p^n$ defined by the condition:

$$\sum_{j=1}^{n} \frac{1}{R_j^2}\left(p_j^2 + x_j^2\right) \leq 1$$

and that its capacity is:

$$\mathrm{Cap}(B(R_1,...,R_n)) = \pi R_1^2$$

(see Subsection 3.8.1). We claim that there exists a periodic orbit on the boundary of $B(R_1,...,R_n)$ which has precisely πR_1^2 as action. The Hamiltonian

$$H = \sum_{j=1}^{n} \frac{1}{R_j^2}\left(p_j^2 + x_j^2\right)$$

has $B(R_1,...,R_n)$ as energy shell, and the functions $x = (x_1,...,x_n)$, $p = (p_1,...,p_n)$ defined by

$$\begin{cases} x_1(t) = x_1' \cos \frac{2t}{R_1} + p_1' \sin \frac{2t}{R_1} \\ p_1(t) = -x_1' \sin \frac{2t}{R_1} + p_1' \cos \frac{2t}{R_1} \\ x_j(t) = p_j(t) = 0 \quad \text{for} \quad j \geq 2 \end{cases}$$

are solutions of Hamilton's equations for H. Choosing for initial conditions x_1' and p_1' such that $x_1'^2 + p_1'^2 = R_1^2$, the trajectory will be the "small circle" $x_1^2 + p_1^2 = R_1^2$ of $B(R_1,...,R_n)$. The corresponding action is the area of this circle, and this area is precisely πR_1^2, proving our claim.

Notice that we could construct, in a similar way, periodic orbits which are any meridian circles with radii $R_2,...,R_n$ of the ellipse $B(R_1,...,R_n)$, but these orbits will have actions $\pi R_2^2,...,\pi R_n^2$ larger than $\pi R_1^2 = \mathrm{Cap}(B(R_1,...,R_n))$.

3.10 Phase Space Quantization and Quantum Blobs

In this Section we propose a quantization scheme based on the property of the symplectic camel. It consists in postulating that no periodic orbits exist, in quantum mechanics, on subsets of phase space with symplectic capacity smaller than $\frac{1}{2}h$. This postulate leads in a rather straightforward way to the Maslov quantization of Lagrangian manifolds (and, in particular, to the correct ground level energies for integrable systems). We then introduce the notion of wave-form on a quantized Lagrangian manifold. These wave-forms are, in a sense, extensions to phase space of the usual wave-functions of quantum mechanics; their definition makes use of the notion of square root of a de Rham form, which is calculated by using the properties of an essential mathematical object, the Leray index. The classical motion of the wave-forms, when projected on configuration space, is just the usual semi-classical mechanics. We begin by reviewing some well-known results from standard quantum mechanics (for details, see the classical treatises [17, 117, 128]).

3.10.1 Stationary States of Schrödinger's Equation

Consider Schrödinger's equation

$$i\hbar\frac{\partial\Psi}{\partial t} = \widehat{H}\Psi$$

associated to a time-independent Hamiltonian function H (in arbitrary dimension). Solving that equation by the method of separation of variables, we find that Ψ is a linear superposition of functions

$$\Psi_N(x,t) = e^{-iE_N t/\hbar}\psi_N(x)$$

where ψ_N is a "stationary state", solution of the eigenvalue problem

$$\widehat{H}\psi = E\psi \tag{3.91}$$

for the eigenvalue $E = E_N$.

Example 76 ***The One-Dimensional Harmonic oscillator.*** *Choose for H the harmonic oscillator Hamiltonian in one dimension:*

$$H = \frac{1}{2m}\left(p^2 + m^2\omega^2 x^2\right).$$

Equation (3.91) is in this case

$$\frac{1}{2m}\left(-\hbar^2\psi'' + m^2\omega^2 x^2\psi\right) = E\psi$$

and has non-zero solutions only if E has the value

$$E_N = \left(N + \tfrac{1}{2}\right)\hbar\omega. \tag{3.92}$$

In particular, the "ground level energy" of the one-dimensional harmonic oscillator is $E_0 = \hbar\omega/2$. The corresponding stationary states are then the functions defined, up to a constant factor, by

$$\psi_N(x) = H_N(\sqrt{\alpha}x)\exp\left(-\alpha x^2/2\right) \tag{3.93}$$

where $\alpha = m\omega/\hbar$ and H_N denotes the N-th Hermite polynomial.

These examples generalize to the Hamiltonian

$$H = \frac{1}{2m}\left(p_x^2 + p_y^2 + p_z^2\right) + \frac{m}{2}\left(\omega_x^2 x^2 + \omega_y^2 y^2 + \omega_z^2 z^2\right)$$

representing anisotropic harmonic oscillations in 3-dimensional physical space. Resolution of the corresponding eigenvalue equation (3.91) leads to the stationary states

$$\psi_{N_x,N_y,N_z}(\mathbf{r}) = \psi_{N_x}(x)\psi_{N_y}(y)\psi_{N_z}(z)$$

where ψ_{N_x}, ψ_{N_y}, ψ_{N_z} are given by (3.93) with ω replaced by ω_x, ω_y, ω_z respectively. The state ψ_{N_x,N_y,N_z} corresponds to the value

$$E_{N_x,N_y,N_z} = E_{N_x} + E_{N_y} + E_{N_z}$$

of the energy, that is:

$$E_{N_x,N_y,N_z} = \left(N_x + \tfrac{1}{2}\right)\hbar\omega_x + \left(N_y + \tfrac{1}{2}\right)\hbar\omega_y + \left(N_z + \tfrac{1}{2}\right)\hbar\omega_z. \tag{3.94}$$

Notice that the ground level energy

$$E_0 = \tfrac{1}{2}\hbar\omega_x + \tfrac{1}{2}\hbar\omega_y + \tfrac{1}{2}\hbar\omega_z \tag{3.95}$$

is the sum of the ground energies of three one-dimensional oscillators with frequencies ω_x, ω_y, ω_z running independently.

These results generalize in a straightforward way to the n-dimensional harmonic oscillator with Hamiltonian

$$H = \sum_{j=1}^{n} \frac{1}{2m_j}\left(p_j^2 + m_j\omega_j x_j^2\right).$$

One finds that the energy levels are in this case

$$E_{N_1,\ldots,N_n} = \sum_{j=1}^{n}\left(N_j + \tfrac{1}{2}\right)\hbar\omega_j \tag{3.96}$$

and the "ground level energy" is thus

$$E_0 = \sum_{j=1}^{n} \tfrac{1}{2}\hbar\omega_j. \tag{3.97}$$

The fact that the ground energy levels are different from zero is often motivated in the physical literature by saying that an observed quantal harmonic oscillator cannot be at rest (that is, one cannot find $x = 0$ and $p = 0$), because this would violate Heisenberg's uncertainty principle. This view is in accord with the Copenhagen interpretation discussed in Chapter 1, and which advocates that the "act of observation" automatically provokes an uncontrolled perturbation of the oscillator. We will see that the non-zero ground energy levels actually have a topological origin, and can be viewed a consequence of the principle of the symplectic camel. (Notice that the energy levels do not depend on the mass of the oscillator; this reflects the fact that m can be eliminated by a convenient symplectic change of variables, and is therefore without influence on the final quantization condition.)

We next use the non-squeezing theorem to introduce a quantization scheme leading to semi-classical mechanics.

3.10.2 Quantum Blobs and the Minimum Capacity Principle

Recall from Subsection 3.8 that we defined the symplectic radius R of a subset $\mathcal{D}$ of phase space as being the radius of the largest ball that can be symplectically embedded in $\mathcal{D}$; the symplectic capacity $\operatorname{Cap}(\mathcal{D})$ is then by definition πR^2. Notice that $\operatorname{Cap}(\mathcal{D})$, which has the dimension of an area, can be any positive number, or ∞.

In quantum thermodynamics and chemistry it is common to "divide" phase space in "cells" with volume having an order of magnitude h^3. We prefer the following definition:

Definition 77 *A* quantum blob *is the image of the phase space ball with radius $R = \sqrt{\hbar}$ by some $s \in \operatorname{Sp}(n)$.*

It immediately follows from this definition that a quantum blob has symplectic capacity $\frac{1}{2}h$. A phase space ball $B(\sqrt{\hbar})$ with radius $\sqrt{\hbar}$ is itself a quantum blob. Notice that while the symplectic capacity of a quantum blob is the same, $\frac{1}{2}h$, in every dimension, its volume

$$\operatorname*{Vol}_{2n} s\left(B(\sqrt{\hbar})\right) = \operatorname*{Vol}_{2n} B(\sqrt{\hbar}) = \frac{h^n}{2^n n!}$$

very quickly decreases as the dimension n increases, since

$$\operatorname*{Vol}_{2n} B(\sqrt{\hbar}) = \frac{h}{2n} \operatorname*{Vol}_{2(n-1)} B(\sqrt{\hbar}).$$

When $n = 3$, corresponding to the case of the phase space of a single particle, the volume of a cell is thus $h^3/48$, but for two particles ($n = 6$) the volume is $h^6/46080$. We begin by discussing the quantization of the harmonic oscillator from the point of view of Theorem 75, and make the following *physical* assumption:

Axiom 78 (Minimum capacity principle) *The only physically admissible periodic orbits of a quadratic Hamiltonian are those lying on an energy shell bounding a quantum cell.*

I view of the "capacity = action" result of Theorem 75, this principle is of course equivalent to:

Axiom 79 (Minimum action principle) *The action of a physically admissible Hamiltonian periodic orbit cannot be inferior to* $\frac{1}{2}h$.

We will call an orbit γ_0 for which equality occurs a *minimal periodic orbit*:

$$\oint_{\gamma_0} pdx = \tfrac{1}{2}h.$$

We are going to see that the minimum capacity/action principle suffices to determine the ground energy levels for the harmonic oscillator in arbitrary dimension n.

3.10.3 Quantization of the N-Dimensional Harmonic Oscillator

We begin by studying the cases $n = 1$ and $n = 2$. Consider the Hamiltonian function

$$H = \frac{1}{2m}(p^2 + m^2\omega^2 x^2)$$

defined on the phase plane $\mathbb{R}_x \times \mathbb{R}_p$. The associated periodic orbits are the ellipses

$$\gamma_E : \frac{1}{2m}(p^2 + m^2\omega^2 x^2) = E;$$

for each positive value of E, γ_E encloses a surface with area $2\pi E/\omega$. These ellipses correspond to the periodic solutions

$$\begin{cases} x = x' \cos \omega t + \dfrac{1}{m\omega} p' \sin \omega t \\ p = -m\omega x' \sin \omega t + p' \cos \omega t \end{cases} \qquad (0 \leq t \leq 2\pi/\omega)$$

of Hamilton's equations, where the initial data satisfy

$$\frac{1}{2m}(p'^2 + m^2\omega^2 x'^2) = E.$$

Since area and capacity coincide for $n = 1$ (see Subsection 3.8.2), the minimum capacity principle, implies that the smallest periodic orbit γ_0, whose energy is denoted by E_0, should satisfy

$$\frac{2\pi E_0}{\omega} = \oint_{\gamma_0} pdx = \tfrac{1}{2}h. \tag{3.98}$$

It follows that $E_0 = \frac{1}{2}\hbar\omega$, which is the "ground energy" predicted by quantum mechanics.

Before we state and prove the most general result, let us study the model case of the two-dimensional oscillator with Hamiltonian

$$H_{x,y} = \frac{1}{2m}\left(p_x^2 + p_y^2 + m^2\omega_x^2 x^2 + m^2\omega_x^2 y^2\right).$$

It is of course no restriction to assume that $\omega_x \geq \omega_y$. The associated orbits $\gamma_{x,y}$ are all periodic if and only if the frequencies ω_x and ω_y are commensurate, that is, if there exist two non-zero integers k and ℓ such that $\omega_x{:}\omega_y = k{:}\ell$. A period is then

$$T = \frac{2k\pi}{\omega_x} = \frac{2\ell\pi}{\omega_y}.$$

However, in all cases there are two distinguished periodic orbits:

$$\gamma_{x,0} : \begin{cases} x = x' \cos \omega_x t + \dfrac{1}{m\omega_x} p'_x \sin \omega_x t \\ p_x = -m\omega_x x' \sin \omega_x t + p'_x \cos \omega_x t \\ y = p_y = 0 \end{cases} \qquad (0 \leq t \leq 2\pi/\omega_x)$$

and

$$\gamma_{0,y}: \begin{cases} x = p_x = 0 \\ y = y' \cos\omega_y t + \dfrac{1}{m\omega_y} p'_y \sin\omega_y t \\ p_y = -m\omega_y y' \sin\omega_y t + p'_y \cos\omega_y t \end{cases} \qquad (0 \leq t \leq 2\pi/\omega_y)$$

with respective actions

$$\mathcal{A}(\gamma_{x,0}) = \frac{\pi}{m\omega_x}(p_x'^2 + m^2\omega_x^2 x'^2)$$

and

$$\mathcal{A}(\gamma_{0,y}) = \frac{\pi}{m\omega_y}(p_y'^2 + m^2\omega_y^2 y'^2).$$

Assume that these orbits lie on the same energy shell $H_{x,y} = E$; then the initial conditions must satisfy

$$\frac{1}{2m}(p_x'^2 + m^2\omega_x^2 x'^2) = \frac{1}{2m}(p_y'^2 + m^2\omega_y^2 y'^2) = E$$

and the corresponding actions are given by

$$\mathcal{A}(\gamma_{x,0}) = \frac{2\pi E}{\omega_x} \quad , \quad \mathcal{A}(\gamma_{0,y}) = \frac{2\pi E}{\omega_y}.$$

Since $\omega_x \geq \omega_y$ we have $\mathcal{A}(\gamma_{x,0}) \leq \mathcal{A}(\gamma_{0,y})$, and a first obvious use of the minimum capacity principle yields $\mathcal{A}(\gamma_{x,0}) = \frac{1}{2}h$. This leads to the value

$$E_0 = \tfrac{1}{2}\hbar\omega_x$$

for the ground energy, which is not the result

$$E_0 = \tfrac{1}{2}\hbar\omega_x + \tfrac{1}{2}\hbar\omega_y$$

predicted by quantum mechanics. This does not however indicate a violation of the minimum capacity principle. Recall, in fact, that the precise statement of that postulate is that there can be no periodic orbits with action less than half Planck's constant, lying on *any* quantum cell, that is, on any convex subset of phase space with capacity inferior to $h/2$. It turns out that the loops $\gamma_{x,0}$ and $\gamma_{0,y}$ lie not only on ellipsoids, but *also on symplectic cylinders* with radius $\sqrt{\hbar}$. To exploit this fact, we reduce the Hamiltonian $H_{x,y}$ to "normal form" by performing the symplectic change of variables

$$\begin{cases} x = \dfrac{x'}{\sqrt{m\omega_x}} \quad , \quad p_x = \sqrt{m\omega_x} p'_x \\ \\ y = \dfrac{y'}{\sqrt{m\omega_y}} \quad , \quad p_y = \sqrt{m\omega_y} p'_y. \end{cases}$$

That change of variable brings $H_{x,y}$ into the form

$$H'_{x,y} = \frac{\omega_x}{2}(p_x^2 + x^2) + \frac{\omega_y}{2}(p_y^2 + y^2)$$

(we are omitting the "primes" on x and y for the sake of notational simplicity). This change of variables preserves the actions, since the form pdx is invariant by the substitution $(x', p') \mapsto (x, p)$. It also preserves the symplectic capacities of subsets of phase space, being symplectic (see Proposition 68). Now, the orbits determined by Hamilton's equations associated to $H'_{x,y}$ are the curves

$$\gamma_{x,y} : \begin{cases} x = x' \cos \omega_x t + p'_x \sin \omega_x t \\ p_x = -x' \sin \omega_x t + p'_x \cos \omega_x t \\ y = y' \cos \omega_y t + p'_y \sin \omega_y t \\ p_y = -y' \sin \omega_y t + p'_y \cos \omega_y t \end{cases}$$

where (x', y', p'_x, p'_y) is the initial point. The energy of such an orbit is

$$E = E(\gamma_{x,y}) = \frac{\omega_x}{2}(p_x'^2 + x'^2) + \frac{\omega_y}{2}(p_y'^2 + y'^2)$$

and the corresponding energy shell is thus the boundary of the ellipsoid

$$\mathcal{E} : \ \frac{\omega_x}{2E}(p_x^2 + x^2) + \frac{\omega_y}{2E}(p_y^2 + y^2) \leq 1$$

whose capacity is $\mathrm{Cap}(\mathcal{E}) = 2\pi E/\omega_x$ (since $\omega_x \geq \omega_y$). Let us now apply the minimum capacity principle to the orbit $\gamma_{x,y}$ (notice that we do not make any assumption of periodicity). We first make the crucial observation that while the curve $\gamma_{x,y}$ belongs to an energy shell of the Hamiltonian $H'_{x,y}$, it also lies on each of the symplectic cylinders

$$Z_x = \{(x, y, p_x, p_y) : x^2 + p_x^2 = R_x^2\}$$

and

$$Z_y = \{(x, y, p_x, p_y) : y^2 + p_y^2 = R_y^2\}$$

where $R_x^2 = p_x'^2 + x'^2$, $R_y^2 = p_y'^2 + y'^2$. These cylinders carry periodic orbits (see Example 74), and their capacities must thus be at least $h/2$. Since by Gromov's non-squeezing theorem the radius and the symplectic radius of a cylinder are equal, we must have both $\pi R_x^2 \geq \frac{1}{2}h$ and $\pi R_y^2 \geq \frac{1}{2}h$, and hence there is a minimal orbit γ_0 such that

$$E(\gamma_0) = \frac{\omega_x}{2}R_x^2 + \frac{\omega_y}{2}R_y^2 = \tfrac{1}{2}\hbar\omega_x + \tfrac{1}{2}\hbar\omega_y$$

which is the correct result.

The discussion hereabove generalizes without difficulty to the case of the Hamiltonian

$$H = \sum_{j=1}^{n} \frac{1}{2m_j}\left(p_j^2 + m_j\omega_j x_j^2\right) \tag{3.99}$$

of the n-dimensional anisotropic harmonic oscillator:

Proposition 80 *The minimum capacity principle implies that the ground energy level of the Hamiltonian (3.99) is*

$$E_0 = \sum_{j=1}^{n} \tfrac{1}{2}\hbar\omega_j. \tag{3.100}$$

Proof. It is a straightforward generalization of the example of the two-dimensional harmonic oscillator: first perform the change of variables

$$(x, p) \longmapsto (Lx, L^{-1}p)$$

where L is the $n \times n$ diagonal matrix with diagonal entries $(m_j\omega_j)^{-1/2}$; this has the effect of changing H into the new Hamiltonian

$$H' = \sum_{j=1}^{n} \frac{\omega_j}{2}\left(p_j^2 + x_j^2\right).$$

The change of variables above being symplectic (we have $(Lx, L^{-1}p) = m_L(x, p)$ with the notations of Subsection 6.2.3), this transformation does not affect the action form pdx, and it does not change the symplectic capacities of sets; we may therefore prove the result with the Hamiltonian H

replaced by H'. Exactly as in the case of the two-dimensional oscillator, we remark that each orbit

$$\gamma : z_j(t) = \begin{pmatrix} x'_j \cos \omega_j t + p'_j \sin \omega_j t \\ -x'_j \sin \omega_j t + p'_j \cos \omega_j t \end{pmatrix} \quad (1 \leq j \leq n)$$

is carried, not only by an energy shell of the Hamiltonian H', but also by each symplectic cylinder

$$Z_j(R_j) = \{z : x_j^2 + p_j^2 = R_j^2\}$$

where $R_j^2 = x_j'^2 + p_j'^2$. However, these cylinders carry periodic orbits (Example 74), and their capacities must satisfy the conditions

$$\operatorname{Cap}(Z_j(R_j)) = \pi R_j^2 \geq \tfrac{1}{2} h$$

in view of the minimum capacity principle. If γ_0 is a minimal periodic orbit, it will thus satisfy

$$E(\gamma_0) = \sum_{j=1}^{n} \frac{\omega_j}{2} R_j^2 = \sum_{j=1}^{n} \tfrac{1}{2} \hbar \omega_j$$

which is the result predicted by ordinary quantum mechanics. ■

3.10.4 Quantum Blobs and the Uncertainty Principle

The usual textbook formulation of Heisenberg's uncertainty principle[1] is

$$\Delta X_j \Delta P_j \geq \tfrac{1}{2} \hbar \quad (1 \leq j \leq n).$$

These inequalities are actually a weak form of the much more precise Schrödinger-Robertson inequalities

$$(\Delta X_j)^2 (\Delta P_j)^2 \geq \Delta(X_j, P_j)^2 + \tfrac{1}{4} \hbar^2 \tag{3.101}$$

where the covariances $\Delta(X_j, P_k)$ (when defined) are the real numbers

$$\Delta(X_j, P_j) = \left\langle \frac{1}{2}(X_j P_j + P_j X_j) \right\rangle_\Psi - \langle X_j \rangle_\Psi \langle P_j \rangle_\Psi \tag{3.102}$$

where the bracket $\langle A \rangle_\Psi$ is the average value of the quantum observable in the (normalized) state Ψ.

[1] It would be better, for ontological reasons, to speak about "quantum indeterminacy", but we will conform to tradition.

The inequalities (3.101) from the following elementary lemma

Proposition 81 *Let A and B be two quantum observables such that $\langle A\rangle_\Psi = \langle B\rangle_\Psi = 0$. We have*

$$(\Delta A)^2(\Delta B)^2 \geq \frac{1}{4}\langle AB + BA\rangle_\Psi^2 + \tfrac{1}{4}\langle AB - BA)\rangle_\Psi^2 . \tag{3.103}$$

Proof. We have $(\Delta A)^2 = \langle A^2\rangle_\Psi$ and $(\Delta B)^2 = \langle B^2\rangle_\Psi$. Using the Cauchy-Schwarz-Bunyakowski inequality we have

$$|\langle\Psi|AB|\Psi\rangle|^2 \leq \langle\Psi|A^2|\Psi\rangle\langle\Psi|B^2|\Psi\rangle$$

that is $|\langle AB\rangle_\Psi|^2 \geq (\Delta A)^2(\Delta B)^2$. Writing

$$AB = \frac{1}{2}(AB + BA) + \frac{1}{2}(AB - BA)$$

we have

$$\langle AB\rangle_\Psi = \frac{1}{2}\langle AB + BA\rangle_\Psi + \frac{1}{2}\langle AB - BA\rangle_\Psi .$$

Now $\langle AB + BA\rangle_\Psi$ is real (because $AB + BA$ is a symmetric operator) and $\langle AB - BA\rangle_\Psi$ is pure imaginary (because $AB - BA$ is antisymmetric), hence

$$|\langle AB\rangle_\Psi|^2 = \frac{1}{4}\langle AB + BA\rangle_\Psi^2 + \tfrac{1}{4}\langle AB - BA)\rangle_\Psi^2$$

and the lemma follows. ■

The Robertson-Schrödinger inequalities (3.101) follow, choosing $A = X_j - \langle X_j\rangle_\Psi$ and $B = P_j - \langle P_j\rangle_\Psi$ and noting that $X_jP_j - P_jX_j = i\hbar$.

Let us introduce the quantum covariance matrix for the state Ψ.

Definition 82 *The quantum covariance matrix in the state Ψ is the real symmetric invertible $2n \times 2n$ matrix*

$$\Sigma = [\Delta(Z_j, Z_k)]_{1\leq j,k\leq 2n} \tag{3.104}$$

where $Z_j = X_j$ for $1 \leq j \leq n$ and $Z_j = P_j$ for $n+1 \leq j \leq 2n$. Here

$$\Delta(Z_j, Z_k) = \frac{1}{2}\langle Z_jZ_k + Z_kZ_j\rangle_\Psi - \langle Z_j\rangle_\Psi \langle Z_k\rangle_\Psi = \Delta(Z_k, Z_j).$$

The associated covariance (or Wigner) ellipsoid *W_Σ is the set of phase space points $z = (x, p)$ such that*

$$\frac{1}{2}\Sigma^{-1}z \cdot z \leq 1. \tag{3.105}$$

The invertibility of the covariance matrix Σ is by no means obvious; to prove this property requires some work; see de Gosson and Luef [71] and the references therein.

Example 83 *When $n = 1$ the covariance matrix is*

$$\Sigma = \begin{bmatrix} \Delta X^2 & \Delta(X,P) \\ \Delta(X,P) & \Delta P^2 \end{bmatrix}$$

and the area of the associated covariance ellipse is $2\pi\sqrt{\det \Sigma}$. In view of the Robertson-Schrödinger inequality (3.101) this area is at least $\pi\hbar = \frac{1}{2}h$.

This example can be generalized to arbitrary dimension n but, not so surprisingly, we have to replace the word "area" with "capacity":

Theorem 84 *The covariance ellipsoid W_Σ has symplectic capacity $\mathrm{Cap}(W_\Sigma) \geq \frac{1}{2}h$. In particular, it contains a quantum blob.*

We will not prove this deep result here, and refer to de Gosson [64, 66] and de Gosson and Luef [71]. The proof is based on the following concise formulation of the Robertson-Schrödinger inequalities well-known in quantum optics:

$$\Sigma + \frac{i\hbar}{2}J \textit{ is Hermitian and nonnegative.}$$

For instance, in the case $n = 1$, this condition reads

$$\begin{bmatrix} \Delta X^2 & \Delta(X,P) + \frac{i\hbar}{2} \\ \Delta(X,P) - \frac{i\hbar}{2} & \Delta P^2 \end{bmatrix} \geq 0;$$

this means that all the eigenvalues of $\Sigma + \frac{i\hbar}{2}J$ are ≥ 0. Since the trace is $\Delta X^2 + \Delta P^2 \geq 0$, this is equivalent to saying that $\det(\Sigma + \frac{i\hbar}{2}J) \geq 0$; now

$$\det\left(\Sigma + \frac{i\hbar}{2}J\right) = \Delta X^2 \Delta P^2 - \Delta(X,P)^2 - \tfrac{1}{4}\hbar^2$$

hence $\det(\Sigma + \frac{i\hbar}{2}J) \geq 0$ if and only if (3.101) holds.

Chapter 4

Action and Phase

Summary. *The gain in action of a system is the integral of its Poincaré-Cartan form along the phase-space trajectory for fixed initial and final times. For completely integrable systems, it is related to the notion of phase of a Lagrangian manifold. This fact yields a geometric interpretation of Hamilton and Jacobi's equation. The amount of action needed to go from one point in physical space to another in a given time is the value of the generating function.*

4.1 Introduction

Physics is where action is (Yu. Manin).

This statement should be taken at face value, because action really *is* one of the most fundamental and ubiquitous quantities in Physics. But *what* is action? To say that it is

$$\textit{energy} \times \textit{time} = \textit{momentum} \times \textit{length}$$

does not leave us any wiser than before. In fact, action is most easily defined in terms of its *variation*: the gain or loss in action of a system moving from a point $z' = (x', p')$ at time t' to a point $z = (x, p)$ at time t is just the integral $\mathcal{A} = \mathcal{A}(z, t; z', t')$ of the Poincaré-Cartan form $\tilde{\lambda}_H = pdx - Hdt$ along the piece of phase space trajectory joining these two points. For short:

$$\mathcal{A} = \int_{z',t'}^{z,t} pdx - Hdt. \tag{4.1}$$

We now ask the following question:

> *Is there any way to determine the amount of action needed to go from (z', t') to $(z, t) = f_{t,t'}(z', t')$ without calculating explicitly the integral above, that is, without first solving Hamilton's equations?*

The answer to that question is that, yes, we can. And there is moreover a bonus:

> *If the time interval $t - t'$ is small enough, then the knowledge of the initial and final positions suffices to determine the action.*

We will see that there is in fact a function with marvelous properties that allows us to do all this. It is called the *generating function* determined by H (or, in somewhat older literature, Hamilton's *characteristic* or *two-point* function).

But, one might object, we have been defining action in terms of a *change*. Is it possible to give an *absolute* definition of action? This subtle question will be discussed at the end of this Chapter, in connection with the notion of *Lagrangian manifold.*

4.2 The Fundamental Property of the Poincaré-Cartan Form

We have seen in Chapter 2 (Remark 26, formula (2.65)) that the contraction of the form

$$\tilde{\Omega}_H = dp \wedge dx - dH \wedge dt$$

with the suspended Hamilton field $\tilde{X}_H$ is zero: $i_{\tilde{X}_H} \tilde{\Omega}_H = 0$. An important consequence of this property is that the integral of the Poincaré-Cartan form

$$\tilde{\lambda}_H = pdx - Hdt$$

along any curve which is shrinkable to a point on the surface of a trajectory tube vanishes. As a consequence, we will recover a classical result from hydrodynamics, known as *Helmholtz's theorem.*

For the benefit of the reader who might be more familiar with the notations of classical vector analysis than with those of intrinsic differential geometry, we begin with the case of a particle moving along the x-axis; the extended phase space is thus here the three-dimensional $\mathbb{R}^3_{x,p,t}$.

4.2.1 Helmholtz's Theorem: The Case $n = 1$

We assume, as usual, that the solutions of the Hamilton equations for H are defined globally and for all times. Thus, every closed curve γ in $\mathbb{R}^3_{x,p,t}$ determines a *trajectory tube* $T_H(\gamma)$: it is the surface swept out by γ under the action of the flow of the suspended vector field $\tilde{X}_H = (\nabla_p H, -\nabla_x H, 1)$. The suspended vector field is tangent to $T_H(\gamma)$, hence at each point (x, p, t) of $T_H(\gamma)$ we must have

$$\tilde{X}_H(x, p, t) \cdot \mathbf{n}(x, p, t) = 0 \tag{4.2}$$

where $\mathbf{n}(x, p, t)$ is the (outward oriented) normal to $T_H(\gamma)$ at (x, p, t).

Theorem 85 *The integral of the Poincaré-Cartan form along any closed curve μ lying on (i.e. not encircling) a trajectory tube $T_H(\gamma)$ is equal to zero:*

$$\int_\mu pdx - Hdt = 0. \tag{4.3}$$

Proof. Introducing the notation $\mathbf{u} = (x, p, t)$, we have

$$\oint_\mu pdx - Hdt = \int_\mu (p, 0, -H) \cdot d\mathbf{u}. \tag{4.4}$$

Let D be the surface in C encircled by μ. By Gauss's formula we have

$$\oint_\mu pdx - Hdt = \iint_D [\nabla_\mathbf{u} \times (p, 0, -H)] \cdot \mathbf{n}dS \tag{4.5}$$

where dS is the area element on C. Now

$$\nabla_\mathbf{u} \times (p, 0, -H) = \tilde{X}_H \tag{4.6}$$

and Eq. (4.3) follows, using Eq. (4.2). ∎

Theorem 85 allows us to give a simple proof of Helmholtz's theorem on conservation of vorticity. (See Westenholz's interesting discussion of this notion in [164], Ch. 13, §4.)

Corollary 86 ***(Helmholtz's theorem)**. Let γ_1 and γ_2 be two arbitrary closed curves encircling a same tube of trajectories $T_H(\gamma)$. The integrals of the Poincaré-Cartan form along γ_1 and γ_2 are equal:*

$$\oint_{\gamma_1} pdx - Hdt = \oint_{\gamma_2} pdx - Hdt. \tag{4.7}$$

Proof. Give the curves γ_1 and γ_2 the same orientation, and let us apply the following "surgery" to the piece of trajectory tube T_{γ_1,γ_2} limited by them: choose a point (z_1, t_1) on γ_1, and a point (z_2, t_2) on γ_2. Let now γ be a curve in T_{γ_1,γ_2} joining these two points, and define the chain

$$\sigma = \gamma_1 + \gamma - \gamma_2 - \gamma.$$

It is a loop encircling T_{γ_1,γ_2} as follows: it starts from (z_1, t_1) and runs (in the positive direction) along γ_1; once it has returned to (z_1, t_1), it runs along γ until it reaches the point (z_2, t_2) and then runs (now in the negative direction) around γ_2 until it is back to (z_2, t_2); finally it runs back to (z_1, t_1) along γ. Obviously σ is contractible to a point, and hence formula (4.3) applies, and the integral of the Poincaré-Cartan form along σ vanishes. Since the contributions from γ and $-\gamma$ cancel, we get

$$\oint_\sigma pdx - Hdt = \oint_{\gamma_1} pdx - Hdt + \oint_{-\gamma_2} pdx - Hdt = 0$$

which proves formula (4.7). ■

4.2.2 Helmholtz's Theorem: The General Case

Everything above can be generalized to the case of the n-dimensional Poincaré-Cartan form:

Theorem 87 *Let $T_H(\gamma)$ be a trajectory tube for the time-dependent flow of X_H. If γ_1 and γ_2 are two homotopic curves (with fixed endpoints) on $T_H(\gamma)$, then we have*

$$\oint_{\gamma_1} pdx - Hdt = \oint_{\gamma_2} pdx - Hdt.$$

Proof. Recall that we are denoting $pdx - Hdt$ by $\tilde{\lambda}_H$. Applying Stoke's theorem to $\sigma = \gamma_1 - \gamma_2$ we have

$$\oint_{\gamma_1} \tilde{\lambda}_H - \oint_{\gamma_2} \tilde{\lambda}_H = \oint_\sigma \tilde{\lambda}_H = \int_D \tilde{\Omega}_H$$

where D is the piece of surface of $T_H(\gamma)$ bounded by γ, so that it is sufficient to prove that the restriction of $\tilde{\Omega}_H$ to $T_H(\gamma)$ is zero:

$$\tilde{\Omega}_H|_{T_H(\gamma)} = 0. \tag{4.8}$$

This amounts to show that for every pair (u, u') of vectors tangent to $T_H(\gamma)$ at a point ξ we have

$$(\tilde{\Omega}_H)_\xi(u, u') = 0. \tag{4.9}$$

Since the suspended vector field $\tilde{X}_H$ is everywhere tangent to the trajectory tube, we can find another tangent vector $\tilde{Y}(\xi)$ such that the pair $(\tilde{X}_H(\xi), \tilde{Y}(\xi))$ is a basis of the tangent space. Writing the vectors u and u' as

$$u = \alpha\tilde{Y}(\xi) + \beta\tilde{X}_H(\xi) \quad , \quad u' = \alpha'\tilde{Y}(\xi) + \beta'\tilde{X}_H(\xi)$$

we get, using the antisymmetry of $\tilde{\Omega}_H$:

$$\begin{aligned}(\tilde{\Omega}_H)_\xi(u, u') &= (\alpha\beta' - \alpha'\beta)\tilde{\Omega}_H(\tilde{X}_H(\xi), \tilde{Y}(\xi)) \\ &= (i_{\tilde{X}_H}\tilde{\Omega}_H)_\xi(\tilde{Y}(\xi)) \\ &= 0\end{aligned}$$

proving (4.9). ■

As a consequence of Theorem 87, we have:

Corollary 88 ***(Helmholtz's theorem)*** ***(1)*** *Let γ_1 and γ_2 be two curves encircling a same trajectory tube $T_H(\gamma)$ in extended phase space. Then*

$$\oint_{\gamma_1} \tilde{\lambda}_H = \oint_{\gamma_2} \tilde{\lambda}_H. \tag{4.10}$$

(2) If in particular γ_1 and γ_2 lie in the parallel planes $t = t_1$ and $t = t_2$, then

$$\oint_{\gamma_1} pdx = \oint_{\gamma_2} pdx. \tag{4.11}$$

Proof. Part **(2)** follows from part **(1)**, since $dt = 0$ on planes with constant t, and **(1)** follows from Lemma 87, using the same "surgery" as in the derivation of the 1-dimensional Helmholtz theorem. ■

Helmholtz's theorem yields an alternative proof of the symplectic invariance of a Hamiltonian flow $(f_{t,t'})$. It goes as follows: let D be a "1-chain" in $\mathbb{R}^n_x \times \mathbb{R}^n_p$, i.e. a two-dimensional piece of surface with boundary γ. Using successively Stoke's theorem and formula (4.11) we have

$$\int_D dp \wedge dx = \oint_\gamma pdx = \oint_{f_{t,t'}(\gamma)} pdx.$$

Applying Stoke's theorem, this time to the last integral, and using the formula of change of variables we get

$$\oint_{f_{t,t'}(\gamma)} pdx = \int_{f_{t,t'}(D)} dp \wedge dx = \int_D f_{t,t'}^*(dp \wedge dx)$$

that is, summarizing, and recalling that $\Omega = dp \wedge dx$:

$$\int_D \Omega = \int_D f_{t,t'}^* \Omega.$$

Since this equality holds for all D, we conclude that $f_{t,t'}^*\Omega = \Omega$, hence $(f_{t,t'})$ is a symplectomorphism.

Helmholtz's theorem can be rephrased by saying that the Poincaré-Cartan form is an *integral invariant*. The subject of integral invariants was initiated in a systematic way by the mathematician Elie Cartan (*b*.1869) in his celebrated *Leçons sur les invariants intégraux* (1922). We refer to Libermann and Marle's treatise [104] for a very clear and complete discussion of the various notions of integral invariants, and of their applications to differential geometry and mechanics.

4.3 Free Symplectomorphisms and Generating Functions

Assume that you want to throw a piece of chalk from a point A to reach some other point B. You decide that there must elapse a time $t-t'$ between the moment the projectile is thrown, and that when it hits its target. Then there will be only one possible trajectory, at least if the time interval $t-t'$ is small enough. (For large $t-t'$ uniqueness is not preserved, because of exotic possibilities, such as the piece of chalk making one, or several, turns around the earth.) That is, the position (x', y', z') of the piece of chalk at time t' and its desired position (x, y, z) at time t will unambiguously determine both the initial and final momenta. This can be stated in terms of the flow $(f_{t,t'})$ determined by the Hamiltonian H of the piece of chalk by saying that $f_{t,t'}$ is a *free symplectomorphism*. More generally:

Definition 89 *A symplectomorphism f defined on (some subset of) phase space is* free *if, given x' and x, the equation*

$$(x, p) = f(x', p')$$

uniquely *determines (p, p').*

Here is a useful criterion:

Lemma 90 *A symplectomorphism $f : (x', p') \longmapsto (x, p)$ defined in some simply connected open subset of phase space is free if and only if*

$$\det \frac{\partial(x, x')}{\partial(p', x')} = \det \frac{\partial x}{\partial p'} \neq 0 \tag{4.12}$$

on that set.

Proof. To prove that f is free if and only if we have

$$\det \frac{\partial x}{\partial p'} \neq 0 \tag{4.13}$$

it suffices to apply the implicit function theorem. In fact, the equation $x = x(x', p')$ can be solved locally in p' if and only if the Jacobian matrix $\partial x / \partial p'$ is non-singular, that is if (4.13) holds. We next note that the equality

$$\det \frac{\partial(x, x')}{\partial(p', x')} = \det \frac{\partial x}{\partial p'}$$

follows from the definition of the Jacobian matrix, because

$$\frac{\partial(x, x')}{\partial(p', x')} = \begin{pmatrix} \frac{\partial x}{\partial p'} & \frac{\partial x}{\partial x'} \\ 0 & I \end{pmatrix}$$

and hence

$$\det \frac{\partial(x, x')}{\partial(p', x')} = \det \left(\frac{\partial x}{\partial p'} \right) \det I = \det \left(\frac{\partial x}{\partial p'} \right)$$

which ends the proof of (4.12). ∎

Example 91 ***A nonlinear free symplectomorphism.*** *Let $n = 1$ and set $H(x, p) = \frac{1}{2} p^2 x^2$. The solutions of the Hamilton equations $\dot{x} = px^2$ and $\dot{p} = -p^2 x$ are given by $x = x' e^{p' x' (t - t')}$ and $p = p' e^{-p' x' (t - t')}$. Let $(f_{t,t'})$ be the corresponding time-dependent flow: $(x, p) = f_{t,t'}(x', p')$. We have*

$$\frac{\partial x}{\partial p'} = e^{p' x' (t - t')} (1 + p'(t - t'))$$

hence $f_{t,t'}$ is free at all points (x', p') such that $p' \neq -1/(t - t')$.

In the linear case, Lemma 90 can be restated in the following pedestrian way:

Lemma 92 *A symplectic matrix* $s = \begin{pmatrix} A & B \\ C & D \end{pmatrix}$ *is free if and only if the two following (equivalent) conditions are satisfied: (**1**)* $\det B \neq 0$*, (**2**)* $s\ell_p \cap \ell_p = 0$*, where* $\ell_p \equiv 0 \times \mathbb{R}^n_p$ *is the p-plane.*

Proof. Conditions (**1**) and (**2**) are evidently equivalent, and (**1**) is equivalent to (4.12). ∎

It turns out that free symplectomorphisms are "generated" by functions defined on twice the configuration space. This important property is studied in the next Subsection.

4.3.1 Generating Functions

Identifying as usual Ω with the differential form

$$dp \wedge dx = dp_1 \wedge dx_1 + \cdots + dp_1 \wedge dx_1$$

to say that a mapping $f : \mathbb{R}^n_x \times \mathbb{R}^n_p \longrightarrow \mathbb{R}^n_x \times \mathbb{R}^n_p$ is a symplectomorphism means that we have

$$dp \wedge dx = dp' \wedge dx' \tag{4.14}$$

where x and p are expressed in terms of the x' and p' by $(x,p) = f(x',p')$. Since by definition of the exterior derivative we have $d(pdx) = dp \wedge dx$ and $d(p'dx') = dp' \wedge dx'$, formula (4.14) is equivalent to

$$d(pdx - p'dx') = 0 \ . \tag{4.15}$$

In view of Poincaré's lemma (the one that says that a closed form on a contractible set is exact), there must exist a function $G = G(x,p;x',p')$ (uniquely defined up to an arbitrary additive constant) such that

$$pdx = p'dx' + dG. \tag{4.16}$$

If we now assume in addition that f is free, then the variables p', p in G are redundant, because they are unambiguously determined by the datum of x' and x, and we can thus define a function W on twice the configuration space by

$$W(x,x') = G(x,x';p(x,x'),p'(x,x')). \tag{4.17}$$

Definition 93 *A function $W : \mathbb{R}^n_x \times \mathbb{R}^n_x \longrightarrow \mathbb{R}$ for which the equivalence (4.18) holds is called a* generating function *for the free symplectomorphism f.*

Here is a basic example:

Example 94 ***The matrix*** J. *The symplectic matrix*

$$J = \begin{pmatrix} 0 & I \\ -I & 0 \end{pmatrix}$$

is obviously free. Let us find a generating function for J. The relation $z = Jz'$ is equivalent to $x = p'$, $p = -x'$ and hence

$$pdx = p'dx' - d(x \cdot x').$$

It follows that a generating function for J is $W(x, x') = -x \cdot x'$.

Notice that the identity operator is not free, and thus does not admit a generating function.

It turns out – and this is one of the most important properties of generating functions – that they allow us to calculate the initial and final momenta given the initial and final positions:

Proposition 95 *Let f be a free symplectomorphism with generating function W. We have*

$$(x, p) = f(x', p') \Longleftrightarrow \begin{cases} p = \nabla_x W(x, x') \\ p' = -\nabla_{x'} W(x, x'). \end{cases} \tag{4.18}$$

Proof. It is elementary: since the differential of W is

$$dW = \nabla_x W dx + \nabla_{x'} W dx'$$

the equivalence (4.18) follows from (4.16). ■

Here is a simple example we will revisit several times in the forthcoming Chapters:

Example 96 ***The tennis player***. *Suppose that a tennis player smashes a ball with his racket at time t', so that its velocity vector is in a given coordinate plane (hereafter called "the x, y-plane"), orthogonal to the ground. We neglect friction or spin effects, and we assume that the ball (identified* *with a material point with mass m) remains forever in that plane. Assuming that the ball is at a point (x', y') at time t', we want to find which initial velocity the tennis player needs to give to the ball so that it reaches another point (x, y) at a prescribed time $t \geq t'$. We set $\mathbf{r} = (x, y)$, $\mathbf{p} = (p_x, p_y)$ and $\mathbf{v} = (v_x, v_y)$, and we choose the y-axis upwards oriented. The Hamiltonian function of the ball is then*

$$H(\mathbf{r}, p) = \frac{1}{2m}\left(p_x^2 + p_y^2\right) + mgy$$

where $g \approx 10\ \mathrm{ms}^{-2}$. The equations of motion are

$$\begin{cases} \dot{x}(t) = v_x(t), \ \ \dot{y}(t) = v_y(t) \\ \dot{v}_x(t) = 0, \ \ \dot{v}_y(t) = -g \end{cases}$$

whose solutions are

$$\begin{cases} x(t) = x' + v_x'(t - t') \\ y(t) = -\frac{g}{2}(t - t')^2 + v_y'(t - t') + y' \\ p_x(t) = p_x' \\ p_y(t) = p_y' - \frac{mg}{2}(t - t'). \end{cases}$$

Keeping x', y', and t' fixed, we want to reach the point (x, y) after time $t - t'$. This requires that we choose the velocity coordinates as

$$v_x' = \frac{x - x'}{t - t'}, \ \ v_y' = \frac{y - y'}{t - t'} + \frac{g}{2}(t - t')$$

and the final velocities are then $v_x = v_x'$, $v_y = v_y' - g(t - t')$.

The following result shows that every function W defined on twice the configuration space whose matrix of second derivatives is non-singular is the generating function of some free symplectomorphism. Recall that the *Hessian* of a function is the determinant of its matrix of second derivatives. (In some texts, it is the matrix of second derivatives itself which is called the Hessian. We will *not* use this convention.)

Proposition 97 *A function $W = W(x, x')$ on $\mathbb{R}^n \times \mathbb{R}^n$ is a generating function for a free symplectomorphism if and only if the Hessian matrix*

$$W''_{x,x'}(x, x') = \left(\frac{\partial^2 W}{\partial x_j \partial x_j'}(x, x')\right)_{1 \leq i,j \leq n}$$

is invertible, that is:

$$W''_{x,x'} \neq 0. \tag{4.19}$$

Proof. Suppose that W is a generating function for some free symplectomorphism f. In view of Lemma 90 we have $\det(\partial x/\partial p') \neq 0$ and hence

$$W''_{x,x'} = \frac{\partial p'}{\partial x} = \left(\frac{\partial x}{\partial p'}\right)^{-1}$$

is invertible. Suppose conversely that $W = W(x, x')$ is a function satisfying (4.19). Let (x', p') be an arbitrary point in phase space, and define x implicitly (and uniquely) by $p' = -\nabla_x W(x, x')$; this is indeed possible in view of the implicit function theorem, and condition (4.19). Next, define p by $p = \nabla_x W(x, x')$, and let f be the mapping $(x', p') \mapsto (x, p)$. We claim that f is a symplectomorphism. In fact,

$$\begin{aligned} pdx - p'dx' &= \nabla_x W(x, x')dx + \nabla_{x'} W(x, x')dx' \\ &= dW(x, x') \end{aligned}$$

and hence $dp \wedge dx = dp' \wedge dx'$. Since $p = \nabla_x W(x, x')$ and $p' = -\nabla_x W(x, x')$, that symplectomorphism is generated by W. ∎

4.3.2 Optical Analogy: The Eikonal

It turns out that the concept of generating function has a straightforward interpretation in terms of the mechanical-optical analogy already sketched in Chapter 3. The optical length of a ray proceeding from x' to x in a medium of index n is here $L = n\ell$ where ℓ is the actual length. Writing

$$L(x, x'; t, t') = n(t - t') + \Delta L(x, x'; t, t') \tag{4.20}$$

where $n(t - t')$ is the optical length of a ray proceeding exactly along the optical axis, the term $\Delta L(x, x'; t, t')$ measures the deviation of the "true" optical length to that of a perfectly coaxial ray; it is called the *eikonal* (from the Greek $\varepsilon\iota\kappa\omega\upsilon$ = image; *cf.* "icon"). In the paraxial approximation, we have

$$\Delta L(x, x'; t, t') = n\frac{(x - x')^2}{2(t - t')} \tag{4.21}$$

which is immediately identifiable with the generating function

$$W(x, x'; t, t') = m\frac{(x - x')^2}{2(t - t')} \tag{4.22}$$

of the free particle.

4.4 Generating Functions and Action

Let $(f_{t,t'})$ be the time-dependent flow determined by a Maxwell Hamiltonian

$$H = \sum_{j=1}^{n} \frac{1}{2m_j} (p_j - A_j(x,t))^2 + U(x,t) \ . \tag{4.23}$$

We begin by showing that $f_{t,t'}$ in fact is always (locally) a free symplectomorphism for sufficiently short time intervals $t - t'$.

4.4.1 The Generating Function Determined by H

It turns out that we have the following very important property: the symplectomorphisms $f_{t,t'}$ determined by a Maxwell Hamiltonian (4.23) are always free if $t - t'$ is small enough (but different from zero: the identity mapping is not free!):

Lemma 98 *For every $z_0 = (x_0, p_0)$ there exists $\varepsilon > 0$ such that $f_{t,t'}$ determined by (4.23) is a free symplectomorphism near z_0 for $0 < |t - t'| < \varepsilon$.*

Proof. Set $z = f_{t,t'}(z')$. The first order Taylor expansion of z at $t = t'$ is

$$z = z' + (t - t')X_H(z') + O((t - t')^2).$$

Denoting by m the mass matrix, the Hamilton vector field associated to H is

$$X_H = \left(m^{-1}(p - A) \, , \, m^{-1}(p - A)\nabla_x A - \nabla_x U\right)$$

and hence

$$\begin{aligned} \frac{\partial x}{\partial p'} &= (t - t')m^{-1} + O_{2n}\left((t - t')^2\right) \\ &= (t - t')m^{-1}\left(I + O_{2n}(t - t')\right) \end{aligned}$$

where $O_{2n}\left((t - t')^k\right)$ $(k = 1, 2)$ is a $2n \times 2n$ matrix whose entries are $O((t - t')^k)$ functions. It follows that

$$\det \frac{\partial x}{\partial p'} \sim (t - t')^{2n} m^{-1}$$

when $t - t' \to 0$, and hence $\partial p / \partial x'$ will be non-singular near x_0 if $|t - t'| \neq 0$ is sufficiently small. In view of Lemma 90, this means that $f_{t,t'}$ is free at x_0 for those values of t, t'. ∎

Remark 99 *It is sufficient, but not necessary, that H be a Maxwell Hamiltonian to determine flows consisting of free symplectomorphisms: see Example 91 above.*

To every free symplectomorphism $f_{t,t'}$ we can associate a generating function $W_{t,t'} = W_{t,t'}(x, x')$, and we have, by Proposition 95:

$$(x,p) = f_{t,t'}(x',p') \Longleftrightarrow \begin{cases} p = \nabla_x W_{t,t'}(x,x') \\ p' = -\nabla_{x'} W_{t,t'}(x,x'). \end{cases} \tag{4.24}$$

We are going to show that for every t' and t we can in fact choose such a family of generating functions having the additional property that $W_{t,t'}$ depends smoothly on (t, t'), and such that the function $(x,t) \longmapsto W_{t,t'}(x,x')$ is, for fixed t', a solution of Hamilton-Jacobi's equation. We will moreover show that $W_{t,t'}(x,x')$ is precisely the action needed to proceed from x' at time t' to x at time t.

Proposition 100 *(1) The function $W_{t,t'}$ defined by*

$$W_{t,t'}(x,x') = \int_{x',t'}^{x,t} pdx - Hds \tag{4.25}$$

where the integral is calculated along the phase space trajectory leading from (x', p', t') to (x, p, t) is a free generating function for the symplectomorphism $f_{t,t'}$; (2) For fixed x' and t', the function $\Phi_{x',t'} : (x,t) \mapsto W(x, x'; t, t')$ is a solution of Hamilton-Jacobi's equation, that is:

$$\frac{\partial}{\partial t}\Phi_{x',t'}(x,t) + H(x, \nabla_x \Phi_{x',t'}, t) = 0. \tag{4.26}$$

Proof. **(1)** Let $W_{t,t'}$ be defined by Eq. (4.25). We have to show that if $(x,p) = f_{t,t'}(x',p')$ then

$$p = \nabla_x W_{t,t'}(x,x') \ \ , \ \ p' = -\nabla_{x'} W_{t,t'}(x,x'). \tag{4.27}$$

It suffices in fact to prove that the first of these identities: since $(f_{t,t'})^{-1} = f_{t',t}$ we have

$$W_{t,t'}(x,x') = -W_{t',t}(x',x) \tag{4.28}$$

and the second formula in (4.27) will hence automatically follow from the first. We set out to prove that

$$p_j = \frac{\partial W_{t,t'}}{\partial x_j}(x,x') \tag{4.29}$$

for $1 \leq j \leq n$. Assuming for notational simplicity that $n = 1$, and giving an increment Δx to x, we set out to evaluate the difference

$$W_{t,t'}(x + \Delta x, x') - W_{t,t'}(x, x').$$

Let γ_1 and γ_2 be the trajectories in joining, respectively, (x', p', t') to (x, p, t) and $(x', p' + \Delta p', t')$ to $(x + \Delta x, p + \Delta p, t)$. The two vectors $p' + \Delta p'$ and $p + \Delta p$ are the new initial and final momenta corresponding to the new final position $x + \Delta x$. We have, by definition of $W_{t,t'}$:

$$W_{t,t'}(x, x') = \int_{\gamma_1} pdx - Hdt$$

and

$$W_{t,t'}(x + \Delta x, x') = \int_{\gamma_2} pdx - Hdt.$$

Let now μ be an arbitrary curve joining the point (x', p') to the point $(x', p' + \Delta p')$ while keeping time constant and equal to t', and $\mu_{t,t'}$ the image of that curve by $f_{t,t'}$: $\mu_{t,t'}$ is thus a curve joining (x, p) to $(x + \Delta x, p + \Delta p)$ in the hyperplane *time* $= t$. In view of Helmholtz's theorem we have

$$\int_{\gamma_2} pdx - Hdt - \int_{\gamma_1} pdx - Hdt = \int_{\mu_{t,t'}} pdx - Hdt - \int_{\mu} pdx - Hdt$$

and hence, since $dt = 0$ on μ and $\mu_{t,t'}$ and $dx = 0$ on μ:

$$W_{t,t'}(x + \Delta x, x') - W_{t,t'}(x, x') = \int_{\mu_{t,t'}} pdx \ .$$

The choice of the curve μ being arbitrary we can assume that $\mu_{t,t'}$ is the line

$$x(s) = x + s\Delta x, \ \ p(s) = p + s\Delta x$$

$(0 \leq s \leq \ 1)$ in which case the integral along $\mu_{t,t'}$ becomes

$$\int_{\mu_{t,t'}} pdx = p\Delta x + \frac{1}{2}(\Delta x)^2$$

and hence

$$\frac{W_{t,t'}(x + \Delta x, x') - W_{t,t'}(x, x')}{\Delta x} = p + \frac{1}{2}\Delta x$$

from which follows that

$$p = \frac{\partial W}{\partial x}(x, x'; t, t')$$

letting $\Delta x \to 0$; we have thus proven (4.29). **(2)** Fixing the starting point x' and the initial time t', x will depend only on t. Writing $\Phi = \Phi_{x',t'}$ we have, by the chain rule:

$$\frac{d\Phi}{dt}(x,t) = \nabla_x\Phi(x,t)\cdot\dot{x} + \frac{\partial\Phi}{\partial t}(x,t) \tag{4.30}$$

that is, since $p = \nabla_x\Phi(x,t)$:

$$\frac{d\Phi}{dt}(x,t) = p\cdot\dot{x} + \frac{\partial\Phi}{\partial t}(x,t). \tag{4.31}$$

On the other hand, by definition of Φ:

$$\Phi(x,t) = \int_{t'}^{t}(p(s)\dot{x}(s) - H(x(s),p(s),s))ds$$

so that

$$\frac{d\Phi}{dt}(x,t) = p\cdot\dot{x} - H(x,p,t). \tag{4.32}$$

Equating the right-hand sides of Eq. (4.31) and Eq. (4.32), we finally get

$$\frac{\partial\Phi}{\partial t}(x,t) + H(x,p,t) = 0$$

which is Hamilton-Jacobi's equation for Φ since $p = \nabla_x\Phi(x,t)$. ∎

Definition 101 *We will call the function $W_{t,t'}$ defined by (4.25) the generating function determined by the Hamiltonian H, and write $W_{t,t'}(x,x') = W(x,x';t,t')$. The function W is thus defined on twice space-time.*

Here are three examples; the first two are classical, but the third is less well-known.

Example 102 ***The free particle.*** *The generating function determined by the free particle Hamiltonian $H = \mathbf{p}^2/2m$ is*

$$W(\mathbf{r},\mathbf{r}';t,t') = m\frac{(\mathbf{r}-\mathbf{r}')^2}{2(t-t')}.$$

Example 103 ***The one-dimensional harmonic oscillator.*** *The generating function determined by the one-dimensional harmonic oscillator Hamiltonian*

$$H = \frac{1}{2m}\left(p^2 + m^2\omega^2x^2\right)$$

is the function

$$W(x,x';t,t') = \frac{m\omega}{2\sin\omega(t-t')}\left((x^2+x'^2)\cos\omega(t-t') - 2xx'\right). \quad (4.33)$$

Notice that W *is defined only for* $t-t' \neq k\pi/\omega$ *(*k *an integer), and that it reduces to the free particle generating function in the limit* $\omega \to 0^+$.

Example 104 ***The nonlinear case.*** *Recall from Example 91 that if* $t-t' \neq -1/p'$ *then the Hamilton equations for* $H(x,p) = \frac{1}{2}p^2x^2$ *determine free symplectomorphisms* $f_{t,t'} : (x',p') \longmapsto (x,p)$ *given by* $x = x'e^{p'x'(t-t')}$ *and* $p = p'e^{-p'x'(t-t')}$*. One easily checks that the corresponding generating function is given by*

$$W(x,x';t,t') = \frac{(Log(x/x'))^2}{2(t-t')}.$$

4.4.2 Action vs. Generating Function

One must be careful to note that the action $\mathcal{A}$ and the generating function W are represented by different mathematical expressions. For instance if $H = p^2/2m$, then

$$\mathcal{A} = p'(x-x') - \frac{p'^2}{2m}(t-t') \quad (4.34)$$

while

$$W(x,x';t,t') = m\frac{(x-x')^2}{2(t-t')}. \quad (4.35)$$

However, inserting the value

$$p' = m\frac{x-x'}{t-t'}$$

in (4.34) yields (4.35). This difference between both notions is much more subtle than it seems at first sight, and has led to many confusions, especially in the literature around Feynman's path integral. A "thumb rule" is that $\mathcal{A}$ depends explicitly on the initial (or final) momentum, while W does not: $\mathcal{A}$ is the *gain* in action as the particle proceeds from x' to x, with initial momentum p', with transit time $t-t'$, while W rather answers the question:

> *"the departure and arrival times* t *and* t' *being given, what amount of action do we need to send the particle from* x' *to* x*?"*

One should remark that the action needed to "go from x to x" (that is, to stay at the point x) is not, in general, zero! For instance, if H is the Hamiltonian of the harmonic oscillator in one spatial dimension (Example 103) then it follows from the expression (4.33) of the generating function that

$$W(x, x; t, t') = -mx^2 \tan \frac{\omega(t - t')}{2}.$$

The generating function determined by a Hamiltonian depends on the gauge in which that Hamiltonian is expressed:

4.4.3 Gauge Transformations and Generating Functions

Although the free generating function determined by a Hamiltonian only depends on the configuration space variables x and x' (and time), it is sensitive to gauge transformations. Consider a Maxwell Hamiltonian

$$H = \sum_{j=1}^{n} \frac{1}{2m_j} (p_j - A_j)^2 + U$$

and a gauge transformation

$$(A, U) \mapsto \left(A + \nabla_x \chi, U - \frac{\partial \chi}{\partial t} \right) \tag{4.36}$$

taking H into the new Hamiltonian

$$H^\chi = \sum_{j=1}^{n} \frac{1}{2m_j} (p_j - A_j - \nabla_x \chi)^2 + U - \frac{\partial \chi}{\partial t}.$$

The following result relates the generating functions of H and H^χ:

Proposition 105 *Let W be the free generating function determined by a Maxwell Hamiltonian H. Then*

$$W^\chi(x, x'; t, t') = W(x, x'; t, t') + \chi(x, t) \tag{4.37}$$

is the free generating function of the transformed Hamiltonian H^χ.

Proof. When one performs the gauge transformation (4.36), the conjugate momentum $p = mv + A$ becomes $p^\chi = p + \nabla_x \chi$, and hence

$$p^\chi = \nabla_x (W + \chi) \quad , \quad p'^\chi = -\nabla_{x'} (W + \chi).$$

It follows that W^χ indeed is a generating function for $f^\chi_{t,t'}$. One immediately verifies that $W^\chi = W + \chi$ solves the Hamilton-Jacobi equation

$$\frac{\partial W^\chi}{\partial t} + H^\chi(x, \nabla_x W^\chi, t) = 0$$

for H^χ, hence W^χ is, as claimed, the generating function determined by H^χ.

■

Example 106 ***Free particle in a non-trivial gauge.*** *Consider the Hamiltonian function*

$$H^\chi = \frac{1}{2m}(p-t)^2 - x$$

which is the free-particle Hamiltonian in the gauge $\chi = xt$. *The free generating function determined by* H *is*

$$W(x, x'; t, t') = m\frac{(x-x')^2}{2(t-t')}$$

and that determined by H^χ *is thus*

$$W^\chi(x, x'; t, t') = m\frac{(x-x')^2}{2(t-t')} + xt$$

in view of formula (4.37).

Generating functions can be used to solve Hamilton's equations:

4.4.4 Solving Hamilton's Equations with W

Let H again be a general Maxwell Hamiltonian

$$H = \sum_{j=1}^{n} \frac{1}{2m_j}(p_j - A_j)^2 + U$$

and W the generating function that it determines. We are going to show that the datum of W allows us to solve explicitly the Hamilton equations

$$\dot{x} = \nabla_p H(x, p, t), \;\; \dot{p} = -\nabla_x H(x, p, t).$$

Proposition 107 *Let W be the generating function determined by a Maxwell Hamiltonian H. The equations*

$$\begin{cases} p = \nabla_x W(x, x', t) \\ p' = -\nabla_{x'} W(x, x', t) \end{cases} \tag{4.38}$$

determine, for given (x', p') and t' two functions

$$t \longmapsto x(x', p', t), \quad t \longmapsto p(x', p', t).$$

These functions are the solutions of Hamilton's equations for H with initial datum $x(t') = x'$ and $p(t') = p'$.

Proof. Differentiating the second equation (4.38) with respect to time yields

$$\frac{dp'}{dt} = -W''_{x,x'}\dot{x} - \nabla_{x'}\left(\frac{\partial W}{\partial t}\right)$$

which can be written, since $dp'/dt = 0$:

$$-W''_{x,x'}\dot{x}(t) + \nabla_{x'}(H(x, \nabla_x W, t)) = 0.$$

Now, by the chain rule,

$$\nabla_{x'}(H(x,p,t)) = \left(\frac{\partial p}{\partial x'}\right)^T (\nabla_p H)(x,p,t) = W''_{x,x'}(\nabla_p H)(x,p,t)$$

so that

$$W''_{x,x'}(\dot{x} - (\nabla_p H)(x,p,t)) = 0.$$

Since the matrix $W''_{x,x'}$ is non-singular (see Proposition 97), this is equivalent to

$$\dot{x} = \nabla_p H(x,p,t)$$

which is the Hamilton equation for x. Similarly, differentiating the first equation (4.38) with respect to t and taking Hamilton-Jacobi's equation into account we get

$$\begin{aligned} \dot{p} &= \nabla_x\left(\frac{\partial W}{\partial t}\right) + W''_{x,x'}\dot{x} \\ &= -\nabla_x(H(x, \nabla_x W)) + W''_{x,x'}\dot{x} \\ &= -(\nabla_x H)(x, \nabla_x W, t) + W''_{x,x'}(\nabla_p H(x, \nabla_x W, t) - \dot{x}) \\ &= -(\nabla_x H)(x, \nabla_x W, t) \\ &= -(\nabla_x H)(x,p,t) \end{aligned}$$

which ends the proof of Proposition 107. ∎

Let us illustrate this result on the harmonic oscillator:

Example 108 ***Again the Harmonic oscillator***. *The generating function at $t' = 0$ determined by*

$$H = \frac{1}{2m}(p^2 + m^2\omega^2 x^2)$$

is

$$W(x, x', t) = \frac{m\omega}{2\sin\omega t}(\cos(x^2 + x'^2) - 2xx').$$

The equations (4.38) are thus

$$p = m\omega\left(\frac{x\cos\omega t - x'}{\sin\omega t}\right) \quad , \quad p' = -m\omega\left(\frac{x'\cos\omega t - x}{\sin\omega t}\right)$$

which, solved for x and p, yield the solutions

$$\begin{cases} x(t) = x'(0)\cos\omega t + p'(0)\frac{1}{m\omega}\sin\omega t \\ p(t) = -m\omega x'(0)\sin\omega t + p'(0)\cos\omega t. \end{cases} \tag{4.39}$$

The flow determined by H thus consists of the symplectic matrices

$$s_t = \begin{pmatrix} \cos\omega t & \frac{1}{m\omega}\sin\omega t \\ -m\omega\sin\omega t & \cos\omega t \end{pmatrix}.$$

We next use the apparatus of generating functions to solve the Hamilton-Jacobi Cauchy problem.

4.4.5 The Cauchy Problem for Hamilton-Jacobi's Equation

We briefly discussed in Chapter 1 the Cauchy problem

$$\begin{cases} \dfrac{\partial\Phi}{\partial t} + H\left(\mathbf{r}, \nabla_{\mathbf{r}}\Phi, t\right) = 0 \\ \Phi(\mathbf{r}, 0) = \Phi_0(\mathbf{r}) \end{cases}$$

for Hamilton-Jacobi's equation, and mentioned that the solution is given by the formula

$$\Phi(\mathbf{r}, t) = \Phi_0(\mathbf{r}_0) + W(\mathbf{r}, \mathbf{r}_0; t, 0) \tag{4.40}$$

where $W(\mathbf{r}, \mathbf{r}_0; t, 0)$ is the action needed to reach the point $\mathbf{r}$ from $\mathbf{r}_0$ after a time t. We are going to prove this claim in arbitrary dimension, by using the properties of the Poincaré-Cartan form.

Proposition 109 *Suppose that there exists ε such that for $0 < |t - t'| < \varepsilon$ the mappings $f_{t,t'}$ are free symplectomorphisms when defined. The Cauchy problem*

$$\begin{cases} \dfrac{\partial \Phi}{\partial t} + H\left(x, \nabla_x \Phi, t\right) = 0 \\ \Phi(x, t') = \Phi'(x) \end{cases} \tag{4.41}$$

then has a unique solution $\Phi = \Phi(x, t)$, defined for $0 < |t - t'| < \varepsilon$, and that solution is given by the formula

$$\Phi(x, t) = \Phi'(x') + \int_{x',t'}^{x,t} pdx - Hds \tag{4.42}$$

where the initial point x' is defined by the condition

$$(x, p) = f_{t,t'}(x', \nabla_x \Phi'(x')). \tag{4.43}$$

Proof. We first note that formula (4.43) really defines x': since $f_{t,t'}$ is free, the datum of x and x' unambiguously determines p and p'; assigning x and $p' = \nabla_x \Phi(x')$ thus also unambiguously determines both p and x'. Now, it is clear that $\lim_{t' \to t} \Phi(x, t') = \Phi'(x)$ since $x' \to x$ as $t' \to t$, so that the Cauchy condition is satisfied. To prove that Φ is a solution of Hamilton-Jacobi's equation one proceeds exactly as in the proof of Proposition 100 (for $n = 1$) to show that

$$\Phi(x + \Delta x, t + \Delta t) - \Phi(x, t) = \int_\beta pdx - Hdt$$

where β is the line in phase space joining (x, p, t) to $(x + \Delta x, p + \Delta p, t + \Delta t)$, p and $p + \Delta p$ being determined by the relations $p = \nabla_x W(x, x'; t, t')$ and

$$p + \Delta p = \nabla_x W(x + \Delta x, x' + \Delta x'; t + \Delta t, t')$$

($\Delta x'$ is the increment of x' determined by Δx). Thus,

$$\begin{aligned} \Phi(x + \Delta x, t + \Delta t) - \Phi(x, t) &= p\Delta x + \frac{1}{2}\Delta p \Delta x \\ &\quad - \Delta t \int_0^1 H(x + s\Delta x, p + s\Delta p, t + s\Delta t) ds \end{aligned}$$

and hence

$$\frac{\Phi(x, t + \Delta t) - \Phi(x, t)}{\Delta t} = - \int_0^1 H(x, p + s\Delta p, t + s\Delta t) ds$$

from which follows that

$$\frac{\partial \Phi}{\partial t}(x,t) = -H(x,p,t) \tag{4.44}$$

since $\Delta p \to 0$ when $\Delta t \to 0$. Similarly,

$$\Phi(x+\Delta x, t) - \Phi(x,t) = p\Delta x + \frac{1}{2}\Delta p \Delta x$$

and $\Delta p \to 0$ as $\Delta x \to 0$ so that

$$\frac{\partial \Phi}{\partial x}(x,t) = p\ . \tag{4.45}$$

Combining (4.44) and (4.45) shows that Φ satisfies Hamilton-Jacobi's equation, as claimed. ■

4.5 Short-Time Approximations to the Action

It is in most cases impossible to solve explicitly Hamilton-Jacobi's Cauchy problem (4.41). We are going to see that it is however rather straightforward to obtain asymptotic solutions for small times of that problem. (In the first edition of this book we were unaware of the fact that Makri and Miller [110, 111] had obtained similar results using somewhat different methods.)

Physicists working on the Feynman integral (about which we will have much to say in Chapter 7) use the following "approximation" for the generating function determined by a Maxwell Hamiltonian in a gauge (A, U):

$$W_{\mathrm{Feyn}}(x,x';t) = m\frac{(x-x')^2}{2t} - (x-x')\cdot A(x'') - U(x')t \tag{4.46}$$

where $x'' = (x'+x)/2$ is the middle of the segment $[x', x]$ (see Schulman [139] for variants of that "midpoint rule"). However, this is really a very crude approximation. To illustrate this, assume that H is, say the harmonic oscillator Hamiltonian

$$H = \frac{1}{2m}(p^2 + m^2\omega^2 x^2).$$

Formula (4.46) then yields

$$W_{Feyn}(x,x';t) = m\frac{(x-x')^2}{2t} - \frac{m\omega^2}{2}x'^2 t.$$

However, we will see (Subsection 4.5.1, Example 110) that the correct formula is

$$W(x,x';t,t') = m\frac{(x-x')^2}{2t} - \frac{m\omega^2}{6}(x^2+xx'+x'^2)t + O(t^2)$$

so that in this elementary case (4.46) is already false at the first order!

We will see that it actually requires little effort to get correct and tractable asymptotic formulae for short-time actions. These formulae have moreover the merit that they allow to understand the apparition of these embarrassing "midpoint rules" in the "Feynman-type" approximations.

We begin with the straightforward situation of a particle moving along the x-axis under the influence of a scalar potential.

4.5.1 The Case of a Scalar Potential

We begin by considering a particle with mass m moving along the x-axis under the influence of a time-independent scalar potential U. The Hamiltonian function is thus

$$H = \frac{p^2}{2m} + U(x).$$

The generating function determined by H satisfies the Hamilton-Jacobi equation

$$\frac{\partial W}{\partial t} + \frac{1}{2m}\left(\frac{\partial W}{\partial x}\right)^2 + U = 0. \tag{4.47}$$

Let us denote by W_f the free-particle generating function:

$$W_f(x,x';t,t') = m\frac{(x-x')^2}{2(t-t')} \tag{4.48}$$

and look for a solution of Hamilton-Jacobi's equation of the form $W = W_f + R$. Inserting $W_f + R$ in (4.47), and expanding the squared bracket, we see that the function $R = R(x,x';t,t')$ has to satisfy the singular partial differential equation

$$\frac{\partial R}{\partial t} + \frac{1}{2m}\left(\frac{\partial R}{\partial x}\right)^2 + U + \frac{1}{t-t'}(x-x')\frac{\partial R}{\partial x} = 0. \tag{4.49}$$

Expanding R to the second order in $t-t'$

$$R = W_0 + W_1(t-t') + W_2(t-t')^2 + O\left((t-t')^3\right)$$

where the W_j ($j = 0, 1, 2$) are smooth functions of x and x', we immediately see that we must have $W_0 = 0$, and that W_1 and W_2 must satisfy the conditions:

$$\begin{cases} W_1 + (x - x')\dfrac{\partial W_1}{\partial x} + U = 0 \\ 2W_2 + (x - x')\dfrac{\partial W_2}{\partial x} = 0. \end{cases}$$

The general solution of the first of these equations is

$$W_1(x, x') = \frac{-1}{x - x'} \int_{x'}^{x} U(x'')dx'' + \frac{k}{x - x'}$$

where k is an arbitrary constant. Since we only want smooth solutions we must choose $k = 0$ and we get

$$W_1 = -\overline{U}$$

where $\overline{U}(x, x')$ is the average value of the potential U on the interval $[x', x]$:

$$\overline{U}(x, x') = \frac{1}{x - x'} \int_{x'}^{x} U(x'')dx'' .$$

Similarly, the only possible choice leading to a smooth W_2 is $W_2 = 0$. It follows that the generating function has the asymptotic form:

$$W = \overline{W} + O\left((t - t')^3\right) \tag{4.50}$$

where

$$\overline{W}(x, x'; t, t') = m\frac{(x - x')^2}{2(t - t')} - \overline{U}(x, x')(t - t'). \tag{4.51}$$

Let us illustrate this on the harmonic oscillator.

Example 110 ***Short-time action for the harmonic oscillator.*** *We suppose that H is given by*

$$H = \frac{1}{2m}\left(p^2 + m^2\omega^2 x^2\right).$$

Formula (4.51) yields the expansion $W = \overline{W} + O\left((t - t')^3\right)$ with

$$\overline{W}(x, x'; t, t') = m\frac{(x - x')^2}{2(t - t')} - \frac{m\omega^2}{6}(x^2 + xx' + x'^2)(t - t'). \tag{4.52}$$

We now consider the n-dimensional case; thus $x = (x_1, ..., x_n)$. We begin by introducing some notation:

Notation 111 *Let $f : \mathbb{R}^n_x \longrightarrow \mathbb{R}$ be a continuous function. We denote by $\bar{f}(x, x')$ its average value on the line segment joining x' to x:*

$$\bar{f}(x, x') = \int_0^1 f(sx + (1-s)x')\, ds \tag{4.53}$$

and we set $\bar{f}(x) = \bar{f}(x, 0)$. If $f = (f_1, ..., f_n) : \mathbb{R}^n_x \longrightarrow \mathbb{R}^n_x$ is a continuous function, then $\bar{f} = (\bar{f}_1, ..., \bar{f}_n)$.

When $n = 1$ formula (4.53) can be rewritten in the familiar form

$$\bar{f}(x, x') = \frac{1}{x - x'} \int_{x'}^{x} f(x'')\, dx''.$$

Note that the average $\bar{f}$ has the following properties:

$$\bar{f}(x, x') = \bar{f}(x', x),\ \bar{f}(x, x) = \bar{f}(x) \tag{4.54}$$

$$\nabla_x \bar{f}(x, x') = \nabla_{x'} \bar{f}(x', x),\ \ \nabla_x \bar{f}(x, x) = \frac{1}{2} \nabla_x f(x). \tag{4.55}$$

The formulas (4.54) are obvious; the first formula in (4.55) is obtained observing that

$$\bar{f}(x, x') = \int_0^1 f(sx + (1-s)x')\, ds$$

and hence

$$\nabla_x \bar{f}(x, x) = \int_0^1 s \nabla_x f(x)\, ds = \frac{1}{2} \nabla_x f(x);$$

the second formula (4.55) is obtained in a similar fashion.

Proposition 112 *Suppose that H is the Hamiltonian of an N-particle system in a scalar potential:*

$$H = \frac{p^2}{2m} + U(x, t).$$

The generating function determined by H satisfies

$$W(x, x'; t, t') = \overline{W}(x, x'; t, t') + O\left((t - t')^2\right)$$

where $\overline{W}$ is defined by the formula

$$\overline{W}(x, x'; t, t') = m \frac{(x - x')^2}{2(t - t')} - \overline{U}(x, x'; t')(t - t').$$

The proof of Proposition 112 is absolutely similar to that of formula (4.51), using the following lemma on singular first-order partial differential equations:

Lemma 113 *For any continuous function* $f : \mathbb{R}^n_x \longrightarrow \mathbb{R}$ *the equation*

$$\left(x - x'\right) \cdot \nabla_x u + u = f \tag{4.56}$$

has $u = \bar{f}(x, x')$ *as only smooth solution defined on all of* $\mathbb{R}^n_x$.

Proof. Let u and v be two solutions of Eq. (4.56). Then $w = v - u$ satisfies the equation $(x - x') \cdot \nabla_x w + w = 0$ whose only smooth solution is $w = 0$ (this is obvious in the case $n = 1$; for general n one can use the method of characteristics). Let us show that $u = \bar{f}$ indeed is a solution of Eq. (4.56). It is sufficient to assume that $x' = 0$ and to show that $u(x) = \bar{f}(x)$ satisfies $x \cdot \nabla_x u + u = f$. We have

$$\begin{aligned}
(x \cdot \nabla_x u + u)(x) &= \int_0^1 ((x \cdot \nabla_x f)(sx)s + f(sx))\, ds \\
&= \sum_{j=1}^{n} \int_0^1 \left(sx_j \frac{\partial f}{\partial x_j}(sx) + f(sx) \right) ds \\
&= \int_0^1 \frac{d}{ds}\left(sf(sx)\right) ds \\
&= f(x)
\end{aligned}$$

hence the result. ∎

4.5.2 One Particle in a Gauge $(\mathbf{A}, U)$

Before we study the general case of a Maxwell Hamiltonian on extended phase space $\mathbb{R}^n_x \times \mathbb{R}^n_p \times \mathbb{R}_t$, we consider a single particle with mass m in a gauge $(\mathbf{A}, U)$. The Hamiltonian function is thus here

$$H = \frac{1}{2m}\left(\mathbf{p} - \mathbf{A}\right)^2 + U \tag{4.57}$$

where $\mathbf{A} = \mathbf{A}(\mathbf{r}, t)$ and $U = U(\mathbf{r}, t)$. As in Chapter 2 we assume that there exists a vector field $\mathbf{B} = \mathbf{B}(\mathbf{r}, t)$ such that $\mathbf{B} = \nabla_{\mathbf{r}} \times \mathbf{A}$. We will use the following result on singular partial differential equations:

Lemma 114 *Let $f = (f_1, ..., f_n) : \mathbb{R}^n_x \longrightarrow \mathbb{R}^n_x$ be a vector-valued continuous function. Then*

$$u(x) = u(x') + (x - x') \cdot \bar{f}(x, x')$$

is the only smooth solution defined on all of $\mathbb{R}^n_x$ of the equation

$$(x - x') \cdot \nabla_x u = (x - x') \cdot f.$$

The proof is similar to that of Lemma 113.

Proposition 115 *The generating function determined by the Hamiltonian function (4.57) is asymptotically given, for $t - t' \to 0$, by $W = \overline{W} + O\left((t-t')^2\right)$ where*

$$\overline{W} = m\frac{(\mathbf{r} - \mathbf{r}')^2}{2(t - t')} + (\mathbf{r} - \mathbf{r}') \cdot \overline{\mathbf{A}}(\mathbf{r}, \mathbf{r}'; t') - \overline{V}(\mathbf{r}, \mathbf{r}'; t')(t - t')$$

where $\overline{V} = \overline{V}(\mathbf{r}, \mathbf{r}'; t')$ is the average on $[\mathbf{r}, \mathbf{r}']$ at time t' of the function $V : \mathbf{r} \longmapsto V(\mathbf{r}, \mathbf{r}'; t')$ defined by

$$V = \frac{1}{2m}\left(\overline{(\mathbf{r} - \mathbf{r}') \times \mathbf{B}}\right)^2 - (\mathbf{r} - \mathbf{r}') \cdot \frac{\partial \mathbf{A}}{\partial t} + U \tag{4.58}$$

(the quantities $\mathbf{B}$, $\partial \mathbf{A}/\partial t$ and U are calculated at time t').

Proof. We look for a solution $W = W(\mathbf{r}, \mathbf{r}'; t, t')$ of Hamilton-Jacobi's equation

$$\frac{\partial W}{\partial t} + H(\mathbf{r}, \nabla_{\mathbf{r}} W, t) = 0$$

such that

$$W(\mathbf{r}, \mathbf{r}'; t, t') \sim m\frac{(\mathbf{r} - \mathbf{r}')^2}{2(t - t')} + \sum_{j=0}^{\infty} W_j(\mathbf{r}, \mathbf{r}'; t')(t - t')^j \ .$$

Insertion of that expression in Hamilton-Jacobi's equation leads to the conditions

$$(\mathbf{r} - \mathbf{r}') \cdot (\nabla_{\mathbf{r}} W_0 - \mathbf{A}_0) = 0 \tag{4.59}$$

and

$$(\mathbf{r} - \mathbf{r}') \cdot (\nabla_{\mathbf{r}} W_1 - \mathbf{A}_1) + W_1 = -\frac{1}{2m}\left(\nabla_{\mathbf{r}} W_0 - \mathbf{A}_0\right)^2 - U_0 \tag{4.60}$$

where $U_0 = U(x, t')$. Equation (4.59) is immediately solved using Lemma 114, and we get

$$W_0 = (\mathbf{r} - \mathbf{r}') \cdot \overline{\mathbf{A}}_0 \tag{4.61}$$

where $\overline{\mathbf{A}}_0 = \overline{\mathbf{A}}(\mathbf{r}, \mathbf{r}'; t')$ is the average of $\mathbf{A}_0 = \mathbf{A}(\mathbf{r}, t')$ evaluated on $[\mathbf{r}', \mathbf{r}]$. To solve Eq. (4.60), it suffices to apply Lemma 113, which yields:

$$W_1 = \overline{-\tfrac{1}{2m}\left(\nabla_{\mathbf{r}} W_0 - \mathbf{A}_0\right)^2 + (\mathbf{r} - \mathbf{r}') \cdot \mathbf{A}_1 - U_0}$$

$$= \overline{-\tfrac{1}{2m}\left(\nabla_{\mathbf{r}} W_0 - \mathbf{A}_0\right)^2} + \overline{(\mathbf{r} - \mathbf{r}') \cdot \mathbf{A}_1} - \overline{U_0}$$

where we have set $\mathbf{A}_1 = \partial \mathbf{A}/\partial t(\mathbf{r}, t')$. To complete the proof it thus suffices to show that

$$\nabla_{\mathbf{r}} W_0 - \mathbf{A}_0 = \overline{(\mathbf{r} - \mathbf{r}') \times \mathbf{B}}$$

(calculated at $(\mathbf{r}, \mathbf{r}'; t')$). In view of the classical formula from vector calculus

$$\nabla_{\mathbf{r}}(\mathbf{f} \cdot \mathbf{g}) = (\mathbf{f} \cdot \nabla_{\mathbf{r}}) \cdot \mathbf{g} + (\mathbf{g} \cdot \nabla_{\mathbf{r}}) \cdot \mathbf{f} + \mathbf{f} \times (\nabla_{\mathbf{r}} \times \mathbf{g}) + \mathbf{g} \times (\nabla_{\mathbf{r}} \times \mathbf{f})$$

we have

$$\begin{aligned}\nabla_{\mathbf{r}} W_0 - \mathbf{A}_0 = &\left((\mathbf{r} - \mathbf{r}') \cdot \nabla_{\mathbf{r}}\right) \overline{\mathbf{A}_0} \\ &+ \left(\overline{\mathbf{A}_0} \cdot \nabla_{\mathbf{r}}\right)(\mathbf{r} - \mathbf{r}') - \overline{\mathbf{A}_0} + (\mathbf{r} - \mathbf{r}') \times \left(\nabla_{\mathbf{r}} \times \overline{\mathbf{A}_0}\right)\end{aligned}$$

that is where we have used the fact that

$$\nabla_{\mathbf{r}} W_0 - \mathbf{A}_0 = \nabla_{\mathbf{r}}\left((\mathbf{r} - \mathbf{r}') \cdot \overline{\mathbf{A}_0}\right) - \overline{\mathbf{A}_0}$$

and $\nabla_{\mathbf{r}} \times (\mathbf{r} - \mathbf{r}') = 0$. Simplifying, we get

$$\nabla_{\mathbf{r}} W_0 - \mathbf{A}_0 = \left((\mathbf{r} - \mathbf{r}') \cdot \nabla_{\mathbf{r}}\right) \overline{\mathbf{A}_0} + \overline{\mathbf{A}_0} + (\mathbf{r} - \mathbf{r}') \times \left(\nabla_{\mathbf{r}} \times \overline{\mathbf{A}_0}\right)$$

that is:

$$\nabla_{\mathbf{r}} W_0 - \mathbf{A}_0 = (\mathbf{r} - \mathbf{r}') \times \left(\nabla_{\mathbf{r}} \times \overline{\mathbf{A}_0}\right).$$

By definition of $\overline{\mathbf{A}_0}$ we have:

$$\nabla_{\mathbf{r}} \times \overline{\mathbf{A}_0}(\mathbf{r}, \mathbf{r}'; t') = \int_0^1 s\left(\nabla_{\mathbf{r}} \times \mathbf{A}\right)(s\mathbf{r} + (1 - s)\mathbf{r}'; t')ds$$

hence, since $\nabla_{\mathbf{r}} \times \mathbf{A} = \mathbf{B}$:

$$\nabla_{\mathbf{r}} \times \overline{\mathbf{A}_0}(\mathbf{r}, \mathbf{r}'; t') = \int_0^1 s\mathbf{B}(s\mathbf{r} + (1 - s)\mathbf{r}'; t')ds$$

yielding

$$(\mathbf{r}-\mathbf{r}')\times\left(\nabla_{\mathbf{r}}\times\overline{\mathbf{A}_0}\right)=\int_0^1 s(\mathbf{r}-\mathbf{r}')\times\mathbf{B}(s\mathbf{r}+(1-s)\mathbf{r}';t')\,ds$$

$$=\overline{(\mathbf{r}-\mathbf{r}')\times\mathbf{B}}.$$

Thus

$$\nabla_{\mathbf{r}}W_0-\mathbf{A}_0=\overline{(\mathbf{r}-\mathbf{r}')\times\mathbf{B}} \tag{4.62}$$

as claimed, and this completes the proof of the Proposition. ∎

Example 116 ***The electron in a uniform magnetic field revisited.*** *We consider, as in Chapter 2, an electron placed in a uniform magnetic field* $\mathbf{B}=(0,0,B_z)$, *and we choose again the "symmetric gauge" defined by*

$$\mathbf{A}=\tfrac{1}{2}(\mathbf{r}\times\mathbf{B}).$$

Proposition 115 then yields the following short-time approximation for W*:*

$$\bar{W}=W_{free}+\frac{eB_z}{2c}(xy'-x'y)+\frac{e^2B_z^2}{4mc^2}(\mathbf{r}+\mathbf{r}')^2(t-t')$$

where W_{free} *is the free-particle generating function:*

$$W_{free}(\mathbf{r},\mathbf{r}';t,t')=m\frac{(\mathbf{r}-\mathbf{r}')^2}{2(t-t')}.$$

The results of this subsection can easily be generalized to many-particle systems:

4.5.3 Many Particle Systems in a Gauge (A,U)

We now assume that H is an N-particle Maxwell Hamiltonian

$$H=\sum_{j=1}^{N}\frac{1}{2m_j}\left(\mathbf{p}_j-\mathbf{A}_j(\mathbf{r}_j,t)\right)^2+U(\mathbf{r},t) \tag{4.63}$$

with the usual notations $\mathbf{r}_j=(x_j,y_j,z_j)$, $\mathbf{p}_j=(p_{x_j},p_{y_j},p_{z_j})$ and $\mathbf{r}=(\mathbf{r}_1,...,\mathbf{r}_N)$. Writing

$$H=\frac{1}{2m}(p-A(x,t))^2+U(x,t) \tag{4.64}$$

where $x=(\mathbf{r}_1,...,\mathbf{r}_N)$, $p=(\mathbf{p}_1,...,\mathbf{p}_N)$, and m is mass matrix, we have:

Proposition 117 *The generating function* W *determined by the Maxwell Hamiltonian (4.63) is asymptotically given by* $W = \overline{W} + O\left((t-t')^2\right)$ *where*

$$\begin{cases} \overline{W} = W_f + (x-x')\cdot\overline{A_0} - \overline{V}(t-t') \\ W_f(x,x';t,t') = m\dfrac{(x-x')^2}{2(t-t')} \end{cases}$$

and $\overline{V} = \overline{V}(x,x';t')$ *is the average on* $[x,x']$ *of*

$$V = \sum_{j=1}^{N}\left(\overline{(\mathbf{r}_j - \mathbf{r}'_j)\times\mathbf{B}_j}\right)^2 - (\mathbf{r}-\mathbf{r}')\cdot\frac{\partial\mathbf{A}}{\partial t} + U.$$

Proof. It is an obvious consequence of Proposition 115. In fact, setting $A_0 = A(x,t')$ Eq. (4.59) becomes

$$(x-x')\cdot(\nabla_x W_0 - A_0) = 0$$

whose solution is $W_0 = (x-x')\cdot\overline{A}_0$ (cf. (4.61)). Similarly, setting $\mathbf{A}_{j,1} = (\partial\mathbf{A}_j/\partial t)(\mathbf{r},t')$, Eq. (4.60) becomes

$$(x-x')\cdot(\nabla_x W_1 - A_1) + W_1 = \frac{1}{m}\left(\nabla_x W_0 - A_0\right)^2 - U_0.$$

Its solution is

$$W_1 = -\frac{1}{2m}\overline{(\nabla_x W_0 - A_0)^2} + \overline{(x-x')\cdot A_1} - \overline{U_0}$$

and Eq. (4.62) thus becomes

$$\nabla_x W_0 - A_0 = \sum_{j=1}^{N}\overline{(\mathbf{r}_j - \mathbf{r}'_j)\times\mathbf{B}_j}$$

in view of the relation

$$\frac{1}{m}\left(\nabla_x W_0 - A_0\right)^2 = \sum_{j=1}^{N}\overline{(\mathbf{r}_j - \mathbf{r}'_j)\times\mathbf{B}_j}.$$

Collecting these results ends the proof of the proposition. (Notice that if $A = 0$ we recover Proposition 112.) ■

4.6 Lagrangian Manifolds

This is a somewhat more technical section. Lagrangian manifolds (whose general definition seem to be due to V.P. Maslov) play an essential role in mechanics, both classical and quantum. As we will see, a Lagrangian manifold is attached to every quantum system, via the phase of the wave function; whereas it only makes sense in classical mechanics for Liouville integrable systems. One of the main properties of Lagrangian manifolds is that one can define a natural notion of phase on them (or, rather, on their universal covering manifolds).

4.6.1 Definitions and Basic Properties

Recall (see Appendix A) that a *Lagrangian plane* in $\mathbb{R}^n_x \times \mathbb{R}^n_p$ is an n-dimensional linear subspace ℓ on which the symplectic form vanishes identically. That is, for every pair (z, z') of vectors of ℓ, we have $\Omega(z, z') = 0$. Every Lagrangian plane can be represented by a system of equations

$$Xx + Pp = 0 \quad \text{with} \quad \begin{cases} P^T X = X^T P \\ \operatorname{rank}(X, P) = n. \end{cases}$$

In particular, a Lagrangian plane is transversal to the vertical plane $\ell = 0 \times \mathbb{R}^n_p$ if, and only if, it has an equation

$$p = Ax \quad \text{with} \quad A = A^T.$$

The set of all Lagrangian planes of $\mathbb{R}^n_x \times \mathbb{R}^n_p$ is denoted by $\operatorname{Lag}(n)$; it is called the *Lagrangian Grassmannian.*

Definition 118 *An n-dimensional submanifold V of $\mathbb{R}^n_x \times \mathbb{R}^n_p$ is called a* Lagrangian manifold *if every tangent space $\ell(z) = T_z V$ is a Lagrangian plane. Thus, V is Lagrangian if and only if the symplectic product of two tangent vectors to V at a same point is zero.*

In the phase plane $\mathbb{R}_x \times \mathbb{R}_p$ every smooth curve is a Lagrangian manifold: two tangent vectors at a point z of a curve are collinear, and their skew-product is thus zero. In higher dimensions an arbitrary manifold is not in general Lagrangian.

A Lagrangian manifold remains Lagrangian if acted upon by a symplectomorphism:

Proposition 119 *Let f be a symplectomorphism and V a Lagrangian manifold. Then the manifold $f(V)$ is also Lagrangian.*

Proof. Suppose that $X(u)$ and $Y(u)$ are two tangent vectors to $f(V)$ at $u = f(z)$. Then there exist vectors $X'(z)$ and $Y'(z)$ tangent to V at z and such that

$$X(u) = Df(z)X'(z), \quad Y(u) = Df(z)Y'(z)$$

where $Df(z)$ is the Jacobian matrix. Hence

$$\Omega(X(u), Y(u)) = \Omega(X'(u), Y'(u)) = 0$$

since $Df(z)$ is symplectic, and V is Lagrangian. It follows that $f(V)$ is also Lagrangian, since it is connected and has the same dimension n as V. ∎

Notice that in general $f(V)$ will not be a graph even if V is: the mapping f can "bend" V in such a way that several points have the same projection on configuration space. The points of $f(V)$ which have a neighborhood which fails to be diffeomorphically mapped on $\mathbb{R}^n_x$ form a set of measure zero, called the *caustic* of $f(V)$. These caustic points are at the origin of the technical difficulties appearing in the traditional "configuration space" descriptions of semiclassical mechanics. More generally:

Definition 120 *Let V be an arbitrary Lagrangian manifold. The* caustic *of V is the set Σ_V of all $z \in V$ who do not have a neighborhood projecting diffeomorphically on $\mathbb{R}^n_x$. Alternatively, Σ_V is the set of points $z \in V$ at which $\ell(z) = T_z V$ is not transversal to the vertical plane $\mathbb{R}^n_p \equiv 0 \times \mathbb{R}^n_p$:*

$$\Sigma_V = \left\{ z \in V : \ell(z) \cap \mathbb{R}^n_p \neq 0 \right\}.$$

The Maslov cycle *is the subset of* $\mathrm{Lag}(n)$ *consisting of all Lagrangian planes ℓ such that $\ell \cap \mathbb{R}^n_p \neq 0$.*

We will see in the next subsection that outside the caustic, we can always find an open set U in $\mathbb{R}^n_x \times \mathbb{R}^n_p$ such that $U \cap V$ can be represented by an equation $p = \nabla_x \Phi_U(x)$ for some smooth function Φ_U (called a *local phase* of V).

One should not forget that caustics have no intrinsic meaning whatsoever: they are just artifacts depending on the linear space on which we are "projecting" the Lagrangian manifold V. For instance, if we rotate V by an angle of $\pi/2$ we obtain new Lagrangian manifold (because this rotation is just multiplication by $-J$), the caustic Σ is replaced by another set of points, which can very well be empty. (This is the case, for instance, when V is represented by an equation $x = \nabla_p \Psi(p)$ where Ψ is some function of p.)

A loop in a manifold V is a continuous mapping $t \longmapsto \gamma(t)$ of the interval $[0, 1]$ (or any other compact interval) into V, and such that $\gamma(0) = \gamma(1)$. If γ is a loop in the Lagrangian manifold V, we will call the real number

$$C(\gamma) = \oint_\gamma pdx \tag{4.65}$$

the *period* of γ.

We have the following result, which extends to arbitrary loops the invariance of the action of periodic orbits:

Proposition 121 *Let $\gamma : [0,1] \longrightarrow V$ be a closed path in the Lagrangian manifold V and $f(\gamma)$ its image by a local symplectomorphism f. Then*

$$C(f(\gamma)) = C(\gamma). \tag{4.66}$$

Proof. Let D be the piece of surface in V bounded by γ; we have, by Stoke's theorem

$$\oint_\gamma pdx = \int_D dp \wedge dx = \int_D \Omega$$

and also

$$\oint_{f(\gamma)} pdx = \int_{f(D)} dp \wedge dx = \int_{f(D)} \Omega.$$

Using the formula of change of variables, together with the fact that $f_*\Omega = \Omega$, we have

$$\oint_\gamma pdx = \int_{f(D)} f_*\Omega = \int_{f(D)} \Omega = \oint_{f(\gamma)} pdx$$

which proves formula (4.66). ∎

Remark 122 *Proposition 121 applies in particular in the following case. Let $(f_{t,t'})$ be the time-dependent flow associated with an arbitrary Hamiltonian function. Choose now a Lagrangian manifold, denote it by $V_{t'}$, and set $V_t = f_{t,t'}(V_{t'})$. If γ' is a loop in $V_{t'}$ and $\gamma = f_{t,t'}(\gamma')$, then*

$$\oint_\gamma pdx = \oint_{\gamma'} pdx \ .$$

In some texts this formula is taken as the definition of adiabatic invariance. *The fallacy of this definition is obvious, as the formula above holds for* all *Hamiltonian flows!*

4.6.2 Lagrangian Manifolds in Mechanics

The simplest example of a Lagrangian manifold of general dimension n is the graph of the gradient of a function. Let in fact $\Phi = \Phi(x)$ be a smooth function defined on an open subset D of $\mathbb{R}^n_x$. We claim that the set

$$V_\Phi = \{(x, \nabla_x \Phi(x)) : x \in D\} \tag{4.67}$$

is a Lagrangian manifold. We first notice that V_Φ is an n-dimensional manifold, since the projection $(x, p) \longmapsto x$ is a diffeomorphism $V_\Phi \longrightarrow D$. Let us next show that, for every $x_0 \in D$, the tangent plane to V_Φ at $z_0 = (x_0, \nabla_x \Phi(x_0))$ is a Lagrangian plane. The definition of V_Φ means that we have

$$(x, p) \in V_\Phi \Longleftrightarrow p_j = \frac{\partial \Phi}{\partial x_j}(x) \quad (1 \leq j \leq n)$$

and hence the tangent space to V_Φ at z_0 is determined by the system of n linear equations

$$p_j = \sum_{i=1}^{n} \frac{\partial^2 \Phi}{\partial x_i \partial x_j}(x_0)\, x_i, \tag{4.68}$$

that is $p = \Phi''(x_0)x$, where $\Phi''(x_0)$ is the matrix of second derivatives of Φ calculated at the point x_0. Since $\Phi''(x_0)$ is symmetric, it follows that the tangent plane to V_Φ at $z_0 = (x_0, p_0)$ is Lagrangian.

We will call Lagrangian manifolds of the type V_Φ *exact Lagrangian manifolds*. The terminology comes from the fact that the 1-form pdx is exact on V_Φ, in fact:

$$pdx = d\Phi \quad \text{on } V_\Phi. \tag{4.69}$$

An exact Lagrangian manifold can be attached to every quantum-mechanical system via its wave function:

Example 123 ***The Lagrangian manifold attached to a wave function.*** *Let Ψ be the wave function of a de Broglie matter wave. Writing Ψ in polar form*

$$\Psi(\mathbf{r}, t) = e^{\frac{i}{\hbar}\Phi(\mathbf{r},t)} R(\mathbf{r}, t)$$

the graph

$$V_t = \{(\mathbf{r}, \nabla_\mathbf{r} \Phi(\mathbf{r})) : \mathbf{r} \in \mathbb{R}^3_\mathbf{r}\}$$

is an exact Lagrangian manifold.

Suppose next H is a Hamiltonian of a system which is Liouville integrable. We are going to see that we can attach a whole family of Lagrangian manifolds to such a system (when the energy shells are compact and connected, these manifolds are the well-known "invariant tori").

Recall from Chapter 2 (Section 2.4) that a Hamiltonian system is said to be Liouville integrable if it has n independent constants of the motion in involution. That is, in addition to the Hamiltonian function $F_1 = H$ itself, there are $n-1$ other independent functions $F_2, ..., F_n$ which are constant along the solution curves to Hamilton's equations, and such that

$$\{F_j, F_k\} = 0 \quad \text{for} \quad 1 \leq j, k \leq n$$

where $\{\cdot, \cdot\}$ are the Poisson brackets (see Chapter 2, Eq. (2.81)). When connected, the manifolds

$$V = \{z : F_j(z) = f_j, 1 \leq j \leq n\} \tag{4.70}$$

are (except for exceptional values of the f_j) symplectomorphic to products of circles and straight lines (if V is moreover compact, it is symplectomorphic to an n-torus).

Proposition 124 *The sets (4.70) are Lagrangian manifolds. In fact, through every point (x', p') of the energy shell $\{z : H(z) = E\}$ passes exactly one Lagrangian manifold V containing the orbit of (x', p') by the flow of X_H.*

Proof. In view of formula (2.82) we have

$$\{F_j, F_k\} = \Omega(X_j, X_k)$$

where $X_j = (\nabla_p F_j, -\nabla_x F_j)$ is the Hamilton vector field associated to F_j, so the involution conditions can be written as

$$\Omega(X_j(z), X_k(z)) = 0$$

for every $z \in V$. Since the functions F_j are independent, the vector fields X_j span V. It follows that for all pairs $Y(z)$, $Y'(z)$ of tangent vectors at $z \in V$, $\Omega(Y(z), Y'(z))$ can be expressed as a linear combination of the terms $\Omega(X_j(z), X_k(z))$ and hence $\Omega(Y(z), Y'(z)) = 0$, so that V is Lagrangian. Let now (x', p') be a point of the energy shell defined by $H(z) = E$, and let $t \longmapsto (x(t), p(t))$ be the orbit of that point. Then $H(x(t), p(t)) = E$ by the law of conservation of energy, so that $(x(t), p(t))$ lies on the energy shell. The $F_j(x(t), p(t))$ being also constant along that orbit, we have $(x(t), p(t)) \in V$ for all t. ∎

4.7 The Phase of a Lagrangian Manifold

What is a phase? According to Webster's New Encyclopedic Dictionary (1994 edition)

> [...a phase is a] *stage or interval in the development of a cycle, or the stage of progress in a regularly recurring motion or a cyclic process (as a wave or vibration) in relation to a reference point.*

I have added the emphasis in the last words to make clear that the phase of a system has to be calculated starting from *somewhere*: the phase is not a quantity intrinsically attached to that system. That problem is usually dodged in physics, because what one is really concerned with when one studies the evolution of a system, is the variation of the phase. We will show in this section that it is possible to define in a notion of phase for Lagrangian manifolds, and that the increment of that phase under the action of a Hamiltonian flow is precisely the integral of the Poincaré-Cartan form along the trajectory.

4.7.1 The Phase of an Exact Lagrangian Manifold

Let V_Φ be an exact Lagrangian manifold, that is:

$$V_\Phi : \quad p = \nabla_x \Phi(x)$$

for some smooth function Φ. We will call Φ a phase of V_Φ. Clearly, two phases of V_Φ differ only by a function $\nabla_x \chi = 0$, that is, by a locally constant function. We will therefore often commit the abuse of language consisting in talking about "the" phase of V_Φ. An immediate property of the phase is that its differential is the action form pdx:

$$d\Phi(x) = pdx \quad \text{on} \quad V_\Phi. \tag{4.71}$$

The notion of phase of an exact Lagrangian manifold allows us to give a nice geometric interpretation of the solutions of Hamilton-Jacobi's equation. Recall from Subsection 4.4.5 (Proposition 109) that the Cauchy problem for Hamilton-Jacobi's equation

$$\begin{cases} \dfrac{\partial \Phi}{\partial t} + H\left(x, \nabla_x \Phi, t\right) = 0 \\ \Phi(x, t') = \Phi'(x) \end{cases}$$

can be (at least locally) solved for short time intervals $0 < |t - t'| < \varepsilon$, and that the solution is

$$\Phi(x,t) = \Phi'(x') + \int_{x',t'}^{x,t} pdx - Hds \tag{4.72}$$

where the point x' is defined by the condition

$$(x,p) = f_{t,t'}(x', \nabla_x \Phi(x')). \tag{4.73}$$

Consider now the Lagrangian manifolds

$$V_{t'} : p' = \nabla_x \Phi(x') \quad , \quad V_t : p = \nabla_x \Phi(x,t).$$

We claim that V_t is just the image of $V_{t'}$ by the symplectomorphism $f_{t,t'}$:

$$V_t = f_{t,t'}(V_{t'}). \tag{4.74}$$

This is obvious, because formula (4.73) says that any point (x,p) of V_t is –by definition!– the image by $f_{t,t'}$ of a point (x',p') of $V_{t'}$. The formula

$$\Phi(x,t) = \Phi'(x') + \int_{x',t'}^{x,t} pdx - Hds$$

yielding the solution of Hamilton-Jacobi's problem (see Proposition 109) is thus just the expression of the phase of V_t in terms of that of $V_{t'}$. This geometric interpretation of the solution of Hamilton-Jacobi's equation immediately makes us understand why the solutions are usually only defined for short time intervals: when $t - t'$ becomes large, $f_{t,t'}$ will make the initial Lagrangian manifold $V_{t'}$ "bend" in such a way that it no longer is a graph, i.e., we can no longer define it by a relation of the type $p = \nabla_x \Phi(x,t)$ because of the appearance of caustic points. However, and this should always be kept in mind, the Lagrangian manifold V_t defined by (4.74) will exist as long as $f_{t,t'}$ is defined. This leads us to wonder whether it would be possible to define the notion of phase for general Lagrangian manifolds. While this question will be answered (affirmatively) in the next subsection, we note that this can already be made in a rather obvious way in the context of the Hamilton-Jacobi equation, whose solution is, as we have seen

$$\Phi(x,t) = \Phi'(x') + \int_{x',t'}^{x,t} pdx - Hds \tag{4.75}$$

where x determines x' by $(x,p) = f_{t,t'}(x',p')$. We can thus define the phase of $V_t = f_{t,t'}(V_{t'})$ as being the function

$$\varphi(z,t) = \Phi'(x') + \int_{z',t'}^{z,t} pdx - Hds \tag{4.76}$$

where the integral is calculated along the extended *phase space* trajectory arriving at (z,t) after time $t-t'$. Formula (4.76) obviously reduces to the formula (4.75) when $t-t'$ is sufficiently small, since we know that in this case the initial and final positions uniquely determine the momenta, hence also the points z' and z.

Observe that formula (4.76) defines the phase of a "moving exact Lagrangian manifold" as a function of $z \in V$ (and of time), and not of x (the latter cannot generally be used as a local variable because of the appearance of the caustics).

Another question which poses itself is whether it is possible to define a tractable notion of phase for arbitrary Lagrangian manifolds. Let us discuss this on the following simple but basic situation.

Consider the unit circle S^1; as every smooth curve, it is a Lagrangian manifold in the phase plane $\mathbb{R}_x \times \mathbb{R}_p$. We want to find some function φ, defined on S^1, and having the property $d\varphi = pdx$. Passing to polar coordinates, we thus require that

$$d\varphi(\theta) = -\sin^2\theta d\theta$$

and we find, by integrating, that

$$\varphi(\theta) = \frac{1}{2}(\sin\theta\cos\theta - \theta). \tag{4.77}$$

Now, there is rub. It comes from the fact that the function φ just obtained is not single valued. For instance, $\varphi(0) = 0$, while $\varphi(2\pi) = -\pi$, but nevertheless 0 and 2π are the angular coordinates of the same point on the circle. In fact, formula (4.77) should be interpreted as defining the phase, not on the circle S^1 itself, but rather on the universal covering $\mathbb{R}_\theta$ of S^1. In fact, that universal covering is defined by the mapping $\pi : \mathbb{R}_\theta \longrightarrow S^1$ given by

$$\pi(\theta) = (\cos\theta, \sin\theta).$$

More generally, the phase of a product of n circles ("the n-torus"):

$$T^n = S^1 \times \cdots \times S^1$$

would be defined on the universal covering $\mathbb{R}_{\theta_1} \times \cdots \times \mathbb{R}_{\theta_n}$ of T^n by:

$$\varphi(\theta_1, ..., \theta_n) = \frac{1}{2}\sum_{j=1}^{n} R_j^2(\sin\theta_j \cos\theta_j - \theta_j).$$

In the following subsections we extend these constructions of the phase to the case where V is an arbitrary Lagrangian manifold. We first need some definitions and properties from the theory of covering spaces.

4.7.2 The Universal Covering of a Manifold*

We begin by noting that, independently of the fact that it is Lagrangian or not, a manifold V always has a universal covering

$$\pi : \check{V} \longrightarrow V.$$

By this we mean that $\check{V}$ is a *simply connected* manifold such that the "projection" π has the following properties:

(1) *π is surjective,* **(2)** *π is a local diffeomorphism.*

Let us shortly describe how the manifold $\check{V}$ is explicitly constructed using homotopy classes of paths. For the reader who is unacquainted with this type of argument, I recommend the lovely little book [98] by Michio Kuga (*b*.1928). (At the time when Kuga wrote this book, in 1968, it got him in trouble with the local mathematical establishment because of the cartoons and the funny examples; it didn't, however, prevent his book from becoming a best-seller!)

Pick a base point $z_0 \in V$, and for every point $z \in V$ consider all the continuous paths joining z_0 to z in V. The set of all these paths is partitioned into equivalence classes by the relation

$$\gamma \sim \gamma' \iff \gamma \text{ and } \gamma' \textit{ are homotopic with fixed endpoints.}$$

We then define $\check{V}$ as being the set of all these equivalence classes. Of course, if we restrict ourselves to loops in V, then we obtain the first homotopy group $\pi_1(V, z_0)$. The projection $\pi : \check{V} \longrightarrow V$ is then defined as being the mapping which to every homotopy class $\check{z}$ associates the endpoint z of a path γ from z_0 to z and representing $\check{z}$. One can prove (but this is noticeably more difficult), that there exists a topology on $\check{V}$ for which $\check{V}$ is a simply connected manifold, and such that π effectively is a covering mapping. (One can in fact show that π is a local diffeomorphism with maximal rank $n = \dim V$, see Godbillon's treatise [51].)

Also observe that the fundamental group $\pi_1(V, z_0)$ acts on $\check{V}$ in the obvious way: if $\gamma_{z_0 z}$ is a path representing $\check{z}$, and $\gamma_{z_0 z_0}$ a loop representing $\gamma_0 \in \pi_1(V, z_0)$, then $\gamma_0 \check{z}$ is the homotopy class of the concatenation $\gamma_{z_0 z_0}\gamma_{z_0 z}$ (i.e. the path $\gamma_{z_0 z_0}$ followed by the path $\gamma_{z_0 z}$).

4.7.3 The Phase: General Case

Let now V be a Lagrangian manifold, which we suppose to be connected (this is no real restriction, because the connected components of an arbitrary Lagrangian manifold are also Lagrangian). We also assume that a "base point" z_0 is chosen once for all.

Definition 125 *The* phase φ *of a Lagrangian manifold is the function*

$$\varphi : \check{V} \longrightarrow \mathbb{R}$$

defined as follows: if $\check{z}$ is the homotopy class of a path $\gamma_{z_0 z}$, then

$$\varphi(\check{z}) = \int_{\gamma_{z_0 z}} p\, dx. \tag{4.78}$$

Let us show that this formula really defines φ as a function $\check{V} \longrightarrow \mathbb{R}$. For this, we have to verify that the integral in Eq. (4.78) only depends on the homotopy class $\check{z}$ of $\gamma_{z_0 z}$. Let $\gamma'_{z_0 z}$ be another path representing $\check{z}$ and set $\sigma = \gamma_{z_0 z} - \gamma'_{z_0 z}$. Since both paths $\gamma_{z_0 z}$ and $\gamma'_{z_0 z}$ are homotopic, the domain Σ in V enclosed by the curve σ is shrinkable to a point. It follows, using Stoke's formula, that

$$\int_\sigma pdx = \int_\Sigma dp \wedge dx = \int_\Sigma \Omega$$

but this is zero, since Ω is zero on V. Hence

$$\int_\sigma pdx = 0 \quad \text{and} \quad \int_{\gamma_{z_0 z}} pdx = \int_{\gamma'_{z_0 z}} pdx$$

as claimed.

We next observe that Definition (4.78) truly extends the definition of the phase of an exact Lagrangian manifold. Suppose in fact that $V = V_\Phi$; in this case

$$\varphi(z) = \int_{\gamma_{z_0 z}} d\Phi = \Phi(x) - \Phi(x_0) \text{ if } z = (x, p)$$

where we have calculated the integral along the projection $\gamma_{x_0 x}$ of $\gamma_{z_0 z}$ on configuration space.

The phase φ has the following obvious property, which extends Eq. (4.71):

$$d\varphi(\check{z}) = pdx \text{ if } \pi(\check{z}) = (x, p)$$

where the differential $d\varphi$ is calculated with respect to the local variable x. Conversely, every function $\varphi : \check{V} \longrightarrow \mathbb{R}$ for which $d\varphi = pdx$ is a phase of V.

Since the phase is defined on $\check{V}$, it does not, in general, make sense to talk about its value at $z \in V$. (Using a somewhat older terminology, one would say that φ is "multiply valued on V".) This "multi-valuedness" is reflected by the formula

$$\varphi(\gamma\check{z}) = \varphi(\check{z}) + \oint_\gamma pdx \tag{4.79}$$

for $\gamma \in \pi_1(V)$. Thus φ is defined on V if and only if all the periods $\int_\gamma p\,dx$ of $p\,dx$ vanish.

4.7.4 Phase and Hamiltonian Motion

We finally study the transformation of the phase of an arbitrary Lagrangian manifold under the action of a Hamiltonian flow (for more on that topic, and complete proofs, see our monograph [59]).

Consider a function $H = H(x, p, t)$ defined on the extended phase space $\mathbb{R}^n_x \times \mathbb{R}^n_p \times \mathbb{R}_t$ (or on an open subset $D \times \mathbb{R}_t$ of it). We do not assume that H has any particular form (for instance that it is a Maxwell Hamiltonian), but only that it is a continuously differentiable function.

Suppose that we are given, at some time t', a Lagrangian manifold $V_{t'}$ on which we select a base point $z_{t'}$. This allows us to define the phase $\varphi'(\check{z}')$ of $V_{t'}$ by formula (4.78), with z_0 replaced by $z_{t'}$, $\check{z}'$ being an element of the universal covering $\check{V}_{t'}$ of $V_{t'}$:

$$\varphi'(\check{z}') = \int_{\gamma_{z_{t'} z'}} p\,dx.$$

The manifold $V_t = f_{t,t'}(V_{t'})$ is also Lagrangian; choosing as base point $z_t = f_{t,t'}(z_{t'})$, the phase of V_t is denoted by $\varphi = \varphi(\check{z})$. Notice that we can identify the universal coverings $\check{V}_t$ and $\check{V}_{t'}$, defining the projection $\pi_t : \check{V}_t \to V_{t'}$ by

$$\pi_t(\check{z}) = z(t) \text{ if } \pi_{t'}(\check{z}) = z(t').$$

We next define a new function $\varphi(\cdot, t) : \check{V}_t \longrightarrow \mathbb{R}$ by

$$\varphi(\check{z}, t) = \varphi'(\check{z}') + \int_{z'}^{z} pdx - Hdt, \tag{4.80}$$

where the integral is calculated along the trajectory $s \to f_{s,t'}(z')$ $(t' \leq s \leq t)$ leading from $z' \in V_{t'}$ to $z \in V_t$. We claim that:

Proposition 126 *(1) The function $\varphi(\cdot, t)$ is a phase of V_t; in fact $d\varphi(\check{z}, t) = pdx$ for fixed t and we have*

$$\varphi(\check{z}, t) = \varphi(\check{z}) + \int_{z_{t'}}^{z_t} p\,dx - H\,dt \tag{4.81}$$

where the integral is calculated along the phase space trajectory leading from the base point $z_{t'} \in V_{t'}$ to its image $z_t \in V_t$. (2) The local expression $\Phi(x, t)$ of that phase satisfies Hamilton-Jacobi's equation

$$\frac{\partial \Phi}{\partial t} + H(x, \nabla_x \Phi, t) = 0 \quad , \quad \Phi(x, t') = \Phi'(x)$$

where Φ' is the local expression of φ'.

Proof. **(1)** Let $\tilde{\lambda}_H = pdx - Hdt$ be the Poincaré-Cartan form. In view of Helmholtz's theorem 85 (see Section 4.2), we have

$$\int_{\gamma_{z_{t'}z'}} \tilde{\lambda}_H + \int_{z'}^{z} \tilde{\lambda}_H = \int_{z_{t'}}^{z_t} \tilde{\lambda}_H + \int_{\gamma_{z_t z}} \tilde{\lambda}_H$$

where $\gamma_{z_{t'}z'}$ is a path representing $\check{z}'$ and $\gamma_{z_t z}$ its image by $f_{t,t'}$ (its homotopy class is thus $\check{z}$). Since $dt' = 0$ on $V_{t'}$ and $dt = 0$ on V_t this equality can be rewritten, in view of (4.78) as:

$$\varphi'(\check{z}') + \int_{z'}^{z} pdx - Hdt = \varphi(\check{z}) + \int_{z_{t'}}^{z_t} pdx - H\,dt \tag{4.82}$$

which is (4.81). Keeping t fixed we thus have

$$d\varphi(\check{z}, t) = d\varphi(\check{z}) = pdx.$$

(2) Follows from the fact that the local expression of φ is

$$\Phi(x,t) = \varphi(\check{z},t) \text{ if } \pi_t(\check{z}) = (x,p)$$

that is

$$\begin{aligned}\Phi(x,t) &= \Phi'(x') + \int_{z',t'}^{z,t} pdx - Hdt \\ &= \Phi'(x') + \int_{x',t'}^{x,t} pdx - Hdt\end{aligned}$$

which is precisely the solution of Hamilton-Jacobi's equation with initial datum Φ' at time t'. ∎

Example 127 ***The harmonic oscillator.*** *Suppose that H is a quadratic homogeneous polynomial (with time-independent coefficients) in the position and momentum coordinates. Applying Euler's identity for homogeneous functions to H, and using Hamilton's equations, we get*

$$\varphi(\check{z},t) = \varphi(\check{z},t') + \frac{1}{2}(px - p'x').$$

4.8 Keller-Maslov Quantization

In physical literature, the Maslov quantization condition is often called the EBK (= Einstein-Brillouin-Keller) or Bohr-Sommerfeld quantization condition. It was actually Keller who was the first to state that condition in a mathematically correct form, in 1958, in connection with the study of WKB approximation. His celebrated paper [95] was the forerunner of Maslov's work [113, 114] on the quantization of Lagrangian manifolds. That work, which made use of an index of Lagrangian loops (the "Maslov index") was further developed and made rigorous by Leray [103, 102, 101]. (We will have more to say about Leray's work in Chapter 5.) However, one already finds a similar quantization condition, though in embryonic form, in an amazingly insightful 1917 article by Einstein (Einstein, [41]).

4.8.1 The Maslov Index for Loops

Let us explain on a very simple situation the idea underlying the Maslov index. Consider an arbitrary smooth loop γ in the phase plane, defined on some closed interval $[a,b]$, say a circle described k times:

$$\gamma(t) = (\cos t, \sin t), \quad -k\pi \leq t < k\pi. \tag{4.83}$$

As t varies from $-k\pi$ to $k\pi$, the tangent $\ell(t)$ to $\gamma(t)$ moves, and will become parallel with the p-axis. Each time this happens we count $+1$, so after k complete turns we have recorded $2k$. This number is, by definition, the Maslov index of the loop γ. This counting procedure extends to arbitrary loops as well: every loop γ in the plane is homotopic to a loop in the circle, and since $\pi_1(S^1) \equiv (\mathbb{Z}, +)$ every loop in the circle is homotopic to a loop (4.83). The Maslov index of γ is, by definition, the even integer

$$m(\gamma) = 2k. \tag{4.84}$$

It is of course here just twice the "winding number" of the circle relatively to the origin (or to any other interior point).

In order to give a general working definition for the Maslov index for loops in an arbitrary Lagrangian manifold, we first recall that we have identified, in Subsection 3.3.1, the unitary group $U(n, \mathbb{C})$ with a subgroup $U(n)$ of $\mathrm{Sp}(n)$. This was done as follows: if $R = A + iB$ is unitary, the real matrices A and B must satisfy the relations

$$\begin{cases} A^T A + B^T B = I \text{ and } AB^T = BA^T \\ AA^T + BB^T = I \text{ and } A^T B = B^T A \end{cases}$$

and the $2n \times 2n$ matrix

$$r = \begin{pmatrix} A & -B \\ B & A \end{pmatrix}$$

is thus symplectic. Now, the natural action of $U(n)$ on $Lag(n)$ induces an action of $U(n, \mathbb{C})$ on $Lag(n)$: if $\ell' \in Lag(n)$, then $\ell = R\ell'$ is the image of the Lagrangian plane ℓ' by the corresponding element r of $U(n) \subset \mathrm{Sp}(n)$. Let now γ be a loop defined on $[a, b]$ in an arbitrary Lagrangian manifold V; we denote by $\ell(t)$ the tangent vector to that curve at the point $\gamma(t)$. The action of $U(n)$ on $Lag(n)$ being transitive (see Appendix A), so is the action of $U(n, \mathbb{C})$, and hence we can find, for every t, a unitary matrix $R(t)$ such that $\ell(t) = R(t)\ell_p$ where $\ell_p = 0 \times \mathbb{R}^n_p$. That matrix $R(t)$ is not uniquely defined, but the product

$$W(t) = R(t)R(t)^T \quad (\ell(t) = R(t)\ell_p) \tag{4.85}$$

is. This is because if $\ell(t) = R_1(t)\ell_p$ and $\ell(t) = R_2(t)\ell_p$ then $R_1(t) = H(t)R_2(t)$ where $H(t)$ is an orthogonal matrix, and hence $R_1(t)R_1(t)^T = R_2(t)R_2(t)^T$. Now, the determinant of $W(t)$ is a complex number with modulus one. As t goes from a to b, $\det W(t)$ makes a certain number of turns around the complex unit circle. By definition, the Maslov index $m(\gamma)$ is that number of turns:

$$m(\gamma) = \frac{1}{2\pi i}\oint_\gamma \frac{d(\det W)}{\det W} \tag{4.86}$$

(if γ is a constant loop, we set $m(\gamma) = 0$). Notice that it immediately follows from this definition that the Maslov index has the following additivity property:

$$m(\gamma * \gamma') = m(\gamma) + m(\gamma') \tag{4.87}$$

where $\gamma * \gamma'$ denotes the loop γ followed by the loop γ'.

The Maslov index is obviously a *homotopy invariant*. In particular, if γ is homotopic to zero (i.e., contractible to a point), then its Maslov index is zero. (The converse of this property is generally not true in higher dimensions.) We note that following result holds:

Proposition 128 *The mapping which to every loop γ in $U(n, \mathbb{C})$ associates the integer*

$$k_\gamma = \frac{1}{2\pi i}\oint_\gamma \frac{d(\det R)}{\det R} \tag{4.88}$$

is an isomorphism $\pi_1(U(n, \mathbb{C})) \xrightarrow{\approx} (\mathbb{Z}, +)$.

(See for instance [79, 101] for a proof.)

Let us check that definition (4.86) of the Maslov index coincides with that given by Eq. (4.84) for loops in the phase plane.

Example 129 ***Maslov index of the circle.*** *The circle S^1 is a Lagrangian submanifold of the phase plane. Consider the loop $\gamma(t) = e^{it}$, $0 \leq t \leq 2\pi$. We have here $u(t) = e^{it}$ and hence*

$$m(\gamma) = \frac{1}{2\pi i}\int_0^{2\pi} \frac{d(e^{2it})}{e^{2it}} = 2.$$

More generally, the Maslov index of a loop homotopic to a loop running k times around S^1 is $2k$, and we thus recover formula (4.84).

More generally:

Proposition 130 *Consider the Lagrangian manifold*

$$V = S_1^1 \times \cdots \times S_k^1 \times \mathbb{R}^{n-k}$$

where S_j^1 is the unit circle in the (x_j, p_j) plane. Then the Maslov index of a loop γ in V is the sum of the projections of the Maslov indices $m(\gamma_j)$ of the projections γ_j of γ on the planes $\mathbb{R}_{x_j} \times \mathbb{R}_{p_j}$. That is, if γ is homotopic to a loop $(\gamma_1, ..., \gamma_k, 0, ..., 0)$ in $\mathbb{R}_x^n \times \mathbb{R}_p^n$ with $\gamma_j(t) \in S^1$, then

$$m(\gamma) = 2\sum_{j=1}^{k} m_j \tag{4.89}$$

where m_j is the number of turns around the circle γ_j.

Proof. Since the first homotopy group of V is

$$\pi_1((S^1)^k \times \mathbb{R}^{n-k}) \equiv \pi_1(S^1)^k \equiv (\mathbb{Z}^k, +)$$

it follows that every loop in V is homotopic to a loop of the type:

$$\gamma(t) = (\gamma_1(t), ..., \gamma_k(t), 0, ..., 0), \ \ 0 \leq t \leq T$$

where γ_j are loops on S^1, i.e., $\gamma_j(0) = \gamma_k(T)$. On the other hand, every loop on S^1 is homotopic to a loop

$$\varepsilon_j(t) = (\cos\omega_j t, \sin\omega_j t), \ \ 0 \leq t \leq T_j$$

so there must exist positive integers $m_1, ..., m_k$ such that

$$m_1 T_1 = \cdots = m_k T_k = T.$$

(In particular, the frequencies $\omega_1, ..., \omega_k$ must be commensurate: $\omega_i : \omega_j$ is rational for all i and j.) We can thus identify γ_j with $m_j\varepsilon_j$, the loop ε_j described "m_j times":

$$m_j\varepsilon_j(t) = (\cos\omega_j t, \sin\omega_j t) \ \ \ 0 \leq t \leq T.$$

The tangent plane $\ell(t) = \dot{\gamma}(t)$ is obtained from the vertical p-plane ℓ_p by $\ell(t) = R(t)\ell_p$ where

$$R(t) = \begin{pmatrix} R^{(k)}(t) & 0 \\ 0 & I_{n-k,n-k} \end{pmatrix}$$

$R^{(k)}(t)$ being the $k \times k$ diagonal matrix

$$R^{(k)}(t) = \begin{pmatrix} e^{i\omega_1 t} & \cdots & 0 \\ \vdots & \ddots & \vdots \\ 0 & \cdots & e^{i\omega_k t} \end{pmatrix}$$

and hence the determinant of $W(t) = R(t)R^T(t)$ is

$$\det W(t) = e^{2i(\omega_1+\cdots+\omega_k)t}.$$

It follows, by definition of the Maslov index, that

$$m(\gamma) = \frac{1}{2\pi i}\int_0^T \frac{d(\det W(t))}{\det W(t)} = \frac{T}{\pi}(\omega_1 + \cdots + \omega_k)$$

which is precisely (4.89). ■

In Proposition 130 the Maslov index is always an even integer. One can in fact prove (see Souriau [150]) that this is always the case when the Lagrangian manifold is orientable. In fact:

Proposition 131 (Souriau) *If V is an oriented Lagrangian manifold then the Maslov index of every loop γ in V is an even integer.*

There are however cases of interest where $m(\gamma)$ can take arbitrary values. This occurs, for instance, when one has problems involving reflections between two walls (see Brack and Bhaduri [25]); these problems are interesting and deserve to be further studied. On the other hand, it may happen that the Maslov index is divisible not only by 2, but also by another number; or it may always be equal to zero (this is the case, for instance, for the plane rotator studied in the next subsection). This, together with Souriau's result motivates following definition:

Definition 132 *A Lagrangian manifold V is said to be q-orientable (q an integer ≥ 1) if we have*

$$m(\gamma) \equiv 0 \mod 2q$$

for every loop in V. That manifold is said to be ∞-orientable if $m(\gamma) = 0$ for every loop in V.

For a detailed study of q-orientability see Dazord [34] or Leray [101] (remark that Dazord says "$2q$-orientable" where we say "q-orientable"). For instance, an orientable Lagrangian manifold is 1-orientable. Any simply connected Lagrangian manifold is ∞-orientable, but the converse is not true, as is shown by the following example:

Example 133 ***Exact Lagrangian manifolds.*** *Recall from Subsection 4.6.2 that V is an exact Lagrangian manifold if it is the graph of a gradient, i.e., if there exists a smooth function $\Phi = \Phi(x)$ such that V has the equation $p = \nabla_x\Phi(x)$. The tangent vectors to V are in this case all transversal to $\ell_p = 0 \times \mathbb{R}^n_p$ and hence $m(\gamma) = 0$ for all loops in V.*

Using the properties of the first homotopy group $\pi_1(V)$ one can show that if V is q-orientable, then it is also q'-orientable if q divides q' (see Dazord [34] or de Gosson [59] for a proof). Notice that since $m(\gamma) = 0$ for all loops, an ∞-orientable is automatically orientable. Since simply connected Lagrangian manifolds are ∞-orientable, one thus has here a direct proof of the fact that a simply connected Lagrangian manifold is orientable (it is a well-known fact from differential geometry that every simply connected manifold is orientable).

Let us next show how Maslov's index intervenes in the quantization of integrable systems.

4.8.2 Quantization of Lagrangian Manifolds

Let V be an arbitrary Lagrangian submanifold of $\mathbb{R}^n_x \times \mathbb{R}^n_p$. Recall from Chapter 2, Subsection 2.4.3 that for almost every E, the energy shells $H = E$ are $(2n-1)$-dimensional submanifolds of $\mathbb{R}^n_x \times \mathbb{R}^n_p$. We also recall that H is said to be completely integrable in the sense of Liouville if there are n independent constants of the motion in involution, and that (see Proposition 37) through every point $z_0 = (x_0, p_0)$ of the energy shell passes a Lagrangian manifold V carrying the orbits passing through z_0. Moreover, when V is connected (which we assume from now on) there exists a symplectomorphism

$$f : V \longrightarrow (S^1)^k \times \mathbb{R}^{n-k} \tag{4.90}$$

where $(S^1)^k$ is the product of k unit circles, each lying in some coordinate plane x_j, p_j (cf. Proposition 130). In particular, if V is compact then it is symplectomorphic to a torus $(S^1)^n$.

Now, the minimum capacity/action principle imposes a condition on the energy shells of any Hamiltonian. That condition is that there should be no periodic orbits with action less than $\frac{1}{2}h$, and that there should exist "minimal periodic orbits" having precisely $\frac{1}{2}h$ as action. The following result is fundamental, because it ties the minimum capacity/action principle to the Maslov index of loops, and thus justifies the "EBK" or "Bohr-Sommerfeld" quantization condition by a purely *topological* argument:

Theorem 134 *Let V be a Lagrangian manifold associated to a Liouville integrable Hamiltonian H and carrying minimal action periodic orbits. Then we have*

$$\frac{1}{2\pi\hbar}\int_\gamma pdx = \frac{1}{4}m(\gamma) \tag{4.91}$$

for every loop γ *on* V.

Proof. Since the actions of loops are symplectic invariants (see Proposition 121), we can use the symplectomorphism (4.90) to reduce the proof to the case $V = (S^1)^k \times \mathbb{R}^{n-k}$. With the notations of the proof of Proposition 130, it follows that the action of any loop γ is

$$\oint_\gamma pdx = \sum_{j=1}^{k} m_j \oint_{\gamma_j} p_j dx_j.$$

By the same argument as that leading to the proof of formula (3.100) in Proposition 80, we must have

$$\oint_{\gamma_j} p_j dx_j = \frac{1}{2}h \quad (1 \leq j \leq k)$$

and hence

$$\oint_\gamma pdx = \frac{1}{2}\left(\sum_{j=1}^{k} m_j\right) h = \frac{m(\gamma)}{4} h$$

which was to be proven. ■

Remark 135 *We urge the reader to note the emphasis on the word* loop *in the end of the statement of Theorem 134: the condition is independent of the existence of periodic orbits on the Lagrangian manifold. It thus applies, in particular, to the case of the* n*-dimensional harmonic oscillator with incommensurate frequencies.*

The important result above motivates the following definition, which was arrived at by other means by Maslov [113], following previous work of Keller [95]:

Definition 136 (Keller-Maslov) *A Lagrangian manifold* V *is said to satisfy the* Keller-Maslov quantization condition, *or to be a* quantized Lagrangian manifold, *if*

$$\frac{1}{2\pi\hbar}\int_\gamma pdx - \frac{1}{4}m(\gamma) \quad \textit{is an integer} \tag{4.92}$$

for every loop γ *in* V.

One easily verifies, by a direct calculation, that in the case of the n-dimensional anisotropic harmonic oscillator with Hamiltonian (3.99), the Lagrangian manifolds singled out by the "selection rule" (4.92) are precisely on which the energy is

$$E_{N_1,\ldots,N_n} = \sum_{j=1}^{n} \left(N_j + \tfrac{1}{2}\right) \hbar\omega_j$$

which are the correct values predicted by quantum mechanics. For more general Hamiltonians, condition (4.92) does not in general yield the correct energy levels. The values obtained are nevertheless asymptotic to the exact values for large quantum numbers (see Brack and Bhaduri [25] for a very complete and detailed study of many cases of genuine physical interest). As we will see later in this Chapter, the Maslov condition (4.92) will allow us to define "semi-classical wave functions"; it is thus the quantization condition leading to *semi-classical mechanics*.

Let us study, as an illustration of the methods introduced above, the plane rigid rotator (which is a crude model for a rotating diatomic molecule).

4.8.3 Illustration: The Plane Rotator

Consider a particle with mass m constrained to move with constant angular velocity ω along the circle $x^2 + y^2 = r^2$ in the x, y plane. Setting $\mathcal{I} = mr^2$ and $\mathcal{L} = xp_y - yp_x = \mathcal{I}\omega$ (it is the angular momentum) and denoting by θ the polar angle, the associated Hamiltonian function is essentially that of the free particle in the coordinates $(\theta, \mathcal{L})$:

$$H(\theta, \mathcal{L}) = \frac{1}{2\mathcal{I}}\mathcal{L}^2. \tag{4.93}$$

The associated Schrödinger equation is

$$i\hbar\frac{\partial\psi}{\partial t} = -\frac{\hbar^2}{2\mathcal{I}}\frac{\partial^2\psi}{\partial\theta^2} \tag{4.94}$$

which we supplement by the periodicity requirement

$$\psi(\theta + 2\pi, t) = \psi(\theta, t). \tag{4.95}$$

(That Eq. (4.95) not necessarily is a realistic condition, is discussed in Schulman [139], page 194, in connection with the occurrence of Bloch waves in a crystal lattice.) The stationary states of equations (4.94)–(4.95) have the form

$$\psi_N(\theta,t)=C_N e^{-iE_N t/\hbar}\cos(N\theta+\theta_0)$$

where the energy levels are given by

$$E_N=\frac{1}{2\mathcal{I}}(N\hbar)^2,\quad N=0,\pm1,\pm2,\ldots\,. \tag{4.96}$$

Let us now apply Maslov quantization to the plane rotator. The solution curves $t\mapsto(\theta(t),\mathcal{L}(t))$ of Hamilton's equations

$$\dot{\theta}=\frac{\partial H}{\partial\mathcal{L}},\quad \dot{\mathcal{L}}=-\frac{\partial H}{\partial\theta}$$

are given by

$$\theta(t)=\omega t+\theta_0,\quad \mathcal{L}(t)=\mathcal{L}_0$$

where $\mathcal{L}_0$ is the constant value of the angular momentum. Returning to the x,y coordinates we get

$$\begin{cases} x(t)=r\cos\theta(t),\quad y(t)=r\sin\theta(t) \\ p_x(t)=-\dfrac{\mathcal{L}_0}{r}\sin\theta(t),\quad p_y(t)=\dfrac{\mathcal{L}_0}{r}\cos\theta(t), \end{cases} \tag{4.97}$$

and the trajectories $t\mapsto(x(t),y(t),p_x(t),p_y(t))$ thus lie on the manifold

$$V:\begin{cases} x^2+y^2=r^2 \\ xp_y-yp_x=\mathcal{L}_0 \end{cases}. \tag{4.98}$$

The functions x^2+y^2 and xp_y-yp_x are constants of the motion in involution:

$$\left\{x^2+y^2,xp_y-yp_x\right\}=0\ \text{ on } V$$

and V is thus a Lagrangian manifold (this can of course also be verified by a direct calculation). Since we are allowed to give arbitrary values to, say, p_y in Eq. (4.98), it follows that V is topologically just the cylinder $S^1\times\mathbb{R}$, so that $\pi_1(V)\equiv(\mathbb{Z},+)$. Let us construct explicitly the universal covering of V. The mapping

$$\pi:\check{V}=\left\{(\theta,\theta'):\left|\theta-\theta'\right|<\tfrac{\pi}{2}\right\}\longrightarrow V$$

defined by

$$\pi(\theta,\theta')=\left(r\cos\theta,r\sin\theta,-\frac{\mathcal{L}_0}{r}\frac{\sin\theta'}{\cos(\theta-\theta')},\frac{\mathcal{L}_0}{r}\frac{\cos\theta'}{\cos(\theta-\theta')}\right)$$

is obviously continuous. It is straightforward to check that it is also surjective, and hence V is connected, being the image of the connected manifold $\check{V}$ by π. The fibers of π are easily calculated; they are the discrete sets

$$\pi^{-1}(x, y, p_x, p_y) = (\theta + 2k\pi, \theta' + 2k\pi) \quad (k \in \mathbb{Z})$$

hence $\pi : \check{V} \longrightarrow V$ is in fact the universal covering of V. The generator γ_V of $\pi_1(V) \equiv (\mathbb{Z}, +)$ is the projection on V of any path γ in $\check{V}$ joining a point (θ, θ') to the point $(\theta + 2\pi, \theta' + 2\pi)$. Performing, if necessary, a re-scaling of the variables, we may assume without loss of generality that $r = \mathcal{L} = 1$, and hence we can choose, taking $\theta = \theta'$,

$$\gamma_V(t) = (\cos t, \sin t, -\sin t, \cos t) \ , \ 0 \leq t \leq 2\pi \ .$$

Let us calculate the corresponding path $\ell(\gamma_V)$ in $Lag(n)$. We first remark that the equations of the tangent plane $\ell(t)$ to V at the point $(\cos t, \sin t, -\sin t, \cos t)$ are

$$\begin{cases} x \cos t + y \sin t = 0 \\ -p_x \sin t + p_y \cos t = 0 \end{cases}$$

so that an orthonormal basis of $\ell(t)$ is

$$\mathcal{B}_t = ((\sin t, -\cos t, 0, 0); (0, 0, \cos t, \sin t)) \ .$$

Setting

$$A(t) = \begin{pmatrix} 0 & \cos t \\ 0 & \sin t \end{pmatrix}, \quad B(t) = \begin{pmatrix} -\sin t & 0 \\ \cos t & 0 \end{pmatrix}$$

one checks that the unitary matrix $R(t) = A(t) + iB(t)$ takes the orthonormal basis $((0, 0, 1, 0), (0, 0, 0, 1))$ of $\ell_p = 0 \times \mathbb{R}^2_p$ to the basis $\mathcal{B}_t$ of $\ell(t)$, and hence $\ell(t) = R(t)\ell_p$. We have

$$W(t) = R(t)R(t)^T = \begin{pmatrix} \cos 2t & \sin 2t \\ \sin 2t & -\cos 2t \end{pmatrix}$$

so $\det W(t) = -1$ and hence, by definition 4.86 of the Maslov index:

$$m(\gamma_V) = \frac{1}{2\pi i} \oint_{\gamma_V} \frac{d(\det W)}{\det W} = 0.$$

Remark 137 *In particular, V is thus ∞-oriented, but is however not simply connected since $\pi_1(V)$ is not trivial.*

Moreover, a straightforward calculation shows that we have

$$\int_{\gamma_V} p\,dr = 2\pi\mathcal{L}$$

hence the quantum condition (4.91) reads $\mathcal{L} = N\hbar$ ($N = 0, 1, ...$). We thus recover the energy levels (4.96) predicted by quantum mechanics.

Chapter 5

Semi-Classical Mechanics

Summary. *The time-evolution of the wave functions of quantum mechanics is essentially the Bohmian motion of half-densities. Semi-classical mechanics is the study of wave forms on Lagrangian manifolds; the definition of these wave forms requires the properties of the Leray index. The shadows on configuration space of wave forms are the usual semi-classical wave functions of Keller and Maslov.*

This Chapter is devoted to the definition and study of semi-classical mechanics in phase space. Except for the first section, it is a rather technical Chapter, and the involved mathematics is not always quite trivial (some working knowledge of covering spaces is certainly helpful at this point). The reader who is more interested in quantum mechanics from the metaplectic viewpoint is advised to proceed directly to Chapter 6, and to come back to the present Chapter following his needs.

In the first section we show, using Bohm's theory of motion, that quantum mechanics, as well as semi-classical mechanics, are mathematically efficiently described by using the notion of half-density. We then define and study the Leray index, which is characterized by two properties, the first of cohomological nature, and the second of topological nature. The Leray index allows us to characterize the orientation (or, lack of) a Lagrangian manifold, and to define the argument of the square root of de Rham forms. This, together with the notion of phase of a Lagrangian manifold leads us to the definition of the wave-forms of the title of this Chapter.

5.1 Bohmian Motion and Half-Densities

We begin by showing that Bohm's approach to quantum mechanics allows us to view the wave function as a half-density on a Lagrangian manifold moving "without caustics", the motion being the quantum motion defined in Chapter 1, Subsection 1.8.2. The idea to include half-densities in quantum mechanics is not new. It is in fact immediately suggested by the fact that it is the square-root of Van Vleck's density that appears in the explicit resolution of Schrödinger's equation as we will see in Chapter 7, but its systematic use apparently goes back to the work of Blattner [16] and Kostant [97]. For detailed accounts of these theories, see Guillemin-Sternberg [79], Woodhouse [169] and the references therein.

5.1.1 Wave-Forms on Exact Lagrangian Manifolds

We will use in this Subsection the word *density* to qualify a function ρ which is a solution of a continuity equation

$$\frac{\partial \rho}{\partial t} + \nabla_x(\rho v) = 0 \tag{5.1}$$

where v is some velocity field on space-time (the notion of density will be somewhat extended later in this Chapter, and studied from the point of view of differential geometry on manifolds). A *half-density* is then the square root of a density. We do not make any special assumption on the nature of the solution ρ of Eq. (5.1): it could be the density of a fluid, or Van Vleck's determinant. It turns out that we have, under adequate smoothness assumptions for ρ that will always be assumed to hold, the following relation between $\rho(x(t), t)$ and $\rho(x, 0)$:

$$\rho(x, 0) = \rho(x(t), t)\frac{dx(t)}{dx} \tag{5.2}$$

(this is a straightforward consequence of Lemma 239 of Chapter 7). Here $t \longmapsto x(t)$ is the trajectory followed by a particle initially located at x and $dx(t)/dx$ is the Jacobian determinant of the "flow mapping" $t \longmapsto x(t)$ that is

$$\frac{dx(t)}{dx} = \det\left(\frac{\partial x_i(t)}{\partial x_j}\right)_{1 \leq i,j \leq n}.$$

Formula (5.2) can be rewritten in differential form as

$$\rho(x(t), t)dx(t) = \rho(x, 0)dx \tag{5.3}$$

showing that the differential form ρdx is constant in time (in fluid mechanics it is just another way of saying that the total mass is conserved during the motion). Consider now a solution

$$\Psi(x,t) = R(x,t)e^{\frac{i}{\hbar}\Phi(x,t)}$$

of Schrödinger's equation

$$i\hbar\frac{\partial \Psi}{\partial t} = \frac{1}{2m}\left(-i\hbar\nabla_x - A\right)^2\Psi + U\Psi.$$

Assuming that Ψ does not vanish, the phase Φ and the amplitude R satisfy the equations

$$\begin{cases} \dfrac{\partial \Phi}{\partial t} + \dfrac{1}{2m}(\nabla_x\Phi - A)^2 + U - \dfrac{\hbar^2}{2m}\dfrac{\nabla_x^2 R}{R} = 0 \\ \dfrac{\partial R^2}{\partial t} + \operatorname{div}\left(\dfrac{1}{m}R^2(\nabla_x\Phi - A)\right) = 0 \end{cases} \tag{5.4}$$

and setting $\rho = R^2$, $v = (\nabla_x\Phi - A)/m$ the second of these equations becomes precisely the continuity equation (5.1), and hence

$$R(x(t),t)^2 dx(t) = R(x,0)^2 dx. \tag{5.5}$$

Now, the second equation (7.38) is Hamilton-Jacobi's equation for the Hamiltonian function

$$H^{\psi} = H - \frac{\hbar^2}{2m}\frac{\nabla_x^2 R}{R}$$

and we know from the general theory of that equation that the solution Φ is such that

$$\Phi(x(t),t) = \Phi(x,0) + \int_{x,0}^{x(t),t} pdx - H^{\psi}dt$$

where $x(t)$ is defined by $(x(t),p(t)) = f_{t,0}^{\psi}(x,p)$, $(f_{t,0}^{\psi})$ being the flow determined by H^{ψ}, and $p(t) = \nabla_x\Phi(x(t),t)$, $p = \nabla_x\Phi(x,0)$. Multiplying both sides of Eq. (5.5) by $\exp(i\Phi(x(t),t)/\hbar)$ we get

$$\Psi(x(t),t)\sqrt{dx(t)} = \exp\left(\frac{i}{\hbar}\int_{x,0}^{x(t),t} pdx - H^{\psi}dt\right)\Psi(x,0)\sqrt{dx}. \tag{5.6}$$

This equation highlights the fact that the evolution of the wave function becomes extremely simple if we view it as the motion of a "wave-form", involving the square root of dx (the argument of that square root being calculated using the Maslov index). Notice that we *immediately* see from Eq. (5.6) that the evolution of Ψ is unitary: taking the moduli of both sides of that equation, squaring, and then integrating yields:

$$\int |\Psi(x(t),t)|^2 \, dx(t) = \int |\Psi(x,0)|^2 \, dx$$

that is

$$\int |\Psi(x(t),t)|^2 \, dx(t) = \int |\Psi(x,0)|^2 \, dx$$

for all t.

Remark 138 *As we have shown in [59], the notion of wave form yields a rather straightforward interpretation of various "geometric phase shifts", including phenomena like "Berry's phase".*

5.1.2 Semi-Classical Mechanics

Semi-classical mechanics is the mechanics we get if we replace, in all the considerations above, the "Bohmian flow" $(f^{\psi}_{t,t'})$ by the flow $(f_{t,t'})$ determined by the classical Hamiltonian H, that is, by neglecting the quantum potential

$$Q = -\frac{\hbar^2}{2m}\frac{\nabla_x^2 R}{R}.$$

It is usually contended that the passage from *bona fide* quantum mechanics to semiclassical mechanics is obtained at the "limit $\hbar \to 0$", but such statements are meaningless. First, one does not see what taking the "limit $\hbar \to 0$" really means –after all, Planck's constant is a constant, not a variable quantity!–, and even if this were justified, it is not because "$\hbar$ is small" that $-\hbar^2\nabla_x^2 R/2mR$ can necessarily be neglected: if one really wants to consider $\hbar$ as a variable parameter, then the form of the equations (5.4) imply that both R and Φ must themselves depend on that parameter, and it could thus, in principle, very well happen that $-\hbar^2\nabla_x^2 R/2mR$ does not tend to zero as $\hbar \to 0$.

The main problem which occurs when one passes to the semiclassical regime is that since the first equation (5.4) is replaced by the ordinary Hamilton-Jacobi equation

$$\frac{\partial \Phi}{\partial t} + \frac{1}{2m}(\nabla_x \Phi - A)^2 + U = 0 \tag{5.7}$$

whose solutions are not, as we know, defined for arbitrary times. At the points where the phase Φ is undefined we will get caustics (see Chapter 4, Subsection 4.6.1). As we will see later in this Chapter, these caustics are actually most easily described using a "cohomological" object, the Leray index, of which the usual Maslov index studied in Chapter 3 is an ancestor, but let us first see what the semi-classical wave functions look like. We begin by noting that in view of Eqs. (5.3) and (5.6) the solution of Schrödinger's equation can be written in the form

$$\Psi(x(t),t) = i^{m(t)} \left| \frac{dx(t)}{dx} \right|^{-1/2} \sqrt{|\rho(x,0)|} e^{\frac{i}{\hbar}\Phi(x(t),t)} \tag{5.8}$$

with $(x(t),p(t)) = f^{\psi}_{t,0}(x,p)$, and $m(t)$ being associated to the argument of the square root of $dx(t)$:

$$\arg dx(t) = (-1)^{m(t)} \arg dx. \tag{5.9}$$

(Formula (5.8), was actually already written down in 1928 by Van Vleck [155].)

If we neglect the quantum potential, Φ is a solution of Eq. (5.7), and is thus not defined (in general) for large values of t. As already discussed in Chapter 4, Subsection 4.7.1, this is related to the fact that the action of the flow (f_t) determined by the classical Hamiltonian H makes the exact Lagrangian manifold

$$V' : p = \nabla_x \Phi(x,0)$$

to "bend", so that caustics appear, and several points z of $f_t(V')$ can have the same projection x on configuration space $\mathbb{R}^n_x$. So what becomes formula (5.8) for large times? Suppose in fact that x is the projection of points $z_1 = (x,p_1), ..., z_N = (x,p_N)$ of the Lagrangian manifold $V_t = f_t(V')$. Each of these points z_j is the image by f_t of a point $z'_j = (x'_j, p'_j)$ of V', so we may define the action at time t along the trajectories leading from z'_j to z_j by

$$\Phi_j(x,t) = \Phi(x,0) + \int_{x'_j,0}^{x,t} pdx - Hdt.$$

One then proves (see Leray [101], Maslov [113], Maslov-Fedoriuk [114]) that the approximate wave function is then given by the formula

$$\Psi(x,t) = \sum_{j=1}^{N} i^{m_j} \left| \frac{dx}{dx'_j} \right|^{-1/2} \sqrt{|\rho(x'_j,0)|} e^{\frac{i}{\hbar}\Phi_j(x,t)} \tag{5.10}$$

where m_j is (up to the sign) the *Morse index* of the trajectory from z'_j to z_j, obtained by counting the number of *conjugate points* along each trajectory (see Arnold [4], Appendix 11, for a discussion of the relationship between the Morse and Maslov indices). For short time intervals t formula (5.10) reduces to (5.8) if one sets $x = x(t)$ and $x' = x$, since $f_t(V')$ remains a graph, so that there is only one trajectory joining z' to z.

The expression in the right-hand side of (5.10) can be viewed as the "shadow" of configuration space of a phase space object, which we will call a *wave form.* These waveforms are defined on quantized Lagrangian manifolds, and play the role of semi-classical phase space wave functions. Let us begin by exposing the idea on a rather elementary example, which nevertheless contains the main ideas that we are going to develop in the forthcoming sections.

5.1.3 Wave-Forms: Introductory Example

Consider the one-dimensional oscillator with Hamiltonian function

$$H = \frac{1}{2}(p^2 + x^2).$$

The associated phase-space trajectories are the circles $S^1_r = \{|z| = r\}$, which carry a natural length element traditionally denoted $ds = rd\vartheta$ where ϑ is the usual polar angle (the notation $d\vartheta$ is standard, but somewhat abusive since $d\vartheta$ is not an exact form).

One looks for stationary solutions of the type

$$\Psi(z) = e^{\frac{i}{\hbar}\varphi(z)}a(z)\sqrt{ds} \tag{5.11}$$

where the phase φ and the amplitude a are real functions, and $\sqrt{ds}$ is supposed to have some well defined meaning. One then immediately encounters two difficulties. First of all, any "reasonable" definition of the phase φ requires that the differential of the phase be the action form:

$$d\varphi = pdx = -r^2 \sin^2 \vartheta \, d\vartheta. \tag{5.12}$$

Unfortunately there exists no such function φ on the circle, because the 1-form pdx is not exact on S^1_r. We can however φ on the *universal covering* of S^1_r, that is on $\mathbb{R}$ together with the projection

$$\begin{cases} \pi : \mathbb{R} \to S^1_r \\ \pi(\theta) = r(\cos\vartheta, \sin\vartheta). \end{cases}$$

One immediately checks (see Subsection 4.7.1, Eq. (4.77)) that the function

$$\varphi(\vartheta) = \frac{r^2}{2}(\sin\vartheta\cos\vartheta - \vartheta) \tag{5.13}$$

satisfies (5.12). We are thus led to consider $\Psi(z, t)$ as being an expression of the type

$$\Psi(\vartheta) = e^{\frac{i}{\hbar}\varphi(\vartheta)}a(\vartheta)\sqrt{r\,d\vartheta} \tag{5.14}$$

where one allows θ to take any real value. However, we then encounter a second more serious obstruction: one does not see how to define unambiguously the square root $\sqrt{ds} = \sqrt{r\,d\vartheta}$. The simplest way out of this difficulty is to decide that one should only consider the (for instance, positive) square root of the *density* $|ds|$, that is to take

$$\Psi(\vartheta) = e^{\frac{i}{\hbar}\varphi(\vartheta)}a(\vartheta)\sqrt{|r\,d\vartheta|}. \tag{5.15}$$

There is however a serious rub with that choice, because it leads to the wrong energy levels for the oscillator: since we are actually interested in *single-valued* objects on S^1_r, we have to impose the condition

$$\Psi(\vartheta + 2\pi) = \Psi(\vartheta) \tag{5.16}$$

to the expression (5.15), that is, the phase must satisfy

$$\varphi(\vartheta + 2\pi) = \varphi(\vartheta) - 2N\pi\hbar \tag{5.17}$$

for some integer N. By definition (5.13), this condition is equivalent to $r^2 = 2N\hbar$ for S^1_r, which leads to the energy levels

$$E_N = \frac{r^2}{2} = N\hbar$$

instead of the physically correct

$$E_N = (N + \tfrac{1}{2})\hbar. \tag{5.18}$$

The way out of these difficulties, and which leads to the correct quantization conditions, is the following: *define* the argument of $d\theta$ by

$$\arg d\vartheta = m(\vartheta)\pi \tag{5.19}$$

where the integer $m(\theta)$ is given by

$$m(\vartheta) = \left[\tfrac{\vartheta}{\pi}\right] + 1 \tag{5.20}$$

(the square brackets $[\cdot]$ mean "integer part of"). This allows us, in turn, to define the square root of $ds = rd\vartheta$ by the formula

$$\sqrt{ds} = i^{m(\vartheta)}\sqrt{|r\, d\vartheta|} \tag{5.21}$$

which leads to the following definition of a wave form on S_r^1: it is the function on $\mathbb{R}_\theta$ defined by

$$\Psi(\vartheta) = e^{\frac{i}{\hbar}\varphi(\vartheta)}a(\vartheta)i^{m(\vartheta)}\sqrt{|r\, d\vartheta|}. \tag{5.22}$$

Notice that this function is discontinuous at the points $\vartheta_k = k\pi$ $(k \in \mathbb{Z})$. With this definition, the single-valuedness condition (5.16) becomes

$$\Psi(\vartheta + 2\pi) = \Psi(\vartheta). \tag{5.23}$$

That condition requires that $r^2 = (2N+1)\hbar$, which yields the energy levels (5.18) predicted by quantum mechanics.

5.2 The Leray Index and the Signature Function*

The Leray index will allow us to construct a Maslov index for arbitrary paths, which we need for the definition of *wave-forms* later in this Chapter.

We begin by introducing some convenient notations and terminology from algebraic topology. The uninitiated reader is urged not to seek any deeper truths or difficulties in this subsection: there are none; we are just introducing a convenient way of expressing combinatorial properties of functions of several variables.

5.2.1 Cohomological Notations

Let X be a set, k an integer ≥ 0, $(G, +)$ an Abelian group. By definition, a *(G-valued) k-cochain* on X (or just *cochain* when the context is clear) is a mapping

$$c : X^{k+1} \longrightarrow G.$$

To every *k-cochain* one associates its *coboundary*: it is the *$(k+1)$-cochain* ∂c defined by

$$\partial c(x_0, ..., x_{k+1}) = \sum_{j=0}^{k+1}(-1)^j c(x_0, ..., \hat{x}_j, ..., x_{k+1}), \tag{5.24}$$

where the cap ˆ suppresses the term it covers. The operator

$$\partial : \{k\text{-cochains}\} \longrightarrow \{(k+1)\text{-cochains}\}$$

defined by (5.24) is called the *coboundary operator* (it would be more appropriate to denote that operator by ∂_k but we will pretend that the subscript k is always implicit). The coboundary operator satisfies the important (but easy to prove) equality

$$\partial^2 c = 0$$

for every cochain c. A cochain c is called a *coboundary* if there exists a cochain m such that $c = \partial m$; a cochain c is called a *cocycle* if

$$\partial c = 0$$

so that any coboundary is a cocycle.

Here is a simple example, which introduces a notation we will use quite often in the forthcoming Subsections:

Example 139 ***The intersection cochain.*** *Choose $X = Lag(n)$ and $G = (\mathbb{Z}, +)$. We will denote by* dim *the* 1*-cochain defined by:*

$$\dim(\ell, \ell') = \dim(\ell \cap \ell').$$

The coboundary of dim *is then the* 2*-cochain defined by:*

$$\partial \dim(\ell, \ell', \ell'') = \dim(\ell, \ell') - \dim(\ell, \ell'') + \dim(\ell', \ell'').$$

One immediately checks that $\partial^2 \dim = 0$.

Let us now define the Leray index in the simplest case $n = 1$. The general case is technically more difficult, and will be treated in Subsection 5.2.3.

5.2.2 The Leray Index: $n = 1$

Recall that $[\cdot]$ is the integer part function: by definition, for any real number α:

$$[\alpha] = \textit{largest integer } \leq \alpha\ .$$

Note that $[\alpha + 1] = [\alpha] + 1$, and that

$$[-\alpha] = \begin{cases} -[\alpha] - 1 & \text{if } \alpha \notin \mathbb{Z} \\ -[\alpha] = -\alpha & \text{if } \alpha \in \mathbb{Z}. \end{cases}$$

We will denote by $[\cdot]_{ant}$ the antisymmetric part of $[\cdot]$, that is

$$[\alpha]_{ant} = \frac{1}{2}([\alpha] - [-\alpha])$$

which we can write, alternatively, as

$$[\alpha]_{ant} = \begin{cases} [\alpha] + \frac{1}{2} & \text{if } \alpha \notin \mathbb{Z} \\ \alpha & \text{if } \alpha \in \mathbb{Z}\ . \end{cases} \tag{5.25}$$

Definition 140 *The* Leray index *on* $\mathbb{R}^2$ *is the* $\mathbb{Z}$*-valued* 1*-cochain on* $\mathbb{R}$ *defined by*

$$\mu(\theta, \theta') = 2\left[\tfrac{\theta-\theta'}{2\pi}\right]_{ant}. \tag{5.26}$$

In view of (5.25), this definition is equivalent to:

$$\mu(\theta, \theta') = \begin{cases} 2\left[\frac{\theta-\theta'}{2\pi}\right] + 1 & \text{if } \theta - \theta' \notin 2\pi\mathbb{Z} \\ 2\left(\frac{\theta-\theta'}{\pi}\right) & \text{if } \theta - \theta' \in 2\pi\mathbb{Z}. \end{cases} \tag{5.27}$$

Notice that the Leray index is obviously antisymmetric:

$$\mu(\theta, \theta') = -\mu(\theta', \theta). \tag{5.28}$$

We now make the following essential observation: for all θ, θ', θ'' the sum

$$\sigma = \mu(\theta, \theta') + \mu(\theta', \theta'') + \mu(\theta'', \theta)$$

will only depend on the classes modulo π of these numbers. Suppose, for instance, that we replace θ by $\theta + 2k\pi$ where k is some integer. Then $\mu(\theta, \theta')$ becomes

$$\mu(\theta + 2k\pi, \theta') = \mu(\theta, \theta') + 2k$$

and, similarly, $\mu(\theta'', \theta)$ becomes $\mu(\theta'', \theta) - 2k$, so that there is no overall change in the sum σ. It turns out that the integer σ has a simple geometric interpretation: consider a line ℓ in the phase plane $\mathbb{R}_x \times \mathbb{R}_p$ passing through the origin, and denote by θ twice the angle modulo 2π of that line with the p-axis (θ is thus *not* the polar angle). We denote that line by $\ell(\bar{\theta})$ (the overbar meaning "class modulo 2π of"). Thus, we may view $\ell(\bar{\theta})$ as the oriented line through the origin, whose polar angle with the oriented x-axis is

$$\vartheta = \pi - 2\theta \mod 2\pi. \tag{5.29}$$

The sum σ only depends on the lines $\ell(\bar{\theta})$, $\ell(\bar{\theta}')$, $\ell(\bar{\theta}'')$ and we therefore denote it by $\sigma(\bar{\theta}, \bar{\theta}', \bar{\theta}'')$. We thus have

$$\mu(\theta, \theta') + \mu(\theta', \theta'') + \mu(\theta'', \theta) = \sigma(\bar{\theta}, \bar{\theta}', \bar{\theta}''). \tag{5.30}$$

Now, it is easy to verify, by exhausting all possible cases, that $\sigma(\bar{\theta}, \bar{\theta}', \bar{\theta}'')$ can only take the values 0 and ± 1. In fact, for every triple (ℓ, ℓ', ℓ'') of lines through the origin, we have

$$\sigma(\bar{\theta}, \bar{\theta}', \bar{\theta}'') = \begin{cases} -1 & \text{if } \ell(\bar{\theta}) \prec \ell(\bar{\theta}') \prec \ell(\bar{\theta}'') \\ +1 & \text{if } \ell(\bar{\theta}') \prec \ell(\bar{\theta}) \prec \ell(\bar{\theta}'') \\ 0 & \text{if } \ell(\bar{\theta}) = \ell(\bar{\theta}') \text{ or } \ell(\bar{\theta}') = \ell(\bar{\theta}'') \\ & \text{or } \ell(\bar{\theta}) = \ell(\bar{\theta}''), \end{cases} \tag{5.31}$$

where the notation $\ell(\bar{\theta}) \prec \ell(\bar{\theta}') \prec \ell(\bar{\theta}'')$ means that the oriented line $\ell(\bar{\theta}')$ lies inside the angular sector determined by the oriented lines $\ell(\bar{\theta})$ and $\ell(\bar{\theta}'')$. The function σ is called the "cyclic order" or the "signature" of the triple of lines $(\ell(\bar{\theta}), \ell(\bar{\theta}'), \ell(\bar{\theta}''))$.

Formula (5.30) has a straightforward cohomological interpretation: using the antisymmetry of the Leray index, we can rewrite that formula as

$$\mu(\theta, \theta') - \mu(\theta, \theta'') + \mu(\theta', \theta'') = \sigma(\bar{\theta}, \bar{\theta}', \bar{\theta}'')$$

and σ thus appears to be a coboundary (and hence a cocycle). In more pedantic cohomological language, one would say that the coboundary of μ "descends to σ". I terms of the coboundary operator ∂ we thus have:

$$\partial\mu(\theta, \theta', \theta'') = \sigma(\bar{\theta}, \bar{\theta}', \bar{\theta}''). \tag{5.32}$$

We will see later in this Chapter that this relation, together with a topological property, is characteristic of the Leray index in any dimension.

5.2.3 The Leray Index: General Case

There are at least two strategies for constructing the complete Leray index, but both make use of the same "final trick" when one wants to extend it to the non-transversal case. That trick is to use the cocycle property of the signature of a triple of Lagrangian planes. In his fundamental work on Lagrangian analysis, J. Leray [101–103] used rather abstract methods from chain intersection theory to define an index $\mu(\ell_\infty, \ell'_\infty)$ when the projections

ℓ and ℓ' were transversal (i.e. when $\ell \cap \ell' = 0$). We then gave a definition of $\mu(\ell_\infty, \ell'_\infty)$ for arbitrary ℓ and ℓ' by using the signature function of triples of Lagrangian planes in our papers [53, 58]. We will use here a different method, already exposed in de Gosson [59], and which consists in identifying the Lagrangian Grassmannian $Lag(n)$ with a set of matrices and to give directly a numerical definition of $\mu(\ell_\infty, \ell'_\infty)$. (The method is due to J.-M. Souriau [149] in the transversal case.) How we arrive at these identifications of $Lag(n)$ with a numerical space is exposed in some detail in Appendix E; the points to retain are the following:

(1) The Lagrangian Grassmannian $Lag(n)$ is identified with the set of all symmetric unitary matrices:

$$Lag(n) \equiv \left\{ W \in U(n, \mathbb{C}) : W = W^T \right\}.$$

Under the identification of $U(n, \mathbb{C})$ with a subgroup $U(n)$ of $\mathrm{Sp}(n)$ (see Chapter 3, Subsection 3.3.1), this identifies the Lagrangian plane ℓ, image of the vertical plane

$$\ell_p = \mathbb{R}^n_p$$

by the matrix

$$u = \begin{pmatrix} A & -B \\ B & A \end{pmatrix} \in U(n),$$

with the product

$$w(\ell) = \begin{pmatrix} A & -B \\ B & A \end{pmatrix} \begin{pmatrix} A^T & -B^T \\ B^T & A^T \end{pmatrix}.$$

(2) The universal covering of $Lag(n)$ is then identified with the set

$$Lag_\infty(n) \equiv \left\{ (W, \theta) : W \in Lag(n) : \det W = e^{i\theta} \right\}$$

the projection $Lag_\infty(n) \longrightarrow Lag(n)$ being defined by $\pi(W, \theta) = W$.

Example 141 ***The case*** $n = 1$. *Let ℓ be the line with equation $x \cos\alpha + p \sin\alpha = 0$ in the plane $\mathbb{R}_x \times \mathbb{R}_p$. It is the line whose angle modulo π with the p-axis is α; its polar angle is thus $\alpha + \frac{\pi}{2}$ $(\mathrm{mod}\, \pi)$. We have*

$$w(\ell) = \begin{pmatrix} \cos 2\alpha & -\sin 2\alpha \\ \sin 2\alpha & \cos 2\alpha \end{pmatrix}$$

and hence $W = e^{2i\alpha}$. In particular, if ℓ is the line in the plane whose polar angle is ϑ, then $w(\ell) = -e^{2i\vartheta}$.

The universal covering of the Lagrangian of all lines through the origin in the plane is easily deduced from the example above:

Example 142 ***The universal covering of*** $Lag(1)$. *The universal covering of* $Lag(1)$ *consists of all pairs* $(e^{i\theta}, \theta)$. *In view of the example above the angle* θ *is twice the angle of the line* $\ell \equiv W$ *with the* p*-axis.*

Here is a very useful characterization of transversality in terms of the matrices W:

Lemma 143 *(1) Two Lagrangian planes* $\ell \equiv W$ *and* $\ell' \equiv W'$ *are transversal if and only if* $W - W'$ *is invertible:*

$$\ell \cap \ell' = 0 \Longleftrightarrow \det(W - W') \neq 0.$$

(2) Equivalently, $\ell \cap \ell' = 0$ *if and only if* -1 *is not an eigenvalue of* $W(W')^{-1}$.

Proof. (1) Let w and w' be the images of W and W' in $U(n) \subset \mathrm{Sp}(n)$; then:

$$\operatorname{rank}(w - w') = 2(n - \dim(\ell \cap \ell'))$$

(see Appendix E) and hence $w - w'$ is invertible if and only if $\dim(\ell \cap \ell') = 0$, that is, if ℓ and ℓ' are transversal. But this is equivalent to saying that $W - W'$ is invertible. **(2)** Since $W(W')^{-1}$ is a unitary matrix, its eigenvalues have modulus one. It thus suffices to show that $+1$ *is not an eigenvalue of* $W(W')^{-1}$. Assume that there exists a vector $z \neq 0$ such that $W(W')^{-1}z = z$, then $(W')^{-1}z = W^{-1}z$ and $(W')^{-1} - W^{-1}$ would not be invertible. But then, in turn,

$$W - W' = WW'\left[(W')^{-1} - W^{-1}\right]$$

would not be invertible. ■

Let us now define the Leray index in the general case:

Definition 144 *Let* $\ell_\infty = (W, \theta)$ *and* $\ell'_\infty = (W', \theta)$ *be two elements of* $Lag_\infty(n)$*; we suppose that their projections* $\ell = W$, $\ell' = W'$ *are transversal:* $\ell \cap \ell' = 0$, *that is* $\det(W - W') \neq 0$. *The* Leray index *of* $(\ell_\infty, \ell'_\infty)$ *is the integer:*

$$\mu(\ell_\infty, \ell'_\infty) = \frac{1}{\pi}\left[\theta - \theta' + i \operatorname{Tr} \operatorname{Log}(-W(W')^{-1})\right] \tag{5.33}$$

where Tr *denotes the trace.*

The logarithm in (5.33) is defined by analogy with the usual logarithm: noting that for every number $m > 0$ we can write

$$\operatorname{Log} m = \int_{-\infty}^{0} \left(\frac{1}{\lambda - m} - \frac{1}{\lambda - 1} \right) d\lambda \tag{5.34}$$

we define

$$\operatorname{Log} M = \int_{-\infty}^{0} \left[(\lambda I - M)^{-1} - (\lambda - 1)^{-1} I \right] d\lambda \tag{5.35}$$

provided that M is invertible and has no negative eigenvalues. One checks without difficulty, by using for instance series expansions, that $\operatorname{Log} M$ has the following properties:

$$\begin{cases} \exp(\operatorname{Log} M) = M \\ \exp(\operatorname{Tr}(\operatorname{Log} M)) = \det M \\ \operatorname{Log}(M^{-1}) = -\operatorname{Log}(M). \end{cases} \tag{5.36}$$

Definition (5.33) indeed makes sense, since $-W(W')^{-1}$ has no negative eigenvalues in view of (**2**) in Lemma 143. Let us check that $\mu(\ell_\infty, \ell'_\infty)$ is really an integer, as claimed in the definition. In view of the second formula (5.36) we have

$$\begin{aligned} \exp \pi i \mu(\ell_\infty, \ell'_\infty) &= \exp(i(\theta - \theta') - \det(-W(W')^{-1})) \\ &= \exp(i(\theta - \theta')) \exp(-\operatorname{Tr} \operatorname{Log}(-W(W')^{-1})) \\ &= \exp(i(\theta - \theta')) \det(-W(W')^{-1})^{-1}. \end{aligned}$$

Now,

$$\det(-W(W')^{-1})^{-1} = (-1)^n \det(W(W')^{-1})^{-1} = \exp(-i(\theta - \theta'))$$

so that $\exp \pi i \mu(\ell_\infty, \ell'_\infty) = (-1)^n$, and hence $\mu(\ell_\infty, \ell'_\infty) \in \mathbb{Z}$.

Remark 145 *Since $(-1)^n = e^{in\pi}$, the argument above actually shows that we have*

$$\mu(\ell_\infty, \ell'_\infty) \equiv n \mod 2 \tag{5.37}$$

when $\ell \cap \ell' = 0$.

Example 146 ***The case*** $n = 1$. *The Leray index defined by formula (5.33) coincides with the function*

$$\mu(\theta, \theta') = 2 \left[\tfrac{\theta - \theta'}{2\pi} \right]_{ant} \tag{5.38}$$

defined in Subsection 5.2.2. In fact, when $n = 1$ we have $(W, \theta) = (e^{i\theta}, \theta)$ and hence, if $\theta \neq \theta'$ $(\operatorname{mod} \pi)$:

$$\mu((W,\theta),(W',\theta')) = \frac{1}{\pi}\left(\theta - \theta' + i\operatorname{Log}(-e^{i(\theta-\theta')})\right).$$

Now, Definition (5.34) of the logarithm yields, by analytic continuation:

$$\operatorname{Log} e^{i\alpha} = i(\alpha - 2k\pi) \quad \text{if} \quad (2k-1)\pi < \alpha < (2k+1)\pi.$$

Since the inequality $(2k-1)\pi < \alpha < (2k+1)\pi$ is equivalent to $k < \frac{\alpha+\pi}{\pi} < k+1$ this can be rewritten as

$$\operatorname{Log}(e^{i\alpha}) = i\left(\alpha - 2\left[\frac{\alpha+\pi}{2\pi}\right]\pi\right)$$

and hence

$$\begin{aligned}\operatorname{Log}(-e^{i(\theta-\theta')}) &= \operatorname{Log}(e^{i(\theta-\theta'+\pi)}) \\ &= i\left(\theta - \theta' + \pi - 2\left[\tfrac{\theta-\theta'+2\pi}{2\pi}\right]\pi\right) \\ &= i\left(\theta - \theta' - \pi - 2\left[\tfrac{\theta-\theta'}{2\pi}\right]\pi\right).\end{aligned}$$

It follows that

$$\begin{aligned}\mu((W,\theta),(W',\theta')) &= \frac{1}{\pi}\left(\theta - \theta' - (\theta - \theta') + \pi + 2\left[\tfrac{\theta-\theta'}{2\pi}\right]\pi\right) \\ &= 2\left[\tfrac{\theta-\theta'}{2\pi}\right] + 1 \\ &= 2\left(\left[\tfrac{\theta-\theta'}{2\pi}\right] + \tfrac{1}{2}\right)\end{aligned}$$

which is the same as (5.38) for $\theta - \theta' \neq 2k\pi$ in view of Eq. (5.25).

As far we have only constructed the Leray index in the transversal case; we will extend its definition to all pairs $(\ell_\infty, \ell'_\infty)$ in a moment. But first we note the following Lemma, due to Souriau [149].

Lemma 147 *For all ℓ_∞, ℓ'_∞, ℓ''_∞ with pairwise transversal projections ℓ, ℓ', ℓ'' the sum $\mu(\ell_\infty, \ell'_\infty) + \mu(\ell'_\infty, \ell''_\infty) + \mu(\ell''_\infty, \ell_\infty)$ only depends on ℓ, ℓ', ℓ''.*

(For a detailed proof, see Guillemin and Sternberg [79], de Gosson [59].) This lemma motivates the following definition:

Definition 148 *Let ℓ_∞, ℓ'_∞, ℓ''_∞ be three elements of $Lag_\infty(n)$ such that $\ell \cap \ell' = \ell' \cap \ell'' = \ell'' \cap \ell = 0$. The integer*

$$\sigma(\ell, \ell', \ell'') = \mu(\ell_\infty, \ell'_\infty) + \mu(\ell'_\infty, \ell''_\infty) + \mu(\ell''_\infty, \ell_\infty) \tag{5.39}$$

is called the signature *of the triple (ℓ, ℓ', ℓ'') of Lagrangian planes.*

Now it turns out (but this is not quite obvious) that the signature $\sigma(\ell, \ell', \ell'')$ can be directly calculated in the following way: consider the quadratic form

$$\mathcal{Q}(z, z', z'') = \Omega(z, z') + \Omega(z', z'') + \Omega(z'', z) \tag{5.40}$$

on the product space $\ell \times \ell' \times \ell''$, that is, in coordinates

$$\mathcal{Q}(z, z', z'') = \sum_{i=1}^{n} \begin{vmatrix} p_i & p'_i \\ x_i & x'_i \end{vmatrix} + \sum_{j=1}^{n} \begin{vmatrix} p'_i & p''_i \\ x'_i & x''_i \end{vmatrix} + \sum_{k=1}^{n} \begin{vmatrix} p''_i & p_i \\ x''_i & x_i \end{vmatrix}.$$

Writing $\mathcal{Q}$ in matrix form

$$\mathcal{Q}(\cdot) = \tfrac{1}{2}(\cdot)^T \mathcal{R}(\cdot)$$

the symmetric matrix $\mathcal{R}$ has σ_+ (resp. σ_-) positive (resp. negative) eigenvalues. Then $\sigma(\ell, \ell', \ell'')$ is just the signature of $\mathcal{R}$:

$$\sigma(\ell, \ell', \ell'') = \sigma_+ - \sigma_-. \tag{5.41}$$

The definition of the signature was first given by Demazure [35] (also see [34]); the definition we are using here is due to Kashiwara, and was originally exposed in Lion–Vergne [105]; also see de Gosson [58, 59]. For a very complete study of the properties of the signature, we refer to Libermann and Marle's treatise [104]. It turns out that the signature has the following property: let ℓ_1, ℓ_2, ℓ_3 and ℓ_4 be four Lagrangian planes; then

$$\sigma(\ell_1, \ell_2, \ell_3) = \sigma(\ell_1, \ell_2, \ell_4) + \sigma(\ell_2, \ell_3, \ell_4) + \sigma(\ell_3, \ell_1, \ell_4) = 0. \tag{5.42}$$

This property will allow us to construct the Leray index $\mu(\ell_\infty, \ell'_\infty)$ for all pairs $(\ell_\infty, \ell'_\infty)$, whether the projections ℓ and ℓ' are transversal, or not. We begin by observing that while the Leray index is, so far, only defined for pairs $(\ell_\infty, \ell'_\infty)$ such that $\ell \cap \ell' = 0$, the signature $\sigma(\ell, \ell', \ell'')$ is defined for *all* triples (ℓ, ℓ', ℓ'') by formula (5.41) above. This observation motivates the following definition:

Definition 149 *Let* $(\ell_\infty, \ell'_\infty)$ *be an arbitrary pair of elements of* $Lag_\infty(n)$. *The Leray index* $\mu(\ell_\infty, \ell'_\infty)$ *is then defined by the formula*

$$\mu(\ell_\infty, \ell'_\infty) = \mu(\ell_\infty, \ell''_\infty) - \mu(\ell'_\infty, \ell''_\infty) + \sigma(\ell, \ell', \ell'') \tag{5.43}$$

where ℓ''_∞ *is chosen so that* $\ell'' \cap \ell = \ell'' \cap \ell' = 0$, *and* $\mu(\ell_\infty, \ell''_\infty)$, $\mu(\ell'_\infty, \ell''_\infty)$ *are calculated using formula (5.33).*

For this definition to make sense one must of course check that formula (5.43) defines unambiguously $\mu(\ell_\infty, \ell'_\infty)$ for all $(\ell_\infty, \ell'_\infty)$, that is, that the right hand side does not depend on the choice of ℓ''_∞. This is however an immediate consequence of the property (5.42) of the signature (see de Gosson [53, 58, 59] for a detailed proof). It is easy to check that the extension (5.43) in the case $n = 1$ yields the value (5.38) in all cases.

5.2.4 Properties of the Leray Index

We begin by stating two elementary properties. The second of these properties generalizes Eq. (5.37):

Proposition 150 *(1) We have* $\mu(\ell_\infty, \ell'_\infty) = -\mu(\ell'_\infty, \ell_\infty)$ *for all* $(\ell_\infty, \ell'_\infty)$; *in particular* $\mu(\ell_\infty, \ell_\infty) = 0$. *(2) Moreover*

$$\mu(\ell_\infty, \ell'_\infty) \equiv n + \dim(\ell, \ell') \mod 2 \tag{5.44}$$

for all $(\ell_\infty, \ell'_\infty)$; *in particular* $\mu(\ell_\infty, \ell_\infty)$ *is an even integer for all* ℓ_∞.

Proof. **(1)** The antisymmetry is obvious from definition (5.33) when $\ell \cap \ell' = 0$. If $\ell \cap \ell' \neq 0$, choose ℓ'' such that $\ell'' \cap \ell = \ell'' \cap \ell' = 0$. In view of (5.43) we then have

$$\mu(\ell_\infty, \ell'_\infty) = \mu(\ell_\infty, \ell''_\infty) + \mu(\ell''_\infty, \ell'_\infty) + \sigma(\ell, \ell', \ell'').$$

Since the signature $\sigma(\ell, \ell', \ell'')$ changes sign under permutation of any two of the planes ℓ, ℓ', ℓ'' we will have

$$\mu(\ell_\infty, \ell'_\infty) = -\mu(\ell''_\infty, \ell_\infty) + \mu(\ell''_\infty, \ell'_\infty) - \sigma(\ell', \ell, \ell'') = -\mu(\ell'_\infty, \ell_\infty)$$

which establishes the antisymmetry in the general case. **(2)** To prove formula (5.44) we begin by noting that it is trivially satisfied when $\ell \cap \ell' = 0$ in view of (5.37). In the general case it follows from definition (5.43) and property (5.52) below of the signature. ∎

Next theorem shows that the Leray index has two essential properties; the first is a topological property, and the second cohomological.

Theorem 151 *The Leray index μ has the following properties:* ***(1)*** $\mu(\ell_\infty, \ell'_\infty)$ *remains constant when ℓ_∞ and ℓ'_∞ move continuously in such a way that the intersection $\ell \cap \ell'$ retains same dimension.* ***(2)*** *We have*

$$\mu(\ell_\infty, \ell'_\infty) + \mu(\ell'_\infty, \ell''_\infty) + \mu(\ell''_\infty, \ell_\infty) = \sigma(\ell, \ell', \ell'') \tag{5.45}$$

for all triples $(\ell_\infty, \ell'_\infty, \ell''_\infty)$ of elements of $Lag_\infty(n)$.

Proof. It is beyond the scope of this book to prove **(1)** (it follows from the topological properties of the signature and of the definition of the Leray index). We refer to de Gosson [53, 58] for a complete proof when $k = 0$, and to de Gosson [59] for the general case. Formula (5.45) follows from (5.43) and Lemma 147. ■

Property **(1)** can be restated by saying that μ is locally constant of the subsets

$$\{(\ell_\infty, \ell'_\infty) : \dim(\ell, \ell') = k\} \tag{5.46}$$

$(0 \leq k \leq n)$ of $(Lag_\infty(n))^2$; In view of the antisymmetry of the Leray index, Property **(2)** can be rewritten as

$$\mu(\ell_\infty, \ell'_\infty) - \mu(\ell_\infty, \ell''_\infty) + \mu(\ell'_\infty, \ell''_\infty) = \sigma(\ell, \ell', \ell''). \tag{5.47}$$

Restated in the cochain notations from Example 139, this means that the coboundary of μ is the signature:

$$\partial\mu(\ell_\infty, \ell'_\infty, \ell''_\infty) = \sigma(\ell, \ell', \ell'')$$

which we will write $\partial\mu = \sigma$, for short.

The Leray index μ is the *only* mapping $(Lag_\infty(n))^2 \longrightarrow \mathbb{Z}$ having the properties **(1)**, **(2)** in Theorem 151. Suppose indeed that μ' is a second mapping having these properties, and set $\nu = \mu - \mu'$. Then

$$\nu(\ell_\infty, \ell'_\infty) = \nu(\ell_\infty, \ell''_\infty) - \nu(\ell'_\infty, \ell''_\infty) \tag{5.48}$$

for all ℓ_∞, ℓ'_∞, ℓ''_∞. But then $\nu(\ell_\infty, \ell'_\infty)$ remains constant when $(\ell_\infty, \ell'_\infty)$ moves in such a way that $\ell \cap \ell'$ keeps a constant dimension. It follows that ν is a locally constant function on $(Lag_\infty(n))^2$; since $Lag_\infty(n)$ is connected ν is actually constant. Taking $\ell_\infty = \ell'_\infty = \ell''_\infty$ in Eq. (5.48), that constant value is 0, hence $\mu = \mu'$. (See de Gosson [53, 58] for details.)

It follows that the properties **(1)**, **(2)** in Theorem 151 are *characteristic* of the Leray index.

Remark 152 *The properties of the Leray index allow the construction of a topological "Lagrangian path intersection index" generalizing the notion of Maslov index to arbitrary paths in a Lagrangian manifold (see de Gosson [61, 62]).*

Let us next study the action of the integer group $(\mathbb{Z},+)$ on the Leray index. This will lead us to an important result that will be used in the definition of wave-forms. Recall that the first homotopy group of a manifold acts on the universal covering of that manifold as a group of "deck transformations". For instance, if $\ell_\infty \in Lag_\infty(n)$ and $\lambda \in \pi_1(Lag(n))$, then $\lambda\ell_\infty$ will be another element of $Lag_\infty(n)$ with the same projection ℓ as ℓ_∞. Also recall that we are identifying $Lag(n)$ with the set $W(n,\mathbb{C})$ of all symmetric matrices, and the universal covering $Lag_\infty(n)$ with

$$Lag_\infty(n) \equiv \left\{(W,\theta) : W \in W(n,\mathbb{C}) : \det W = e^{i\theta}\right\} \tag{5.49}$$

where the projection $Lag_\infty(n) \longrightarrow Lag(n)$ is the mapping $\pi(W,\theta) = W$. The fibers $\pi^{-1}(W)$ are all discrete, in fact, if $\det W = e^{i\theta}$ then

$$\pi^{-1}(W) = \{(W,\theta + 2k\pi) : k \in \mathbb{Z}\}$$

so that we may identify the first homotopy group $\pi_1(Lag(n))$ with the integer group $(\mathbb{Z},+)$, the action of that group on $Lag_\infty(n)$ being then given by

$$k * \ell_\infty \equiv k * (W,\theta) \equiv (W,\theta + 2k\pi). \tag{5.50}$$

Of course $k * \ell_\infty$ and ℓ_∞ have same projection ℓ on $Lag(n)$:

$$\pi(k * \ell_\infty) = \pi(\ell_\infty) = \ell.$$

It immediately follows from (5.50) and the definition of the Maslov index that (see de Gosson [59]):

Proposition 153 *The isomorphism $\pi_1(Lag(n)) \cong (\mathbb{Z},+)$ described above associates to every $\gamma \in \pi_1(Lag(n))$ the Maslov index $m(\gamma)$ of that loop.*

Using these identifications, the action of the integer group on the Leray index can be described as follows:

Proposition 154 *The action of $\pi_1(Lag(n)) \equiv (\mathbb{Z},+)$ on the universal covering $Lag_\infty(n)$ is given by:*

$$\mu(k * \ell_\infty, k' * \ell'_\infty) = \mu(\ell_\infty, \ell'_\infty) + 2k - 2k' \tag{5.51}$$

for all integers k and k'.

Proof. Suppose first that $\ell \cap \ell' = 0$. Then formula (5.51) immediately follows from Eq. (5.50), by definition (5.33) of the Leray index in the transversal case. If $\ell \cap \ell' \neq 0$, write

$$\mu(\ell_\infty, \ell'_\infty) = \mu(\ell_\infty, \ell''_\infty) - \mu(\ell'_\infty, \ell''_\infty) + \sigma(\ell, \ell', \ell'')$$

where ℓ''_∞ is chosen so that $\ell'' \cap \ell = \ell'' \cap \ell' = 0$ (see (5.43)). Then, since $k \cdot \ell_\infty$ and $k' \cdot \ell'_\infty$ have projections $k * \ell$ and $k' * \ell'$:

$$\mu(k * \ell_\infty, k' \cdot \ell'_\infty) = \mu(k * \ell_\infty, \ell''_\infty) - \mu(k * \ell'_\infty, \ell''_\infty) + \sigma(\ell, \ell', \ell'')$$

hence, taking the first case into account:

$$\begin{aligned}
\mu(k * \ell_\infty, k' * \ell'_\infty) &= \mu(k * \ell_\infty, \ell''_\infty) - \mu(k * \ell'_\infty, \ell''_\infty) + \sigma(\ell, \ell', \ell'') \\
&= \mu(\ell_\infty, \ell''_\infty) + 2k - \mu(\ell'_\infty, \ell''_\infty) - 2k' + \sigma(\ell, \ell', \ell'') \\
&= \mu(\ell_\infty, \ell'_\infty) + 2k - 2k'
\end{aligned}$$

which proves the proposition in all cases. ∎

5.2.5 More on the Signature Function

The signature function σ has the following properties:

(1) *Antisymmetry*: $\sigma(\ell, \ell', \ell'')$ changes sign when any two of the Lagrangian planes ℓ, ℓ', ℓ'' are swapped; in particular $\sigma(\ell, \ell', \ell'') = 0$ if $\ell = \ell'$, $\ell' = \ell''$ or $\ell'' = \ell$;

(2) $\mathrm{Sp}(n)$-*invariance*: $\sigma(s\ell, s\ell', s\ell'') = \sigma(\ell, \ell', \ell'')$ *for all* $s \in \mathrm{Sp}(n)$;

(3) *Value modulo* 2: *for all Lagrangian planes* ℓ, ℓ', ℓ'':

$$\sigma(\ell, \ell', \ell'') \equiv n + \dim(\ell \cap \ell') + \dim(\ell \cap \ell'') + \dim(\ell' \cap \ell'') \mod 2 \tag{5.52}$$

(4) *Locally constant*: $\sigma(\ell, \ell', \ell'')$ remains constant when ℓ, ℓ', ℓ'' move continuously in $Lag(n)$ in such a way that $\dim(\ell \cap \ell')$, $\dim(\ell \cap \ell'')$, $\dim(\ell' \cap \ell'')$ do not change.

While properties **(1)** and **(2)** are immediate (**(1)** follows from the antisymmetry of Ω and **(2)** from the definition of a Lagrangian plane), the proofs of **(3)** and **(4)** are somewhat lengthy exercises in linear algebra (see Marle and Libermann [104], or Lion–Vergne [105]). We remark that Property 5.52 can be rewritten, in cohomological notation, as

$$\sigma \equiv n + \partial \dim \mod 2.$$

The signature of a triple of Lagrangian planes can be viewed as a device measuring the relative positions of these planes, which for $n = 1$ reduces to the "cyclic order" of three lines. To see this, we note one-dimensional Lagrangian planes are just the straight lines through the origin in the phase plane, and the quadratic form (5.40) here reduces to the expression

$$\mathcal{Q}(z, z', z'') = \begin{vmatrix} p & p' \\ x & x' \end{vmatrix} + \begin{vmatrix} p' & p'' \\ x' & x'' \end{vmatrix} + \begin{vmatrix} p'' & p \\ x'' & x \end{vmatrix}.$$

Choosing for ℓ and ℓ'' the "coordinate planes" $\mathbb{R}_x$ and $\mathbb{R}_p$, and for ℓ' the line $\ell_a : p = ax$, we have

$$\mathcal{Q} = -axx' - p''x' + p''x$$

which can be put, after diagonalization, in the form

$$\mathcal{Q} = Z^2 - (X^2 + \operatorname{sign}(a)Y^2)$$

and hence:

$$\sigma(\mathbb{R}_x, \ell_a, \mathbb{R}_p) = \begin{cases} -1 & \text{if } a > 0 \\ 0 & \text{if } a = 0 \\ +1 & \text{if } a < 0\,. \end{cases} \tag{5.53}$$

The signature $\sigma(\ell, \ell', \ell'')$ is thus 0 if any two of the lines coincide, -1 if the line ℓ' lies "between" ℓ and ℓ'' (the plane being oriented in the usual way), and $+1$ if it lies outside. An essential observation is that we would get the same values for an arbitrary triple ℓ, ℓ', ℓ'' of lines having the same relative positions as $\mathbb{R}_x$, ℓ_a, $\mathbb{R}_p$ because one can always reduce the general case to that of the triple $(\mathbb{R}_x, \ell_a, \mathbb{R}_p)$, by using a matrix with determinant one. More generally, in higher dimensions:

Lemma 155 *Let ℓ_A be the Lagrangian plane with equation $p = Ax$ (A a symmetric matrix). Let ℓ_x and ℓ_p be, respectively, the x- and p-planes $\mathbb{R}^n_x \times 0$ and $0 \times \mathbb{R}^n_p$. Then*

$$\sigma(\ell_x, \ell_A, \ell_p) = -\operatorname{sign}(A) \tag{5.54}$$

where $\operatorname{sign}(A)$ *is the number of* > 0 *eigenvalues of* A *minus the number of* < 0 *eigenvalues of* A.

(The proof is based on elementary calculations in linear algebra; since it is somewhat lengthy we do not reproduce it here and refer to de Gosson [59].)

5.2.6 The Reduced Leray Index

The concept of index of inertia of a triple of pairwise Lagrangian planes is due to Leray [101] (Ch. I, §2,4). The definition we give here is a generalization of that which was given in de Gosson [53, 58, 59].

Definition 156 *The* index of inertia *of a triple* (ℓ, ℓ', ℓ'') *of Lagrangian planes is*

$$\operatorname{Inert}(\ell, \ell', \ell'') = \frac{1}{2}(n + \partial \dim(\ell, \ell', \ell'') + \sigma(\ell, \ell', \ell''))$$

where $\partial \dim$ *is the coboundary of the cochain* $\dim(\ell, \ell') = \dim(\ell \cap \ell')$.

Writing explicitly

$$\operatorname{Inert}(\ell, \ell', \ell'') = \frac{1}{2}(n + \dim(\ell \cap \ell') - \dim(\ell \cap \ell'') + \dim(\ell' \cap \ell'') + \sigma(\ell, \ell', \ell''))$$

the claim that $\operatorname{Inert}(\ell, \ell', \ell'')$ is an integer immediately follows from formula (5.52) of the last subsection. One can prove that (see de Gosson [53, 58]) when

$$\ell \cap \ell' = \ell' \cap \ell'' = \ell'' \cap \ell = 0$$

then $\operatorname{Inert}(\ell, \ell', \ell'')$ is the common index of inertia of the quadratic forms

$$z \mapsto \Omega(z, z') \ , \ z' \mapsto \Omega(z', z'') \ , \ z'' \mapsto \Omega(z'', z)$$

defined on, respectively ℓ, ℓ', ℓ'' when $z + z' + z'' = 0$. (This is Leray's original definition of the index of inertia in [101].)

Let us next introduce the reduced Leray index:

Definition 157 *The* reduced Leray index *is the function* $m : Lag^2_\infty(n) \longrightarrow \mathbb{Z}$ *defined by*

$$m(\ell_\infty, \ell'_\infty) = \frac{1}{2}\left(\mu(\ell_\infty, \ell'_\infty) + n + \dim(\ell, \ell')\right) \tag{5.55}$$

where ℓ *and* ℓ' *are the projections of* ℓ_∞ *and* ℓ'_∞.

That $m(\ell_\infty, \ell'_\infty)$ is an integer is an immediate consequence of Eq. (5.44) in Proposition 150, which implies that $(\mu(\ell_\infty, \ell'_\infty) + n + \dim(\ell, \ell'))$ is an even integer. Notice that when $\ell_\infty \equiv (W, \theta)$ and $\ell'_\infty \equiv (W', \theta')$ have transversal projections, then the reduced Leray index is given by

$$m(\ell_\infty, \ell'_\infty) = \frac{1}{2\pi}\left[\theta - \theta' + i \operatorname{Tr}(\operatorname{Log}(-W(W')^{-1}))\right] + \frac{n}{2} \tag{5.56}$$

in view of definition (5.33) of $\mu(\ell_\infty, \ell'_\infty)$.

Example 158 ***The case*** $n = 1$. *We have seen in Subsection 5.2.3 that*

$$\mu(\theta, \theta') = 2\left[\tfrac{\theta-\theta'}{2\pi}\right]_{ant}.$$

It follows that

$$m(\theta, \theta') = \left[\tfrac{\theta-\theta'}{2\pi}\right]_{ant} + \frac{1}{2}$$

that is, by formula (5.25):

$$m(\theta, \theta') = \left[\tfrac{\theta-\theta'}{2\pi}\right] + 1 \tag{5.57}$$

in all cases.

It follows from Theorem 151 that m enjoys the same topological property as μ (see (1) in Theorem 151), and that it satisfies

$$m(\ell_\infty, \ell'_\infty) - m(\ell_\infty, \ell''_\infty) + m(\ell'_\infty, \ell''_\infty) = \text{Inert}(\ell, \ell', \ell'') \tag{5.58}$$

which we can write $\partial m = \text{Inert}$ for short. This formula allows us to calculate $m(\ell_\infty, \ell'_\infty)$ for arbitrary pairs $(\ell_\infty, \ell'_\infty)$ by the same method as we used to extend the Leray index in Subsection 5.2.3. Also observe that the action of $\pi_1(Lag(n)) \equiv (\mathbb{Z}, +)$ on the reduced Leray index is given by

$$m(k * \ell_\infty, k' * \ell'_\infty) = m(\ell_\infty, \ell'_\infty) + k - k' \tag{5.59}$$

in view of formula (5.51) in Proposition 154. It follows, using Proposition 153, that we have the following relation between the Maslov index and the reduced Leray index:

$$m(\gamma) = m(\gamma\ell_\infty, \ell'_\infty) - m(\ell_\infty, \ell'_\infty)$$

for every $\gamma \in \pi_1(V)$; $\gamma\ell_\infty$ is the element of $Lag_\infty(n)$ obtained by concatenating a representant of γ and one of ℓ_∞.

5.3 De Rham Forms

Roughly speaking, a de Rham form –also called "twisted", or "pseudo" form in the literature– is to ordinary volume forms what a pseudo-vector is to a vector. More precisely, a de Rham form is an orientation-dependent differential form: its sign depends on the (local) orientation. (The notion was defined by Charles de Rham [134] in the late 1950's.)

5.3.1 Volumes and their Absolute Values

To grasp the difference between de Rham and "ordinary" forms, let us consider the following example. Suppose we want to calculate the volume of the parallelepiped spanned by three vectors $\mathbf{r} = (x, y, z)$, $\mathbf{r}' = (x', y', z')$, $\mathbf{r}'' = (x'', y'', z'')$. For this purpose we can use either the formula

$$\nu(\mathbf{r}, \mathbf{r}', \mathbf{r}'') = \det(\mathbf{r}, \mathbf{r}', \mathbf{r}'') = \begin{vmatrix} x & x' & x'' \\ y & y' & y'' \\ z & z' & z'' \end{vmatrix}, \tag{5.60}$$

which involves the determinant, or the formula

$$\nu(\mathbf{r}, \mathbf{r}', \mathbf{r}'') = (\mathbf{r} \times \mathbf{r}') \cdot \mathbf{r}''. \tag{5.61}$$

However, the latter only gives the same result as (5.60) once we have given $\mathbb{R}^3_{\mathbf{r}}$ the usual "positive" orientation: if we change that orientation, then the right hand side of (5.61) changes sign (one sometimes says that $\mathbf{r} \times \mathbf{r}'$ is not a true vector, but rather a "pseudo-vector"). In the language of differential-forms, (5.60) is the same thing as

$$\nu(\mathbf{r}, \mathbf{r}', \mathbf{r}'') = dx \wedge dy \wedge dz(\mathbf{r}, \mathbf{r}', \mathbf{r}'')$$

whereas (5.61) is

$$(\mathbf{r} \times \mathbf{r}') \cdot \mathbf{r}'' = \pm dx \wedge dy \wedge dz(\mathbf{r}, \mathbf{r}', \mathbf{r}'')$$

where we choose the sign $+$ if we have oriented $\mathbb{R}^3_{\mathbf{r}}$ positively, and $-$ if we oriented it negatively. Now, there is a third option: if we feel puzzled by the fact that formula (5.61) is orientation-dependent, we have the option to define the volume by taking the absolute value of the right-hand side of (5.61):

$$\nu(\mathbf{r}, \mathbf{r}', \mathbf{r}'') = |(\mathbf{r} \times \mathbf{r}') \cdot \mathbf{r}''| \tag{5.62}$$

but we then lose linearity: while the properties of the vector product imply that

$$(\mathbf{r} \times \mathbf{r}') \cdot (\mathbf{r}''_1 + \mathbf{r}''_2) = (\mathbf{r} \times \mathbf{r}') \cdot \mathbf{r}''_1 + (\mathbf{r} \times \mathbf{r}') \cdot \mathbf{r}''_2$$

we have, in general

$$|(\mathbf{r} \times \mathbf{r}') \cdot (\mathbf{r}''_1 + \mathbf{r}''_2)| \neq |(\mathbf{r} \times \mathbf{r}') \cdot \mathbf{r}''_1| + |(\mathbf{r} \times \mathbf{r}') \cdot \mathbf{r}''_2|.$$

Now, a "naive" remark. We can restore linearity in Eq. (5.62) by doing the following: replace $|(\mathbf{r} \times \mathbf{r}') \cdot \mathbf{r}''|$ by 0 if the vectors $\mathbf{r}$, $\mathbf{r}'$, $\mathbf{r}''$are dependent, and by

$$(-1)^{m(\mathbf{r},\mathbf{r}',\mathbf{r}'')}|(\mathbf{r} \times \mathbf{r}') \cdot \mathbf{r}''|$$

if they are independent; $m(\mathbf{r}, \mathbf{r}', \mathbf{r}'')$ is an integer equal to 0 if the basis $(\mathbf{r}, \mathbf{r}', \mathbf{r}'')$ is positively oriented, and 1 if it is negatively oriented. This remark is naive, because we have done nothing else than reconstructing $(\mathbf{r} \times \mathbf{r}') \cdot \mathbf{r}''$ itself:

$$(\mathbf{r} \times \mathbf{r}') \cdot \mathbf{r}'' = (-1)^{m(\mathbf{r},\mathbf{r}',\mathbf{r}'')} |(\mathbf{r} \times \mathbf{r}') \cdot \mathbf{r}''| \tag{5.63}$$

and one can of course wonder what is the point with this apparently cumbersome reconstruction. It is of course totally useless in this precise situation, but we will see that it is the key to the construction of de Rham forms on non-orientable manifolds, on which the notion of volume form does not make sense.

For the purposes of semi-classical mechanics, the introduction of de Rham forms is essential, because one can define unambiguously their square roots, using the properties of the Leray index (more precisely the Lagrangian path index deduced from the Leray index).

The introduction of de Rham forms in physics should not be too surprising after all. It is well-known that many phenomena really exhibit this dependence on orientation, the most elementary example of this phenomenon being the magnetic field, which is a "pseudo-vector" in physicists terminology. For more about the use of de Rham forms in physics, I highly recommend Frankel's book [48].

One can define, more generally, a de Rham form on $\mathbb{R}^n_x$ in the following way: if $\nu = a(x)dx_1 \wedge ... \wedge dx_n$ is a differential n-form or $\mathbb{R}^n_x$ then the associated de Rham form is $\nu^{\pm} = \operatorname{Or}(\mathbb{R}^n_x)\nu$, where $\operatorname{Or}(\mathbb{R}^n_x)$ is $\pm$ depending on whether $\mathbb{R}^n_x$ is positively or negatively oriented.

Remark 159 *A de Rham form on $\mathbb{R}^n_x$ can thus be viewed as an ordinary differential form defined on the union of two copies of $\mathbb{R}^n_x$. On a more sophisticated level, a de Rham form on a manifold V is actually just an ordinary differential form, but defined on the oriented double-covering $\tilde{V}$ of that manifold.*

We will however proceed a little bit differently in the forthcoming subsections. It turns out that there is another (equivalent) approach, constructing directly de Rham forms from densities on a manifold. We will then specialize our constructions to the case of Lagrangian manifolds.

5.3.2 Construction of de Rham Forms on Manifolds

By definition, an *s-density* $(s > 0)$ on a vector space L is a mapping ρ associating to every n-tuple $(\xi_1, ..., \xi_n)$ of vectors of L a number (real or complex), and such that

$$\rho(A\xi_1, ..., A\xi_n) = |\det(A)|^s \, \rho(\xi_1, ..., \xi_n) \tag{5.64}$$

for every invertible $n \times n$ matrix A. We will often write this formula in the more compact form

$$\rho(A\xi) = |\det(A)|^s \, \rho(\xi)$$

where we are thus using the notations

$$\xi = (\xi_1, ..., \xi_n) \quad , \quad A\xi = (A\xi_1, ..., A\xi_n)$$

where $\xi_j \in L$.

Definition 160 *An s-*density *ρ on a manifold V is the datum, for every $z \in V$ of an s-density $\rho(z)$ on the tangent space T_zV at z, and depending smoothly on z. A 1-density is called a* density*; a $\frac{1}{2}$-density is called a* half-density.

One can always construct s-densities on a vector space: it suffices to take

$$\rho(\xi_1, ..., \xi_n) = |\det(\xi_1, ..., \xi_n)|^s \, . \tag{5.65}$$

One can moreover prove that this is essentially the only example, because all s-densities are proportional, and hence proportional to $|\det|^s$ (see Appendix D). It follows that an s-density has the property that

$$\rho(\xi_1, ..., \lambda\xi_j, ..., \xi_n) = |\lambda|^s \rho(\xi_1, ..., \xi_n) \tag{5.66}$$

for every scalar $\lambda \neq 0$. (This property follows at once from (5.64), choosing A such that $A\xi_j = \lambda\xi_j$ and $A\xi_k = \xi_k$ if $k \neq j$; alternatively, one can note that since $\rho(\xi_1, ..., \lambda\xi_j, ..., \xi_n)$ and $\rho(\xi_1, ..., \xi_n)$ are both s-densities, they must be proportional; using Eq. (5.64) one finds that the proportionality constant is precisely $|\lambda|^s$.)

Let now V be a manifold (we do not suppose that V is Lagrangian at this point). It follows from the discussion above that every s-density ρ on V is locally of the type

$$\rho(z)(\xi) = |a(x)dx_1 \wedge \cdots \wedge dx_n(\xi)|^s$$

where a is some smooth function of the coordinates $x = (x_1, ..., x_n)$, and $\xi = (\xi_1, ..., \xi_n)$ is a tangent vector to V at z. If f is a diffeomorphism of a manifold V' onto V, the pull-back $f^*\rho$ of an s-density ρ on V by f is defined by

$$f^*\rho(z')(\xi) = \rho(f(z'))(f'(z')\xi)$$

where we have set

$$f'(z')\xi = (f'(z')\xi_1, ..., f'(z')\xi_n);$$

$f'(z')$ is the Jacobian matrix of f at z'.

Formula (5.63) relating a volume to its absolute value suggests the following definition:

Definition 161 *Let ρ be a density on the manifold V. The two de Rham forms μ and $-\mu$ associated to ρ are defined by*

$$\mu(z)(\xi) = \begin{cases} 0 & \text{if the vectors } \xi_1, ..., \xi_n \text{ are dependent;} \\ (-1)^{m(z;\xi)}\rho(z)(\xi) & \text{if they are independent} \end{cases} \tag{5.67}$$

where $m(z;\xi)$ is an integer equal to 0 if the vectors $\xi_1, ..., \xi_n$ determine the positive orientation of the tangent space T_zV, and to 1 if these vectors determine the negative orientation.

The following result connects this definition with the definition given at the end of Subsection:

Proposition 162 *A de Rham form is an antisymmetric linear form at each $z \in V$:*

$$\begin{cases} \mu(z)(\xi_1, ..., \lambda\xi_j, ..., \xi_n) = \lambda\mu(z)(\xi) \\ \mu(z)(\xi_1, ..., \xi_j + \xi_j', ..., \xi_n) = \mu(z)(\xi) + \mu(z)(\xi') \end{cases} \tag{5.68}$$

for all $\lambda \in \mathbb{R}$ and $\xi = (\xi_1, ..., \xi_j, ..., \xi_n)$, $\xi' = (\xi_1, ..., \xi_j', ..., \xi_n)$ in T_zV. Moreover,

$$\mu(z)(A\xi) = (\det A)\mu(z)(\xi) \tag{5.69}$$

for every invertible $n \times n$ matrix A.

Proof. Let us begin by proving formula (5.69). If the vectors $\xi_1, ..., \xi_n$ are linearly dependent, then both $\mu(z)(A\xi)$ and $\mu(z)(\xi)$ are zero, so that (5.69) trivially holds in this case. Suppose next that $\xi_1, ..., \xi_n$ are independent, and that they determine the positive orientation of T_zV. If $\det A > 0$, the vectors $A\xi_1, ..., A\xi_n$ are also linearly independent, and determine the same orientation, so that

$$\begin{aligned}\mu(z)(A\xi) &= \rho(z)(A\xi) \\ &= (\det A)\rho(z)(\xi) \\ &= (\det A)\mu(z)(\xi).\end{aligned}$$

If $\det A < 0$, then A reverses the orientation, and hence

$$\begin{aligned}\mu(z)(A\xi) &= |\det A|\rho(z)(\xi) \\ &= -|\det A|\mu(z)(\xi) \\ &= \det A\mu(z)(\xi).\end{aligned}$$

The case where $\xi_1, ..., \xi_n$ is a negatively oriented basis is similar. It immediately follows from (5.69) that the function $\xi \mapsto \mu(z)(\xi)$ is antisymmetric. The first formula (5.68) follows from property (5.66) of densities: suppose that $\lambda > 0$ and let $\xi_1, ..., \xi_n$ be a positively oriented basis; then so is $\xi_1, ..., \lambda\xi_j, ..., \xi_n$ and we thus have

$$\mu(z)(\xi_1, ..., \lambda\xi_j, ..., \xi_n) = \lambda\mu(z)(\xi_1, ..., \xi_n).$$

If $\lambda < 0$, then $\xi_1, ..., \lambda\xi_j, ..., \xi_n$ is negatively oriented and thus

$$\begin{aligned}\mu(z)(\xi_1, ..., \lambda\xi_j, ..., \xi_n) &= -(-\lambda\mu(z)(\xi_1, ..., \xi_n)) \\ &= \lambda\mu(z)(\xi_1, ..., \xi_n).\end{aligned}$$

The case where $\xi_1, ..., \xi_n$ is negatively oriented dealt with in the same way. Let us now show that $\mu(z)$ is additive, that is that we have

$$\mu(z)(\xi + \xi') = \mu(z)(\xi) + \mu(z)(\xi').$$

In view of the antisymmetry, it actually suffices to show that we have, say

$$\mu(z)(\xi_1 + \xi_1', ..., \xi_n) = \mu(z)(\xi_1, ..., \xi_n) + \mu(z)(\xi_1', ..., \xi_n)$$

(we are using the shorter notation $(\xi_1+\xi_1', ..., \xi_n)$ in place of $(\xi_1+\xi_1', \xi_2, ..., \xi_n)$). In view of the definition of a de Rham form we have

$$\mu(z)(\xi_1 + \xi_1', ..., \xi_n) = (-1)^{m(z;\xi_1+\xi_1',...,\xi_n)}\rho(z)(\xi_1 + \xi_1', ..., \xi_n)$$

that is, writing $\xi_1' = \sum_j a_j\xi_j$:

$$\mu(z)(\xi_1 + \xi_1', ..., \xi_n) = (-1)^{m(z;\xi_1+\xi_1',...,\xi_n)}\rho(z)(A\xi)$$

where A is the upper triangular matrix:

$$A = \begin{pmatrix} 1+a_1 & a_2 & \cdots & a_n \\ 0 & 1 & \cdots & 0 \\ \vdots & \vdots & \vdots & \vdots \\ 0 & 0 & \cdots & 1 \end{pmatrix}.$$

Since $\det A = 1 + a_1$ we have

$$\mu(z)(\xi_1 + \xi_1', ..., \xi_n) = (-1)^{m(z,\xi_1+\xi_1',...,\xi_n)} |1 + a_1| \rho(z)(\xi).$$

Suppose now, for instance, that $1 + a_1 > 0$; then $\det A > 0$ and $\xi_1 + \xi_1', \xi_2, ..., \xi_n$ thus has the same orientation as $\xi_1, \xi_2, ..., \xi_n$. It follows that

$$m(z, \xi_1 + \xi_1', ..., \xi_n) = m(z, \xi_1, ..., \xi_n)$$

and hence

$$\mu(z)(\xi_1 + \xi_1', ..., \xi_n) = (-1)^{m(z,\xi_1,...,\xi_n)}(1 + a_1)\rho(z)(\xi).$$

Assume now $a_1 > 0$; then

$$\begin{aligned} a_1\rho(z)(\xi) &= \rho(z)(a_1\xi_1, ..., \xi_n) \\ &= \rho(z)(a_1\xi_1, ..., \xi_n) \\ &= \rho(z)(\Sigma_j a_j \xi_j, ..., \xi_n) \\ &= \rho(z)(\xi_1', ..., \xi_n) \end{aligned}$$

so that we have

$$\mu(z)(\xi + \xi') = \mu(z)(\xi) + \mu(z)(\xi')$$

proving the linearity when $a_1 > 0$. The cases $a_1 < 0$ and $1 + a_1 < 0$ are treated quite similarly (see page 487 in Abraham *et al.* [1] for details; Lang [100] gives a very nice geometrical proof). ■

In the next subsection we specialize our study to the case where V is a Lagrangian manifold; we will see that the de Rham form associated to a density on such a manifold can be expressed using the Leray index.

5.3.3 De Rham Forms on Lagrangian Manifolds

Consider a line ℓ through the origin in the plane: it is an element of the Lagrangian Grassmannian $Lag(1)$, and can be identified with the complex number $e^{i\theta}$, where θ is here twice the angle α of ℓ with the p-axis. Taking *twice* α is just a way to say that we do not care about the orientation of the line: if we rotate ℓ by an angle π we get the same line, and this is reflected by the relation $e^{i(\theta+2\pi)} = e^{i\theta}$. Suppose now that we *do* care about the orientation of ℓ, that is that we want to be able to distinguish between two "arrows" ℓ^+ and ℓ^- having the same support ℓ. We can then identify these "arrows", or oriented lines, by writing $\ell^+ = (e^{i\theta}, \theta)$ and $\ell^- = (e^{i\theta}, \theta + 2\pi)$. Since each of these oriented lines is invariant by a rotation of 2π, we could actually write as well $\ell^+ = (e^{i\theta}, \theta + 4k\pi)$ and $\ell^- = (e^{i\theta}, \theta + 2\pi + 4k\pi)$ where k is an arbitrary integer. The best choice is thus to write

$$\ell^+ = (e^{i\theta}, \tilde{\theta}) \text{ and } \ell^- = (e^{i\theta}, \widetilde{(\theta + 2\pi)})$$

where the tilde $\sim$ means here "class modulo 4π of". Since the universal covering $Lag_\infty(1)$ is identified with the set of all $(e^{i\theta}, \theta + 2k\pi)$ it follows that ℓ^+ and ℓ^- are just elements of the double covering space

$$Lag_2(1) = Lag(1)/4\pi\mathbb{Z}$$

of $Lag(1)$. In higher dimensions, the situation is quite similar:

Proposition 163 *(1) Let $Lag_\infty(n) = \{(W, \theta) : \det W = e^{i\theta}\}$ be the universal covering of $Lag(n)$. The elements of the double covering*

$$Lag_2(n) = Lag(n)/4\pi\mathbb{Z}$$

of $Lag(n)$ are the oriented Lagrangian planes. (2) The first homotopy group of $Lag_2(n)$ is the subgroup

$$\pi_1(Lag(n)) = \{\bar{0}, \bar{1}\}$$

of $\mathbb{Z}_2 = \mathbb{Z}/2\mathbb{Z}$. That group acts on $Lag_2(n)$ by the law

$$\bar{0} * (W, \tilde{\theta}) = (W, \tilde{\theta}) \quad \text{and} \quad \bar{1} * (W, \tilde{\theta}) = (W, \widetilde{(\theta + 2\pi)}) \tag{5.70}$$

(cf. the definition (5.50) of $k * (W, \theta)$).

(All the statements in this proposition are obvious in view of the standard theory of covering spaces; see for instance Frankel [48], Godbillon [51], or Kuga [98].)

We will denote the generic elements of $Lag_2(n)$ by $\ell^{\pm}$, $\ell'^{=}$, etc. The choice of what we call the "positive" orientation of an oriented Lagrangian plane $\ell^{\pm}$ is of course arbitrary. Proposition 163 shows that "$*$-multiplication" by $\bar{1}$ reverses the orientation of an oriented Lagrangian plane:

$$\bar{1} * \ell^{+} = \ell^{-}, \ \ \bar{1} * \ell^{-} = \ell^{+}.$$

We now proceed to study de Rham forms on a Lagrangian manifold V. We begin by noting that we can always assign to every $z \in V$ an oriented Lagrangian plane $\ell^{\pm}(z)$ by choosing (arbitrarily, see the remark above) an orientation on the tangent plane $\ell(z) = T_z V$. *Assume now that V is oriented.* Then we can choose once and for all a positive orientation on one tangent plane $\ell(z_0)$ and induce the same orientation on all $\ell(z)$ by letting z vary smoothly in V. That is, we can construct a continuous mapping $z \mapsto \ell^{+}(z)$ from V to $Lag_2(n)$ such that the projection of $\ell^{+}(z)$ on $Lag(n)$ is precisely $\ell(z)$. Since the choice of what we call "positive orientation" is purely a matter of taste, we can actually construct *two* such mappings:

$$z \mapsto \ell^{+}(z), \ \ z \mapsto \ell^{-}(z). \tag{5.71}$$

We can even do a little bit more. Consider two paths $\gamma_{z_0 z}$ and $\gamma'_{z_0 z}$ in V joining a point z_0 to a point z. If we transport the tangent plane $\ell(z)$ from z_0 to z along any of theses curves, we will end up at z with the same orientation (that this property does not hold in non-oriented manifolds is easily seen by making such "transports" in, say, the Möbius strip). Recalling from Subsection 4.7.2 in Chapter 4 that the universal covering $\check{V}$ of a manifold V is the set of all homotopy classes $\check{z}$ (with fixed endpoints) of continuous paths in V originating from some "base point" z_0, the mappings (5.71) induce mappings

$$\check{z} \mapsto \ell^{+}_{\infty}(\check{z}), \ \ \check{z} \mapsto \ell^{-}_{\infty}(\check{z}) \tag{5.72}$$

of $\check{V}$ in $Lag_{\infty}(n)$. These mappings $\ell^{\pm}_{\infty}(\cdot)$ are constructed as follows: consider a continuous path $\gamma_{z_0 z}$: $[a, b] \longrightarrow V$, and let $\ell^{+}(\gamma_{z_0 z})$ be its image by the mapping $\ell^{+}(\cdot) : V \longrightarrow Lag_2(n)$ that to every point z of V associates the oriented tangent plane $\ell^{+}(z)$. Thus, $\ell^{+}(\gamma_{z_0 z})$ is the path in $Lag_2(n)$ defined by

$$\ell^{+}(\gamma_{z_0 z})(t) = \ell^{+}(\gamma_{z_0 z}(t)), \ \ a \leq t \leq b.$$

If we choose $\ell^{+}(z_0)$ as base point ℓ^{+}_0 in $Lag_2(n)$, then the image of the homotopy class $\check{z}$ by the mapping $\ell^{+}(\cdot)$ becomes the homotopy class of $\ell^{+}(\gamma_{z_0 z})$

in $Lag_2(n)$, in fact an element of $Lag_\infty(n)$, since $Lag(n)$ and $Lag_2(n)$ have the same universal covering. It is this element of $Lag_\infty(n)$ that we denote by $\ell^+_\infty(\check{z})$. The mapping $\ell^-(\cdot)$ is constructed in a similar way.

Summarizing, we have two commutative diagrams

$$\begin{array}{ccc} \check{V} & \overset{\ell^+_\infty(.)}{\longrightarrow} & Lag_\infty(n) \\ \pi\downarrow & & \downarrow\pi \\ V & \overset{\ell^+(.)}{\longrightarrow} & Lag_2(n) \end{array} \quad \text{and} \quad \begin{array}{ccc} \check{V} & \overset{\ell^-_\infty(.)}{\longrightarrow} & Lag_\infty(n) \\ \pi\downarrow & & \downarrow\pi \\ V & \overset{\ell^-(.)}{\longrightarrow} & Lag_2(n) \end{array} \tag{5.73}$$

where π is a collective notation for the covering projections.

We will use in the sequel the following notation: ξ^+ is the datum of a positively oriented basis $\xi = (\xi_1, ..., \xi_n)$ of $\mathbb{R}^n$ (identified with $\ell(z)$), and ξ^- the datum of a negatively oriented basis.

To each $\ell_{\alpha,\infty} \in Lag_\infty(n)$ we associate two integer valued functions

$$m^+_\alpha, m^-_\alpha : \check{V} \longrightarrow \mathbb{Z}$$

by the formula

$$m^\pm_\alpha(\check{z}) = m(\ell_{\alpha,\infty}, \ell^\pm_\infty(\check{z})). \tag{5.74}$$

Lemma 164 *(1) For every $\gamma \in \pi_1(V)$ we have*

$$m^\pm_\alpha(\gamma\check{z}) = m(\ell_{\alpha,\infty}, \ell^\pm_\infty(\check{z})) = m^\pm_\alpha(\check{z}) \mod 2 \tag{5.75}$$

and the functions m^+_α, m^-_α are thus defined modulo 2 on the oriented Lagrangian manifold V. (2) These functions are constant on each connected subset of $V \setminus \Sigma_\alpha$ where

$$\Sigma_\alpha = \{z \in V : \ell(z) \cap \ell_\alpha \neq 0\}$$

is the caustic of V relative to ℓ_α.

Proof. **(1)** Since V is oriented, the Maslov index $m(\gamma)$ is even; formula (5.75) then follows from the fact that

$$\begin{aligned} m^\pm_\alpha(\gamma\check{z}) &= m(\ell_{\alpha,\infty}, \ell^\pm_\infty(\gamma\check{z})) \\ &= m(\ell_{\alpha,\infty}, m(\gamma) * \ell^\pm_\infty(\check{z})) \\ &= m(\ell_{\alpha,\infty}, \ell^\pm_\infty(\check{z})) - m(\gamma) \end{aligned}$$

where we have used (5.59) to obtain the last equality. **(2)** In view of the topological property of the Leray index (property **(1)** in Theorem 151),

$m(\ell_{\alpha,\infty}, \ell_{\infty}^{\pm}(\check{z}))$ is locally constant outside the caustic Σ_α, and hence constant on the connected components of $V \setminus \Sigma_\alpha$. ∎

The properties of the functions $m_\alpha^{\pm}$ allow us to construct de Rham forms on oriented Lagrangian manifolds:

Proposition 165 *Let ρ be a density on the oriented Lagrangian manifold V, and $\ell_\alpha \in Lag(n)$.* ***(1)*** *The de Rham forms μ and $-\mu$ associated with ρ are given, outside the caustic $\Sigma_\alpha = \{z \in V : \ell(z) \cap \ell_\alpha \neq 0\}$ by*

$$\mu_\alpha(z)(\xi^{\pm}) = (-1)^{m_\alpha^{\pm}(\check{z})}\rho(z)(\xi) \tag{5.76}$$

where the sign $\pm$ in $m_\alpha^{\pm}(\check{z})$ is determined by the sign $\pm$ of the basis $\xi = (\xi_1, ..., \xi_n)$ of the tangent plane $\ell(z)$. (If the vectors $\xi_1, ..., \xi_n$ are linearly dependent, then $\mu(z)(\xi) = 0$.) ***(2)*** *Two such expressions μ_α and μ_β, corresponding to the choices $\ell_{\alpha,\infty}$ and $\ell_{\beta,\infty}$ are related by*

$$\mu_\alpha(z) = (-1)^{-m(\ell_{\alpha,\infty},\ell_{\beta,\infty})+\mathrm{Inert}(\ell_\alpha,\ell_\beta,\ell(z))}\mu_\beta(z). \tag{5.77}$$

Proof. (1) In view of Lemma 164, the right-hand side of (5.76) is really defined on V. In view of the same lemma $(-1)^{m_\alpha^{\pm}(\check{z})}$ keeps a constant value on each connected subset of $V \setminus \Sigma_\alpha$ once an orientation is chosen, and changes sign when the orientation is reversed. **(2)** Formula (5.77) immediately follows from the cohomological property (5.58) of the reduced Leray index, since

$$\begin{aligned} m_\alpha^{\pm}(\check{z}) - m_\beta^{\pm}(\check{z}) &= m(\ell_{\alpha,\infty}, \ell_{\infty}^{\pm}(\check{z})) - m(\ell_{\beta,\infty}, \ell_{\infty}^{\pm}(\check{z})) \\ &= -m(\ell_{\alpha,\infty}, \ell_{\beta,\infty}) + \mathrm{Inert}(\ell_\alpha, \ell_\beta, \ell(z)) \end{aligned}$$

where $\ell(z)$ is the projection of $\ell_{\infty}^{\pm}(\check{z})$. ∎

The following example shows that Proposition 165 contains as a particular case the construction of the de Rham form on the circle in the introductory Subsection 5.1.3:

Example 166 ***The circle***. *Consider the density $r|d\vartheta|$ on the circle S_r^1 with radius r, where ϑ is the usual polar angle. The positively oriented line ℓ^+ whose polar angle is ϑ is thus identified with $-e^{-2i\vartheta}$, and hence*

$$\ell_{\infty}^{+} \equiv (-e^{-2i\vartheta}, \pi - 2\vartheta)$$

is an element of $Lag_\infty(1)$ *with projection* ℓ^+. *Choose now for* ℓ_α *the* x-*axis, that is* $\ell_\alpha \equiv -1$, *and* $\ell_{\alpha,\infty} \equiv (-1,\pi)$. *It follows, using formula (5.57) for the reduced Leray index, that*

$$m_\alpha(\vartheta) = \left[\tfrac{\pi-(\pi-2\vartheta)}{2\pi}\right] + 1 = \left[\tfrac{\vartheta}{\pi}\right] + 1$$

which is just formula (5.20) in Subsection 5.1.3. The de Rham forms associated with $r|d\vartheta|$ *are thus* $\pm\mu$ *where*

$$\mu = (-1)^{\left[\frac{\vartheta}{\pi}\right]+1} r|d\vartheta|.$$

The non-orientable case. Until now we have supposed the manifold oriented. To deal with the non-orientable case, where we can no longer define mappings $\ell^\pm(\cdot)$, it suffices to use the following very simple trick. We begin by noting that in Proposition 165 the de Rham form μ is defined on V by the expression

$$\mu_\alpha(z)(\xi^\pm) = (-1)^{m_\alpha^\pm(\check{z})}\rho(z)(\xi)$$

involving the variable $\check{z}$ of the universal covering; the fact that μ_α is really defined on V coming from the fact that the Maslov index of every loop in an orientable Lagrangian manifold is even. This suggests that in the non-orientable case we *define* the de Rham form μ on the universal covering $\check{V}$, rather than on V itself, by the formula

$$\mu_\alpha(\check{z})(\xi^\pm) = (-1)^{m_\alpha^\pm(\check{z})}\rho(z)(\xi) \tag{5.78}$$

where $\rho(z)$ is viewed as a density on $\check{V}$ (this amounts to identify the density ρ on V with its pullback $\pi^*\rho$ to $\check{V}$). The functions $m_\alpha^\pm(\check{z})$ are defined as in (5.74):

$$m_\alpha^\pm(\check{z}) = m(\ell_{\alpha,\infty}, \ell_\infty^\pm(\check{z})) \tag{5.79}$$

$\ell_\infty^+(\check{z})$ and $\ell_\infty^-(\check{z})$ corresponding to two orientations of $\check{V}$. ($\check{V}$ is orientable, being simply connected.) If we now define a de Rham form on a Lagrangian manifold by (5.78) in all cases, Proposition 165 holds *mutatis mutandis* for all Lagrangian manifolds, whether they are orientable, or not!

Remark 167 *A Lagrangian plane* ℓ_α *being given, the choice of the element* $\ell_{\alpha,\infty}$ *with projection* ℓ_α *is arbitrary. We will therefore assume that such an element is fixed once and for all. If we change* $\ell_{\alpha,\infty}$ *into another element* $\ell'_{\alpha,\infty}$ *with same projection* ℓ_α, *we will have*

$$m(\ell'_{\alpha,\infty}, \ell_\infty^\pm(\check{z})) = m(\ell_{\alpha,\infty}, \ell_\infty^\pm(\check{z})) + k_\alpha$$

where k_α is an integer only depending on the pair $(\ell_{\alpha,\infty}, \ell'_{\alpha,\infty})$, and the effect on μ_α will simply be a change of overall sign. In order to avoid this ambiguity one can for example decide to choose, once and for all, the base point of $Lag_\infty(n)$ as being ℓ_α, and to take for $\ell_{\alpha,\infty}$ the constant loop.

These constructions motivate the following definition of the argument of a de Rham form on a Lagrangian manifold:

Definition 168 *Let μ be a de Rham form on $\check{V}$, and $\ell_{\alpha,\infty} \in Lag_\infty(n)$. Set $\ell_\infty^\pm(\check{z}) = \ell_\infty(\check{z})$. The argument of μ relative to $\ell_{\alpha,\infty}$ is*

$$\arg_\alpha \mu(\check{z}) \equiv m_\alpha(\check{z})\pi \mod 2\pi \tag{5.80}$$

where the integer $m_\alpha(\check{z})$ is defined by:

$$m_\alpha(\check{z}) = m(\ell_{\alpha,\infty}, \ell_\infty(\check{z})). \tag{5.81}$$

Notice that this definition cannot in general be used to define a global argument for a de Rham form on the manifold V itself. For instance, the de Rham form $d\vartheta = \pm |d\vartheta|$ on the circle S^1 has no well-defined argument (but it has, if we view it as a form on $\mathbb{R}_\vartheta$, the universal covering of S^1). However, if we denote by μ_α the restriction of μ to the set $V \setminus \Sigma_\alpha$ ($\Sigma_\alpha$the caustic of V relative to ℓ_α), then it makes sense to define

$$\arg \mu_\alpha(z) \equiv m_\alpha(\check{z})\pi \mod 2\pi. \tag{5.82}$$

We are now, at last, able to define the wave forms in the general case.

5.4 Wave-Forms on a Lagrangian Manifold

Recall from Subsection 5.1.1 that we called "wave form" an expression of the type $\Psi(x)\sqrt{dx}$ involving the square root of dx, and where $\Psi(x) = e^{\frac{i}{\hbar}\Phi(x)}R(x)$ was a wave function. In this section we show that similar objects can be defined on arbitrary quantized Lagrangian manifolds, and relate our constructions to the usual formulae of semi-classical mechanics.

5.4.1 Definition of Wave Forms

We begin by defining the square roots of a de Rham form on a Lagrangian manifold. This is made possible by Definition 168 above. Let μ be a de

Rham form on the universal covering $\check{V}$ of the Lagrangian manifold V. For each $\ell_{\alpha,\infty} \in Lag_\infty(n)$ we define

$$\sqrt{\mu_\alpha}(\check{z})(\xi^\pm) = i^{m_\alpha^\pm(\check{z})}\sqrt{\rho(z)(\xi)}. \tag{5.83}$$

Using the short-hand conventions of Definition 168 this can be written simply as

$$\sqrt{\mu_\alpha}(\check{z}) = i^{m_\alpha(\check{z})}\sqrt{\rho(z)}. \tag{5.84}$$

The following straightforward lemma relates the different square roots of a de Rham form:

Lemma 169 *(1) Two square roots of a de Rham form corresponding to different choices $\ell_{\alpha,\infty}$, $\ell_{\beta,\infty}$ are related by the formula*

$$\sqrt{\mu_\alpha}(\check{z}) = i^{m_{\alpha\beta}(z)}\sqrt{\mu_\beta}(\check{z}) \tag{5.85}$$

where the function $m_{\alpha\beta}$ is given by

$$m_{\alpha\beta}(z) = m(\ell_{\alpha,\infty}, \ell_{\beta,\infty}) - \operatorname{Inert}(\ell_\alpha, \ell_\beta, \ell(z)). \tag{5.86}$$

(2) The function $m_{\alpha\beta}$ is constant on each connected subset U of V such that $U \cap \Sigma_\alpha = U \cap \Sigma_\beta = \emptyset$ (Σ_α, Σ_β the caustics of V relative to ℓ_α and ℓ_β, respectively).

Proof. **(1)** Formula (5.86) follows from the cohomological property (5.58) of the reduced Leray index. Property **(2)** follows from the fact that $\operatorname{Inert}(\ell_\alpha, \ell_\beta, \ell(z))$ is locally constant when $\ell(z)$ moves in such a way that it remains transversal to ℓ_α and ℓ_β (as follows from Property **(4)** of the signature in Subsection 5.2.5). ∎

Recall from Chapter 4, Subsection 4.7.3, that the phase of a Lagrangian manifold V is the function φ defined on the universal covering $\check{V}$ of V by the formula

$$\varphi(\check{z}) = \int_{\gamma_{z_0 z}} p\,dx$$

where $\check{z}$ is the homotopy class of the path $\gamma_{z_0 z}$ joining the base point z_0 to z in V. That phase φ satisfies:

$$d\varphi(\check{z}) = pdx \quad \text{if} \quad \pi(\check{z}) = (x, p)$$

where the differential $d\varphi$ is calculated with respect to the local variable x.

Definition 170 *(1) Let μ be a de Rham form on $\check{V}$ and $\ell_{\alpha,\infty} \in Lag_\infty(n)$. The wave-form associated to the pair (μ, ℓ_α) is the expression*

$$\widetilde{\Psi}_\alpha(\check{z}) = e^{\frac{i}{\hbar}\varphi(\check{z})}\sqrt{\mu_\alpha}(\check{z}). \tag{5.87}$$

(2) The set $\widetilde{\Psi} = (\widetilde{\Psi}_\alpha)_\alpha$ of all wave-forms (5.87) determined by the same density ρ is called the catalogue *of ρ, and the $\widetilde{\Psi}_\alpha$ are called the* pages *of the catalogue $\widetilde{\Psi}$. The set of all catalogues on V is denoted by $Cat(V)$.*

In terms of the density ρ associated to μ we thus have

$$\widetilde{\Psi}_\alpha(\check{z}) = e^{\frac{i}{\hbar}\varphi(\check{z})}i^{m_\alpha(\check{z})}\sqrt{\rho}(z). \tag{5.88}$$

The pages of a given catalogue are easily deduced from one another. Let in fact ℓ_α, ℓ_β be two Lagrangian planes; in view of formula (5.85) we have:

$$\widetilde{\Psi}_\alpha(\check{z}) = i^{m_{\alpha\beta}(z)}\widetilde{\Psi}_\beta(\check{z}), \tag{5.89}$$

where

$$m_{\alpha\beta}(z) = -m(\ell_{\alpha,\infty}, \ell_{\beta,\infty}) + \operatorname{Inert}(\ell_\alpha, \ell_\beta, \ell(z)).$$

Every page of a catalogue thus contains all the information about all the other pages of that catalogue.

A wave-form is a priori defined on the universal covering $\check{V}$, and not V itself. However, if we impose to the manifold V to be quantized, then $\widetilde{\Psi}_\alpha$ is single-valued:

Proposition 171 *A wave form is defined on the manifold V if, and only if, Maslov's quantization condition*

$$\frac{1}{2\pi\hbar}\int_\gamma pdx - \frac{1}{4}m(\gamma) \in \mathbb{Z} \tag{5.90}$$

holds for all loops γ in V.

Proof. To say that $\widetilde{\Psi}(\check{z})$ is single-valued amounts to say that it keeps the same value if we replace $\check{z}$ with another element of $\check{V}$ with same projection z on V. Every such element being of the type $\gamma\check{z}$ where $\gamma \in \pi_1(V, z_0)$, it suffices to show that if Eq. (5.90) holds, then $\widetilde{\Psi}(\gamma\check{z}) = \widetilde{\Psi}(\check{z})$ for every $\gamma \in \pi_1(V, z_0)$. Now

$$\varphi(\gamma\check{z}) = \int_{\gamma_{z_0 z}} p\,dx + \int_\gamma p\,dx$$

and

$$m_\alpha(\gamma\check{z}) = m(\ell_{\alpha,\infty}, \ell_\infty(\gamma\check{z})) = m_\alpha(\gamma\check{z}) + m(\gamma)$$

hence, in view of the expression (5.88) of the wave form:

$$\widetilde{\Psi}(\gamma\check{z}) = \exp\left[i\left(\frac{1}{\hbar}\int_\gamma p\,dx + \frac{\pi}{2}m(\gamma)\right)\right]\widetilde{\Psi}(\check{z}).$$

It follows that $\widetilde{\Psi}(\gamma\check{z}) = \widetilde{\Psi}(\check{z})$ for all γ if and only if the exponential on the right hand side is one for all γ; but this is equivalent to Maslov's condition (5.90). ■

5.4.2 The Classical Motion of Wave-Forms

Let H be a Hamiltonian function (not necessarily of Maxwell type) and (f_t) the time-dependent flow it determines. Let $\widetilde{\Psi} = (\widetilde{\Psi}_\alpha)$ be a catalogue on a Lagrangian manifold V. We define the action of f_t on

$$\widetilde{\Psi}_\alpha(\check{z}) = e^{\frac{i}{\hbar}\varphi(\check{z})}\sqrt{\mu_\alpha}(\check{z}) = e^{\frac{i}{\hbar}\varphi(\check{z})}i^{m_\alpha(\check{z})}\sqrt{\rho}(z) \tag{5.91}$$

by the formula:

$$f_t\widetilde{\Psi}_\alpha(\check{z}) = e^{\frac{i}{\hbar}\varphi(\check{z},t)}i^{m_\alpha(\check{z},t)}(f_t)_*\sqrt{\rho}(z) \tag{5.92}$$

where $(f_t)_*\sqrt{\rho}$ is the push-forward to $V_t = f_t(V)$ of $\sqrt{\rho}$:

$$(f_t)_*\sqrt{\rho}(z) = \sqrt{(f_t)_*\rho}(z).$$

The functions $\varphi(\check{z}, t)$, $m_\alpha(\check{z}, t)$ are defined as follows: $\varphi(., t)$ is the phase of $V_t = f_t(V)$, that is

$$\varphi(\check{z}, t) = \varphi(\check{z}) + \int_z^{z(t)} pdx - Hds \tag{5.93}$$

where the integration is performed along the trajectory going from the point $z \in V$ to $z(t) = f_t(z) \in V_t$. (See Proposition 126 of Section 4.7.) To define the integer $m_\alpha(\check{z}, t)$ we have to introduce a few supplementary notations. Let $s_t(z)$ be the Jacobian matrix of $f_t(z)$ at the point z:

$$s_t(z) = f_t'(z).$$

Since f_t is a symplectomorphism, we have $s_t(z) \in \mathrm{Sp}(n)$ for every z, and

$$\ell(f_t(z)) = s_t(z)\ell(z).$$

The image $f_t(\check{z})$ of $\check{z} \in \check{V}$ is an element of $\check{V}_t$. Denoting by $\ell_\infty(f_t(\check{z}))$ the image of $\ell_\infty(\check{z})$ by $s_t(z)$, $z = \pi(\check{z})$, that is:

$$\ell_\infty(f_t(\check{z})) = s_t(z)\ell_\infty(\check{z})$$

$m_\alpha(\check{z}, t)$ is then the integer

$$m_\alpha(\check{z}, t) = m(\ell_{\alpha,\infty}, \ell_\infty(f_t(\check{z}))) \tag{5.94}$$

($\ell_\infty(\cdot)$ being here defined on the Lagrangian manifold $\check{V}_t = f_t(V)$). Notice that we have $m_\alpha(\check{z}, 0) = m_\alpha(\check{z})$. Formula (5.92) defines $f_t\widetilde{\Psi}_\alpha$ as a wave-form on V_t equipped with the phase (5.93). We will denote the corresponding catalogue by $\widetilde{\Psi}(., t)$.

That the image of a single-valued wave form by f_t also is single-valued follows from:

Lemma 172 *If the Lagrangian manifold V is quantized, then so is $V_t = f_t(V)$.*

Proof. It is obvious, since the Keller-Maslov condition is in essence a condition on the universal covering of V, which is identical to that of V_t (alternatively one can use the symplectic invariance of the action along loops together with the fact that $m_\alpha(\gamma\check{z}, t) = m_\alpha(\check{z}, t) + m(\gamma)$ for all t). ■

It turns out that the mapping

$$f_t : \mathrm{Cat}(V) \to \mathrm{Cat}(V_t)$$

defined by (5.92) is an isomorphism in the following sense: suppose we choose the phase $\varphi(., t)$ of V_t as being given by (5.93). Then, if

$$\widetilde{\Psi}_\beta(\check{z}, t) = e^{\frac{i}{\hbar}\varphi(\check{z},t)} i^{m_\beta(\check{z})}\mu(z)$$

for some half-density μ on V_t, we can determine *uniquely* $\widetilde{\Psi}_\alpha$ such that $f_t\widetilde{\Psi}_\alpha = \widetilde{\Psi}_\beta$ by choosing $\ell_{\alpha,\infty} = s_t\ell_{\beta,\infty}$ and $\rho = (f_t)^*\mu$. (See de Gosson [57].) We will denote the catalogue on V thus defined by $(f_t)^{-1}\widetilde{\Psi}$. For $\widetilde{\Psi} \in \mathrm{Cat}(V_{t'})$ and arbitrary t, t' we define

$$f_{t,t'}\widetilde{\Psi}(\check{z}, t') = f_t(f_{t'})^{-1}\widetilde{\Psi}(\check{z}, t').$$

In view of the discussion here above it is clear that the Chapman-Kolmogorov law

$$f_{t,t'}f_{t',t'}\widetilde{\Psi} = f_{t,t''}\widetilde{\Psi}$$

holds for all t, t', t'' and $\widetilde{\Psi} \in \mathrm{Cat}(V_{t''})$. In fact, we have the somewhat stronger result

$$f_{t,t'}f_{t',t''}\widetilde{\Psi}_\alpha = f_{t,t''}\widetilde{\Psi}_\alpha$$

which shows that the Chapman-Kolmogorov law holds even for the individual pages of a catalogue (see de Gosson [57, 60] for details).

5.4.3 The Shadow of a Wave-Form

We are now going to see why, and how, these abstract geometrical constructions are related to the semi-classical solutions (5.10) of Schrödinger's equation.

In what follows we assume that the Lagrangian manifold V is quantized:

$$\frac{1}{2\pi\hbar}\int_\gamma pdx - \frac{1}{4}m(\gamma) \in \mathbb{Z}$$

and we choose for ℓ_α the vertical plane $\ell_p = 0 \times \mathbb{R}^n_p$.

We assume that for every x there exist at most a *finite* number of points $z_j = (x, p_j)$ in V. Denoting by χ the projection of V on $\mathbb{R}^n_x$, this means that $\chi^{-1}(x)$ is always a finite (or empty) set: $\chi^{-1}(x) = \emptyset$ or

$$\chi^{-1}(x) = \{z_1, ..., z_N\}.$$

Let $\widetilde{\Psi}_p$ be a wave form on V associated to ℓ_p and a density ρ:

$$\widetilde{\Psi}_p(\check{z}) = e^{\frac{i}{\hbar}\varphi(\check{z})}i^{m_p(\check{z})}\sqrt{\rho}(z).$$

For a point $z = (x,p)$ of V outside the caustic $\Sigma = \Sigma_p$ the local expression of $\widetilde{\Psi}_p$ is

$$\widetilde{\Psi}_p(x) = e^{\frac{i}{\hbar}\Phi(x)}i^{m(x)}a(x)\,|dx|^{1/2}$$

where $a(x)\,|dx|^{1/2}$ is the local expression of the half-density $\sqrt{\rho}$ and $\Phi(x)$, $m(x)$ are defined as follows:

$$\Phi(x) = \varphi(\check{z}), \quad m(x) = m_p(\check{z})$$

where $\check{z}$ is any point of the universal covering $\check{V}$ whose projection on V is z. The choice of that $\check{z}$ is irrelevant because V is quantized: if we change $\check{z}$ into another point $\check{z}'$ with the same projection z on V, then $\check{z}' = \gamma\check{z}$ for some $\gamma \in \pi_1(V)$ and thus

$$\varphi(\gamma\check{z}) = \varphi(\check{z}) + \int_\gamma p\,dx \quad , \quad m_\alpha(\gamma\check{z}) = m_\alpha(\gamma\check{z}) + m(\gamma)$$

which causes no overall change in $\widetilde{\Psi}_p(x)$.

Definition 173 *The* shadow of the wave form $\widetilde{\Psi}_p$ *is the function* $\Psi = \Sigma\widetilde{\Psi}_p$ *of* x *defined by:* $\Psi(x) = 0$ *if there are no points* z_j *in* V *such that* $z_j = (x, p_j)$, *and by*

$$\Psi(x) = \Sigma\widetilde{\Psi}_p(x) = \sum_j e^{\frac{i}{\hbar}\Phi_j(x)} i^{m_j(x)} a_j(x) \tag{5.95}$$

otherwise. The functions $a_j(x)$, $\Phi_j(x)$ *and* $m_j(x)$ *in (5.95) are defined as follows:* ***(1)*** $a_j(x)|dx|^{1/2}$ *is the local expression of the density* $\rho(x)$ *near the point* $z_j = (x, p_j) \in V$*;* ***(2)*** *the functions* Φ_j *and* m_j *are given by*

$$\Phi_j(x) = \varphi(\check{z}_j), \quad m_j(x) = m_p(\check{z}_j) \tag{5.96}$$

where, for each j, $\check{z}_j$ *is any point of* $\check{V}$ *with projection* z_j. *(See de Gosson [57, 60] for the definition of shadows on an arbitrary Lagrangian plane.)*

That the choice of the $\check{z}_j$ in the formulas (5.96) is unimportant is again a consequence of the fact that V is quantized.

Here is the essential result of this section; it shows that the semi-classical wave functions are just the shadows of our wave forms. We assume that the wave form is defined, as in Subsection 5.1.1, on an exact Lagrangian manifold

$$V : p = \nabla_x\Phi(x).$$

The local expression of $\widetilde{\Psi}_p$ on configuration space is thus globally defined and can be written in the form:

$$\widetilde{\Psi}_p(x) = e^{\frac{i}{\hbar}\Phi(x)} a(x)|dx|^{1/2}.$$

We assume, for simplicity, that V is simply connected, but the result remains valid in the general case (see our articles de Gosson [57, 60]). Since we then have $\check{V} = V$ we can identify the variables $\check{z}$ and z.

Theorem 174 *Let (f_t) be the time-dependent flow determined by a Hamiltonian function H. Let $\widetilde{\Psi}_p$ be a wave form on the exact Lagrangian manifold V. The shadow of $\widetilde{\Psi}_p(\cdot,t) = f_t\widetilde{\Psi}_p$ at a point $x(t)$ such that there exists $p_j(t)$ with $(x(t),p_j(t)) \in V_t$ is given by the formula*

$$\Sigma\widetilde{\Psi}_p(x(t),t) = \sum_{j=1}^{N} i^{m_j(x(t))} e^{\frac{i}{\hbar}\Phi_j(x(t),t)} a(x_j) \left|\frac{dx(t)}{dx_j}\right|^{-1/2} \tag{5.97}$$

where the $dx(t)/dx_j$ are the Jacobian determinants of the diffeomorphisms $x_j \longmapsto x(t)$ defined in a neighborhood of the points x_j such that $(x(t),p_j(t)) = f_t(x_j,p_j)$. The functions Φ_j are given by

$$\Phi_j(x(t),t) = \Phi(x_j,t) + \int_{x_j,0}^{x(t),t} pdx - Hdt' \tag{5.98}$$

and the functions m_j by:

$$m_j(x(t)) = m(\ell_{p,\infty}, s_t(z_j)\ell_\infty(z_j)). \tag{5.99}$$

Proof. We set $z(t) = f_t(z)$ in the proof. We first note that we have

$$\widetilde{\Psi}_\alpha(z,t) = e^{\frac{i}{\hbar}\mathcal{A}(z(t),t)} i^{-m(z(t),t)} (e^{\frac{i}{\hbar}\Phi(x)}(f_t)_*\sqrt{\rho}(z)) \tag{5.100}$$

where

$$\mathcal{A}(z(t),t) = \int_z^{z(t)} pdx - Hdt' \tag{5.101}$$

and

$$m(z(t),t) = m_\alpha(\ell_{\alpha,\infty}, s_t(z)\ell_{\alpha,\infty}). \tag{5.102}$$

(Formula (5.101) is obvious, and (5.102) follows from the fact that $\ell(z(t)) = s_t(z)\ell(z)$.) The local expressions of $\mathcal{A}(z(t),t)$ and of $m(z(t),t)$ near x_j are given by, respectively and

$$\mathcal{A}(x_j,t) = \int_{x_j,0}^{x(t),t} pdx - Hdt'$$

(5.99). Writing the local expression of ρ near $(x(t),p_j(t))$ as

$$\rho_j(x(t)) = a_j(x(t))|dx(t)|$$

we have

$$a_j(x(t)) = a(x_j)\left|\frac{dx(t)}{dx_j}\right|^{-1/2}$$

by the transformation properties of densities (see Appendix D), so that we see that the local expression of $\widetilde{\Psi}_\alpha(z,t)$ is

$$\widetilde{\Psi}_p(x(t)) = e^{\frac{i}{\hbar}\Phi_j(x(t),t)}i^{m_j(x(t)))}a(x_j)|dx_j|^{1/2}.$$

The theorem follows, by definition (5.95) of the shadow of a wave form. ∎

We thus recover the semi-classical formula (5.10), changing $x(t)$ in x and x in x'_j in (5.97).

We refer the reader interested in various extensions of Theorem 174 to our articles [57, 60], where we have used a more cohomological approach than here. In these articles the relation between the function

$$m(z(t),t) = m_\alpha(\ell_{\alpha,\infty}, s_t(z)\ell_{\alpha,\infty})$$

(see (5.102) and the Maslov index on the metaplectic group $\mathrm{Mp}(n)$ (which we shall study in the next Chapter)) is also made explicit. In [59] we analyzed geometric phase shifts ("Berry's phase") in terms of wave forms on quantized Lagrangian manifolds.

Chapter 6

Metaplectic Group, Maslov Index, and Quantization

Summary. *The symplectic group has a connected double covering which can be realized as a group of unitary operators on* $L^2(\mathbb{R}^n)$*, the metaplectic group* $\mathrm{Mp}(n)$*. Each element of* $\mathrm{Mp}(n)$ *is the product of two "quadratic Fourier transforms". This property allows the definition of the Maslov index on* $\mathrm{Mp}(n)$*, which can be expressed in terms of the Leray index.*

Here we touch one of the central themes of this book, the metaplectic representation of the symplectic group. It is a deep and fascinating subject of mathematics, unfortunately unknown to most physicists. It is however essential to the understanding of the relationship between classical and quantum mechanics. For the readers who do not want to absorb all the technicalities underlying the construction of the metaplectic group and the Maslov index, I suggest to only read the Introduction to this Chapter, and then to proceed directly to the next Chapter, which is devoted to the Schrödinger equation.

6.1 Introduction

6.1.1 Could Schrödinger have Done it Rigorously?

No doubt that this question will provoke strong reactions, going from stupor to horror and indignation among many readers. You see, it is often sustained that there is no way, whatsoever, to *derive* quantum mechanics from classical mechanics. And this is right, no doubt, because there is no loophole for introducing Planck's constant in Newtonian mechanics, which is a "self-sufficient" theory. However, I claim that the answer to the question in the title of this subsection is:

> "*Schrödinger in a sense* did *it rigorously, because he used arguments that could have led him to discover a deep mathematical property, the metaplectic representation of the symplectic group*"

In order to understand this statement, let us first review what Schrödinger did (for detailed arguments, see for instance, Jammer [92], Messiah [117], or Park [128]).

6.1.2 Schrödinger's Idea

Remember that Schrödinger was desperately looking for an equation governing the time evolution of de Broglie's matter waves. He did not finally arrive to his equation by a rigorous argument, but reasoned as follows (see e.g. [92, 132]). Elaborating on Hamilton's mechanical-optical analogy (which are discussed in Arnold [4] or Park [128], §46), Schrödinger made the assumption that this analogy remains valid even for de Broglie's matter waves. Using Hamilton-Jacobi's theory as guideline, he postulated the equation

$$\nabla_{\mathbf{r}}^2\psi + \frac{8\pi^2 m}{h^2}(E-U)\psi = 0 \tag{6.1}$$

which is of course just the same thing as

$$\hat{H}\psi = E\psi \tag{6.2}$$

where $\hat{H}$ is the quantum operator

$$\hat{H} = -\frac{\hbar^2}{2m}\nabla_{\mathbf{r}}^2 + U.$$

Realizing that Eq. (6.1) is an eigenvalue problem, and that the energy E thus in general only takes a discrete sequence of values, and consequently must not occur in the wave equation, Schrödinger eliminated E by setting

$$\psi(\mathbf{r}) = \Psi(\mathbf{r}, t)\exp\left(-\frac{2\pi i}{h}Et\right)$$

and finally obtained the equation

$$-\frac{h^2}{2m}\nabla_{\mathbf{r}}^2\Psi + U\Psi = \frac{h}{2\pi i}\frac{\partial\Psi}{\partial t}$$

which is the same thing as

$$i\hbar\frac{\partial\Psi}{\partial t} = \hat{H}\Psi. \tag{6.3}$$

While it is true that Schrödinger's argument was not rigorous (it was rather a "sleepwalker" argument[1]), all the mathematically "forbidden" steps he took ultimately lead him to his famous equation (6.3). But it all worked so well, because what he was discovering, using rudimentary and awkward mathematical methods, was a property of pure mathematics. He in fact discovered *the metaplectic representation of the symplectic group*, to which this Chapter is devoted.

6.1.3 Sp(n)'s "Big Brother" Mp(n)

The metaplectic group Mp(n) has, as such, a rather recent history (it goes back to the 1950's) although its implicit appearance can probably be traced back to Fresnel's and Gouy's work in optics around 1820; see the historical account in Folland [46] or Guillemin–Sternberg [79, 80]. The first rigorous constructions of Mp(n) as a group seem to have been initiated by the work of I. Segal [142] and L. van Hove [154]. D. Shale [144] remarked about a decade later that the metaplectic representation of the symplectic group is to bosons what the spin representation of $SO(2n,\mathbb{R})$ is to fermions. The study of the metaplectic group was generalized by A. Weil [162] to arbitrary fields in connection with C. Siegel's work on number theory. Historically it seems that V.P. Maslov was the first to observe in 1965, following the work of V.C. Buslaev [27], the role played by the metaplectic group in the theory of asymptotic solutions to partial differential equations depending on a small parameter, in particular WKB theory (Maslov actually considered the subgroup of Mp(n) generated by the partial Fourier transforms in his work [113, 114]). Maslov's theory was clarified and improved by J. Leray [101] who used the properties of Mp(n) to define a new mathematical structure, Lagrangian Analysis. For a slightly different approach, together with many interesting applications (for instance the Fock-Bargmann complex representation or R. Howe's "oscillator group") we refer to Folland's book [46]; also

[1] As described in Arthur Koestler's book "Sleepwalkers".

see Dubin *et al.* [40] which addresses the metaplectic group from a somewhat different point of view. We also refer to A. Voros [157, 158] for an interesting discussion of the metaplectic group applied to semi-classical expansions of the wave function.

Let us begin by very briefly discussing things from an abstract point of view.

The symplectic group has covering groups of all orders $q = 1, 2, ..., +\infty$. By this we mean that there exists *connected* groups $\mathrm{Sp}_2(n)$, $\mathrm{Sp}_3(n)$,..., $\mathrm{Sp}_\infty(n)$ together with homomorphisms ("projections")

$$\Pi_q : \mathrm{Sp}_q(n) \longrightarrow \mathrm{Sp}(n)$$

having the following properties: if $q < +\infty$ then:

(1) π_q is onto and q-to-one: every $s \in \mathrm{Sp}(n)$ is the image of exactly q elements $S_1, ..., S_q$ elements of $\mathrm{Sp}_q(n)$; equivalently $\Pi_q^{-1}(I)$ consists of exactly q elements;

(2) Π_q is continuous, it is in fact a local diffeomorphism of $\mathrm{Sp}_q(n)$ onto $\mathrm{Sp}(n)$: every $s \in \mathrm{Sp}(n)$ has a neighborhood $\mathcal{U}$ such that the coimage $\Pi_q^{-1}(\mathcal{U})$ is the disjoint union of neighborhoods $\mathcal{U}_1, ..., \mathcal{U}_k$ of $S_1, ..., S_q$.

In the case $q = +\infty$:

(3) $\mathrm{Sp}_\infty(n)$ is simply connected (i.e. contractible to its identity element) and $\Pi_q^{-1}(I) \equiv (\mathbb{Z}, +)$ (the integer group).

The *existence* of the covering groups $\mathrm{Sp}_q(n)$ follows from a standard argument from algebraic topology. That argument goes as follows: since $\mathrm{Sp}(n)$ is topologically the product $U(n) \times \mathbb{R}^{n(n+1)}$, the first homotopy group $\pi_1(\mathrm{Sp}(n))$ is isomorphic to $\pi_1(U(n, \mathbb{C})) = (\mathbb{Z}, +)$. It follows that $\mathrm{Sp}(n)$ has covering groups $\mathrm{Sp}_q(n)$ which are in one-to-one correspondence with the quotient groups $\mathbb{Z}/q\mathbb{Z}$ for $q < \infty$, and $\mathrm{Sp}_\infty(n)$ is simply connected.

It turns out that one of these "companion groups" can be realized as a group of unitary operators:

Among all these covering groups, there is one which plays a privileged role for us. It is the double covering $\mathrm{Sp}_2(n)$, and it is the *only* covering group of $\mathrm{Sp}(n)$ that can be represented as a group of unitary operators acting on the space $L^2(\mathbb{R}_x^n)$ of square integrable functions on configuration space. This "realization" of $\mathrm{Sp}_2(n)$ is called the *metaplectic group*, and it is denoted by

Mp(n). We will see that Mp(n) is generated by a set of operators that are closely related to the Fourier transform

$$F\Psi(x) = \left(\frac{1}{2\pi i}\right)^{n/2} \int e^{-ix\cdot x'}\Psi(x')\,d^n x'$$

where the product $x \cdot x'$ is replaced by non-degenerate quadratic forms in the variables (x, x'), which are the generating function of free linear symplectomorphisms. In fact (see Subsection 6.4.2) we could, for this purpose, as well use any Fourier transform

$$F^{\varepsilon}\Psi(x) = \left(\frac{1}{2\pi i\varepsilon}\right)^{n/2} \int e^{-\frac{i}{\varepsilon}x\cdot x'}\Psi(x')\,d^n x'$$

where ε is a positive parameter (for instance $\hbar$, as in the applications to quantum mechanics).

In terms of representation theory Mp(n) is thus a unitary representation of the double cover $\mathrm{Sp}_2(n)$ in the square integrable functions. One can show that this representation is reducible, but that its sub-representations on the spaces $L^2_{odd}(\mathbb{R}^n)$ and $L^2_{even}(\mathbb{R}^n)$ are irreducible (see Folland [46], Chapter 4).

6.2 Free Symplectic Matrices and their Generating Functions

In this section we complement our discussion of the symplectic group of Chapter 3. In particular, we study thoroughly the free symplectic matrices and their generating functions.

We will write as usual symplectic matrices in the block-form

$$s = \begin{pmatrix} A & B \\ C & D \end{pmatrix} \tag{6.4}$$

where the entries A, B, C, D are subject to the equivalent constraints:

$$\begin{cases} A^T C,\ D^T B \ \text{ symmetric, } A^T D - C^T B = I \\ AB^T,\ DC^T \ \text{ symmetric, } \ DA^T - CB^T = I \\ AC^T,\ DB^T \ \text{ symmetric, } \ AD^T - BC^T = I \end{cases} \tag{6.5}$$

(see (3.4) in Chapter 3). Also, when P is a symmetric $n \times n$ matrix we will use the shorthand notation Px^2 for the quadratic form $x^T Px$ and for an arbitrary $n \times n$ matrix L we will write $Lx \cdot x'$ for $x'^T Lx$.

6.2.1 Free Symplectic Matrices

Recall from Subsection 4.3, Lemma 92, that a symplectic matrix (6.4) is free if and only if its right upper corner is invertible:

$$\det B \neq 0 \ . \tag{6.6}$$

This condition is equivalent to

$$s\ell_p \cap \ell_p = 0 \tag{6.7}$$

where we are using the shorthand notation ℓ_p for the vertical plane $0 \times \mathbb{R}^n_p$.

We next study the generating functions of free symplectic matrices.

Proposition 175 *(1) Suppose that s is a free symplectic matrix. Then a generating function for s is the quadratic form*

$$W(x, x') = \frac{1}{2}DB^{-1}x^2 - B^{-1}x \cdot x' + \frac{1}{2}B^{-1}Ax'^2 \tag{6.8}$$

(the matrices DB^{-1} and $B^{-1}A$ are symmetric in view of (6.5)). (2) If, conversely, W is a quadratic form of the type

$$\begin{cases} W(x, x') = \dfrac{1}{2}Px^2 - Lx \cdot x' + \dfrac{1}{2}Qx'^2 \\ P = P^T, \quad Q = Q^T, \quad \det L \neq 0 \end{cases} \tag{6.9}$$

then the matrix

$$s_W = \begin{pmatrix} L^{-1}Q & L^{-1} \\ PL^{-1}Q - L^T & PL^{-1} \end{pmatrix} \tag{6.10}$$

is a free symplectic matrix whose generating function is given by Eq. (6.9). (3) If $(x, p) = s(x', p')$, then that generating function is given by the formula

$$W(x, x') = \frac{1}{2}(p \cdot x - p' \cdot x'). \tag{6.11}$$

Proof. **(1)** Formula (6.8) is obtained by a routine calculation using the relations

$$p = \nabla_x W(x, x') \ , \ p' = -\nabla_{x'} W(x, x').$$

(2) Using the expression (6.8) for W, we see that

$$(x, p) = s_W(x', p') \Longleftrightarrow \begin{cases} p = Px - L^T x' \\ p' = Lx - Qx \end{cases}$$

and since $\det L \neq 0$, we can solve explicitly these equations in (x, p); this yields (6.10) after a few straightforward calculations. That the matrix s_W is symplectic can be checked using for instance the conditions (6.5). **(3)** Formula (6.11) is just Euler's formula for homogeneous functions applied to the quadratic polynomial W. ∎

There is thus a one-to-one correspondence between the quadratic forms (6.8), and free symplectic matrices. In fact, to every such quadratic form W one can associate the free symplectic matrix and, conversely, every free symplectic matrix can be written in this form, and thus determines W.

Corollary 176 *The inverse $(s_W)^{-1}$ is the free symplectic matrix given by*

$$(s_W)^{-1} = s_{W^*} \quad , \; W^*(x, x') = -W(x', x). \tag{6.12}$$

Proof. To prove (6.12), it suffices to note that the inverse of the symplectic matrix s_W is

$$(s_W)^{-1} = \begin{pmatrix} (L^{-1})^T P & -(L^{-1})^T \\ -Q(L^{-1})P + L^T & Q(L^{-1})^T \end{pmatrix}$$

which shows that the inverse matrix $(s_W)^{-1}$ is associated to the quadratic form W^* obtained from W by changing the triple of matrices (P, L, Q) into the new triple $(-Q, -L^T, -P)$. ∎

Notation 177 *We will use the shorthand notation $W = (P, L, Q)$ for quadratic forms (6.9).*

The fact that there is a one-to-one correspondence between the set of all $W = (P, L, Q)$ and the set $\mathrm{Sp}_0(n)$ of free matrices can be used to count "how many" free matrices there are in $\mathrm{Sp}(n)$. In fact:

Proposition 178 *The subset $\mathrm{Sp}_0(n)$ of $\mathrm{Sp}(n)$ consisting of all free symplectic matrices is a submanifold with dimension $(n+1)(2n-1)$ of $\mathrm{Sp}(n)$. Thus $\mathrm{Sp}_0(n)$ has codimension 1 in $\mathrm{Sp}(n)$ (and has therefore measure zero).*

Proof. Topologically $\mathrm{Sp}_0(n)$ and the set of all $W = (P, L, Q)$ are identical. The latter being essentially the product

$$Sym(n, \mathbb{R}) \times GL(n, \mathbb{R}) \times Sym(n, \mathbb{R}) \equiv \mathbb{R}^{(n+1)(2n-1)}$$

($\mathrm{Sym}(n,\mathbb{R})$ the real symmetric $n\times n$ matrices), it follows that the free symplectic matrices form a submanifold of dimension $(n+1)(2n-1)$ in the symplectic group $\mathrm{Sp}(n)$ and hence, since $\mathrm{Sp}(n)$ has dimension $n(2n+1)$:

$$\dim \mathrm{Sp}(n) - \dim \mathrm{Sp}_0(n) = 1$$

as claimed. ∎

It follows from Proposition 178 that free symplectic matrices are the overwhelming majority in $\mathrm{Sp}(n)$, and that those who are not are the exception!

6.2.2 The Case of Affine Symplectomorphisms

Recall from Subsection 3.5 that $\mathrm{ISp}(n)$ denotes the inhomogeneous symplectic group, that is, the group of all symplectomorphisms

$$\tau(z_0)\circ s = s\circ\tau(s^{-1}z_0)$$

where $s\in\mathrm{Sp}(n)$ and $\tau(z_0)$ is the phase space translation $z\longmapsto z_0$. We identified $\mathrm{ISp}(n)$ with the group of all matrices

$$\langle s, z_0\rangle \equiv \begin{pmatrix} s & z_0 \\ 0_{1\times 2n} & 1 \end{pmatrix}.$$

The following result characterizes the generating functions of the free elements of $\mathrm{ISp}(n)$:

Proposition 179 *An affine symplectomorphism $\langle s, z_0\rangle$ is free if and only if s is free. A free generating function of $f=\tau(z_0)\circ s_W$ ($z_0=(x_0,p_0)$) is the inhomogeneous quadratic polynomial*

$$W_{z_0}(x,x') = W(x-x_0,x') + p_0\cdot x \tag{6.13}$$

where W is a free generating function for s. Conversely, if W is the generating function of a symplectic transformation s, then any polynomial

$$W_{z_0}(x,x') = W(x,x') + \alpha\cdot x + \alpha'\cdot x' \tag{6.14}$$

($\alpha,\alpha'\in\mathbb{R}^n_x$) is a generating function of an affine symplectic transformation, the translation vector $z_0=(x_0,p_0)$ being

$$(x_0,p_0) = (B\alpha, Da+\beta) \tag{6.15}$$

when s is written in block-matrix form (6.4).

Proof. Let W_{z_0} be defined by (6.13), and set $(x', p') = s(x'', p'')$, $(x, p) = \tau(z_0)(x', p')$. We have

$$\begin{aligned} pdx - p'dx' &= (pdx - p''dx'') + (p''dx'' - p'dx') \\ &= (pdx - (p - p_0)d(x - x_0) + dW(x'', x') \\ &= d(p_0 \cdot x + W(x - x_0, x')) \end{aligned}$$

which shows that W_{z_0} is a generating function. Finally, formula (6.15) is obtained by a direct computation, expanding $W(x - x_0, x')$. ■

Corollary 180 *Let $f = \langle s_W, z_0 \rangle$ be a free affine symplectic transformation, and set $(x, p) = f(x', p')$. The function Φ_{z_0} defined by*

$$\Phi_{z_0}(x, x') = \tfrac{1}{2}(px - p'x') + \tfrac{1}{2}\Omega(z, z_0) \tag{6.16}$$

(Ω the symplectic form) is also a free generating function for f; in fact:

$$\Phi_{z_0}(x, x') = W_{z_0}(x, x') + \tfrac{1}{2}p_0 \cdot x_0. \tag{6.17}$$

Proof. Setting $(x'', p'') = s(x, p)$, the generating function W satisfies

$$W(x'', x') = \tfrac{1}{2}(p'' \cdot x'' - p' \cdot x')$$

in view of (6.11). Let Φ_{z_0} be defined by formula (6.16); in view of (6.13), we have

$$W_{z_0}(x, x') - \Phi_{z_0}(x, x') = \tfrac{1}{2}p_0 \cdot x - \tfrac{1}{2}p \cdot x_0 - \tfrac{1}{2}p_0 \cdot x_0$$

which is (6.17); this proves the corollary since all generating functions of a symplectomorphism are equal up to an additive constant. ■

6.2.3 The Generators of Sp(n)

We next study the relationship between free symplectic matrices and the generators of the symplectic group.

Lemma 181 *Let s_W and $s_{W'}$ be two free symplectic matrices, associated to $W = (P, L, Q)$ and $W' = (P', L', Q')$. Their product $s_W s_{W'}$ is a free symplectic matrix $s_{W''}$ if and only if*

$$\det(P' + Q) \neq 0 \tag{6.18}$$

in which case we have $W'' = (P'', L'', Q'')$ *with*

$$\begin{cases} P'' = P - L^T(P' + Q)^{-1}L \\ L'' = L'(P' + Q)^{-1}L \\ Q'' = Q' - L'(P' + Q)^{-1}L'^T \end{cases} . \tag{6.19}$$

Proof. In view of Eq. (6.10) the product $s_W s_{W'}$ is given by

$$\begin{pmatrix} L^{-1}Q & L^{-1} \\ PL^{-1}Q - L^T & PL^{-1} \end{pmatrix} \begin{pmatrix} L'^{-1}Q' & L'^{-1} \\ P'L'^{-1}Q' - L'^T & P'L'^{-1} \end{pmatrix}$$

and performing the matrix multiplication, the right upper-corner of that product is $L^{-1}(P' + Q)L'^{-1}$ which is invertible if and only if (6.18) holds. If it holds, set

$$s_{W''} = \begin{pmatrix} L''^{-1}Q'' & L''^{-1} \\ P''L''^{-1}Q'' - L''^T & P''L''^{-1} \end{pmatrix}$$

and solve successively for P'', L'', Q''. ■

Let P and L be two $n \times n$ matrices, P symmetric and L invertible. It immediately follows from the conditions (6.5) that the following matrices are symplectic:

$$v_P = \begin{pmatrix} I & 0 \\ -P & I \end{pmatrix}, \quad m_L = \begin{pmatrix} L^{-1} & 0 \\ 0 & L^T \end{pmatrix}. \tag{6.20}$$

We will prove that the set of all matrices v_{Pp}, m_L, together with the matrix

$$J = \begin{pmatrix} 0 & I \\ -I & 0 \end{pmatrix}$$

generates the symplectic group $\mathrm{Sp}(n)$. To my knowledge, there are at least four proofs of this fact. One can either use a topological argument (see for instance de Gosson [59] or Wallach [160]) or elementary linear algebra, as in the first Chapter of Guillemin-Sternberg [80] (but the calculations are then rather complicated). One can also use methods from the theory of Lie groups (see for instance Mneimé and Testard [118]). We are going to present a fourth method, which consists of using the properties of free symplectic matrices we have developed above. This approach has the advantage of giving a rather straightforward factorization of an arbitrary symplectic matrix. We begin with two preparatory results:

Lemma 182 *Every free symplectic matrix s_W can be (uniquely) written as a product*

$$s_W = v_{-P} m_L J v_{-Q} \tag{6.21}$$

where v_{-P}, v_{-Q}, and m_L are defined by (6.20).

Proof. Performing the product on the right-hand side of Eq. (6.21) we get

$$s_W = \begin{pmatrix} L^{-1}Q & L^{-1} \\ PL^{-1}Q - L^T & PL^{-1} \end{pmatrix}$$

(*cf.* (6.10)). Writing s_W in the usual block-matrix form (6.4), we get the following equations for P, L and Q:

$$\begin{cases} A = L^{-1}Q, \ \ B = L^{-1} \\ C = PL^{-1}Q - L^T, \ \ D = PL^{-1}. \end{cases}$$

Since B is invertible, we get $L = B^{-1}$, $P = DB^{-1}$, $Q = B^{-1}A$. ■

The following important Lemma can, in principle, be proven by a direct calculation. But we will rather give a neat geometrical proof, using the notion of Lagrangian plane (see Appendix A).

Lemma 183 *Every symplectic matrix is the product of two free symplectic matrices.*

Proof. Set $\ell = 0 \times \mathbb{R}^n_p$ and choose ℓ' transversal to both ℓ and $s\ell$:

$$\ell' \cap \ell = \ell' \cap s\ell = 0.$$

Since $\mathrm{Sp}(n)$ acts transitively on pairs of transversal Lagrangian planes (see Appendix A), there exists $s_1 \in \mathrm{Sp}(n)$ such that $(s\ell, \ell') = s_1(\ell', \ell)$, and we can find $s_2' \in \mathrm{Sp}(n)$ such that $\ell' = s_2'\ell$. Hence $s\ell = s_1 s_2' \ell$, and we have $s = s_1 s_2' h$ for some $h \in \mathrm{Sp}(n)$ such that $h\ell = \ell$. Now, s_1 and $s_2 = s_2' h$ satisfy

$$s_1\ell \cap \ell = \ell' \cap \ell = 0 \ \ , \ \ s_2\ell \cap \ell = s_2'\ell \cap \ell = \ell' \cap \ell = 0$$

and are hence free symplectic matrices. ■

Combining the two Lemmas above we get:

Proposition 184 *The matrices v_P, m_L and J generate the symplectic group* $\mathrm{Sp}(n)$.

Proof. In view of Lemma 183, every $s \in \mathrm{Sp}(n)$ can be written as a product $s_W s_{W'}$ and hence, by formula (6.21) in Lemma 182,

$$s = v_{-P} m_L J v_{-(P'+Q)} m_{L'} J v'_{-Q} \tag{6.22}$$

which shows that s is a product matrix of the type v_P, m_L and J. ∎

6.3 The Metaplectic Group $\mathrm{Mp}(n)$

We begin by defining the "quadratic Fourier transforms" associated with free symplectic matrices (or, rather, their generating functions).

6.3.1 Quadratic Fourier Transforms

The unitary Fourier transform F is defined by

$$F\psi(x) = \left(\tfrac{1}{2\pi i}\right)^{n/2} \int e^{ixx'} \psi(x')\, d^n x' \tag{6.23}$$

(ψ in the Schwartz space $\mathcal{S}(\mathbb{R}^n_x)$; the integral is calculated over $\mathbb{R}^n_x$). In this formula the argument of i is $\pi/2$, so the normalizing factor in front of the integral should be interpreted as

$$\left(\tfrac{1}{2\pi i}\right)^{n/2} = \left(\tfrac{1}{2\pi}\right)^{n/2} e^{-in\pi/4}. \tag{6.24}$$

The inverse F^{-1} of F is given by

$$F^{-1}\psi(x) = \left(\tfrac{i}{2\pi}\right)^{n/2} \int e^{-ixx'} \psi(x')\, d^n x' \tag{6.25}$$

that is

$$F^{-1}\psi = (F\psi^*)^* \tag{6.26}$$

where the star * denotes complex conjugation. Both F and F^{-1} are unitary Fourier transforms, in the sense that

$$||F\psi||_{L^2} = ||\psi||_{L^2} \tag{6.27}$$

where $||\cdot||_{L^2}$ is the usual norm on $L^2(\mathbb{R}^n_x)$, defined by

$$||\psi||^2_{L^2} = \int \psi(x)\psi^*(x) d^n x.$$

Let now s_W be a free symplectic matrix, with $W = (P, L, Q)$. Recall that P and Q are symmetric and that

$$\operatorname*{Hess}_{x,x'}(-W) = \det L \neq 0 \ .$$

To s_W we associate the *two* "quadratic Fourier transforms" defined, for ψ in the Schwartz space $\mathcal{S}(\mathbb{R}^n_x)$, by the formula

$$S_{W,m}\psi(x) = \left(\frac{1}{2\pi i}\right)^{n/2} \Delta(W) \int_{\mathbb{R}^n_x} e^{iW(x,x')}\psi(x')\, d^n x' \tag{6.28}$$

where

$$\Delta(W) = i^m \sqrt{|\det L|}. \tag{6.29}$$

The integers m are defined by the condition

$$\arg\det L = m\pi \mod 2\pi \tag{6.30}$$

Thus

$$\begin{cases} m \text{ is even if } \det L > 0 \\ m \text{ is odd if } \det L < 0. \end{cases} \tag{6.31}$$

Formula (6.30) can be written as

$$m = \arg \operatorname*{Hess}_{x,x'}(-W). \tag{6.32}$$

The formulae above motivate the following definition:

Definition 185 *A choice of an integer m satisfying (6.31) is called a* Maslov index *of the quadratic form $W = (P, L, Q)$. Thus, exactly two Maslov indices modulo 4 are associated to each W, namely m and $m+2$. The Maslov index of a quadratic Fourier transform $S_{W,m}$ is then, by definition, the integer m; we write $m(S_{W,m}) = m$.*

In particular,

$$m(F) = 0 \ , \ m(F^{-1}) = n \tag{6.33}$$

since we obviously have $F = S_{(0,I,0),0}$ and $F^{-1} = S_{(0,-I,0),n}$.

We will see later in this Chapter the relation between the Maslov index as defined above, and the Maslov index for Lagrangian paths.

Notation 186 *We will denote by $\aleph$ the set of all pairs (W, m), where m is one of the two integers modulo 4 defined by (6.30).*

It is time now to define the metaplectic group:

Definition 187 *The metaplectic group* Mp(n) *is the set of all products*

$$S = S_{W_1,m_1} \cdots S_{W_k,m_k}$$

of a finite number of quadratic Fourier transforms.

Notice that it is not at all clear from this definition that Mp(n) really is a group! That Mp(n) is a semi-group is clear (it is closed under multiplication, and it contains the identity operator I, because $I = FF^{-1}$), but one does not immediately see why the inverse of a product of two quadratic Fourier transforms should also be such a product.

We will actually see that:

(1) Mp(n) is a Lie group;

(2) There exists a "two-to-one" and "onto" homomorphism $\Pi : \mathrm{Mp}(n) \longrightarrow \mathrm{Sp}(n)$ ("the projection") whose kernel $\Pi^{-1}(I)$ consists of the two elements $\pm I$ of Mp(n).

Remark 188 *It turns out that* Mp(n) *is also connected, but the proof of this property is beyond the scope of this book; the reader who wants to find out how this is done is referred to Leray [101] or de Gosson [59].*

6.3.2 The Operators $M_{L,m}$ and V_P

We defined in Subsection 6.2.3 (formulae (6.20)) the symplectic matrices m_L and v_P, and we proved that these matrices, together with J, generate Sp(n). We are going to see that, similarly, Mp(n) is generated by operators $M_{L,m}$ and V_P together with the Fourier transform J.

The operators $S_{W,m}$ are essentially Fourier transforms as the denomination "quadratic Fourier transform" is intended to suggest. To make this statement more precise, we define operators $M_{L,m}$ and V_{-P} acting on functions ψ by

$$M_{L,m}\psi(x) = i^m \sqrt{|\det L|}\psi(Lx) \tag{6.34}$$

where L is an invertible $n \times n$ matrix, the integer m being defined, as in (6.30), by

$$\arg \det L = m\pi \mod 2\pi$$

and

$$V_P\psi(x) = \exp\left(-\frac{i}{2}Px^2\right)\psi(x) \tag{6.35}$$

where P is a symmetric $n \times n$ matrix. The operators $M_{L,m}$ and V_P have the obvious group properties

$$M_{L,m}M_{L',m'} = M_{L'L,m+m'}, \quad (M_{L,m})^{-1} = M_{L^{-1},-m} \tag{6.36}$$

(beware the ordering in the first formula: it is $L'L$, and *not* LL', that appears on the right-hand side), and

$$V_PV_{P'} = V_{P+P'}, \quad (V_P)^{-1} = V_{-P} \ . \tag{6.37}$$

Remark 189 *The group $M\ell(n) = \{M_{L,m} : \det L \neq 0\}$ is sometimes called the "metalinear group" in the literature. It plays an important role in various areas of geometric quantization. (See Guillemin-Sternberg [79], Woodhouse [169] and the references therein.)*

In addition to (6.36), (6.37) the operators $M_{L,m}$, V_P satisfy the following intertwining relations (the proofs are omitted, because they are pedestrian):

$$\begin{cases} M_{L,m}V_P = V_{L^TPL}M_{L,m} \\ V_PM_{L,m} = M_{L,m}V_{(L^T)^{-1}PL^{-1}} \end{cases} \quad \begin{cases} FM_{L,m} = M_{(L^T)^{-1},m}F \\ F^{-1}M_{L,m} = M_{(L^T)^{-1},m}F^{-1}. \end{cases} \tag{6.38}$$

These formulae are very useful when one has to perform products of quadratic Fourier transforms. They will, for instance, allow us to find very easily the inverse of $S_{W,m}$. Let us first prove the following factorization result (*cf.* Lemma 182):

Lemma 190 *Let $W = (P, L, Q)$ be a quadratic form, and m an integer defined by*

$$\arg\det L = m\pi \mod 2\pi \tag{6.39}$$

(that is $(W, m) \in \aleph$, with Notation 186). We have the factorization

$$S_{W,m} = V_{-P}M_{L,m}FV_{-Q}. \tag{6.40}$$

Proof. Since $V_PV_{-P} = I$ we have, by definition of $S_{W,m}$:

$$V_PS_{W,m}V_Q\psi(x) = \left(\tfrac{1}{2\pi i}\right)^{n/2}\sqrt{\det L}\int_{\mathbb{R}^n_x} e^{-iLx\cdot x'}\psi(x')\,d^nx'$$

that is

$$V_P S_{W,m} V_Q \psi = M_{L,m} F \psi.$$

It follows that we have $V_P S_{W,m} V_Q = M_{L,m} F$, hence Eq. (6.40) since $(V_P)^{-1} = V_{-P}$ and $(V_Q)^{-1} = V_{-Q}$. ∎

Lemma 190 allows us to prove very easily that the inverse of a quadratic Fourier transform is also a quadratic Fourier transform:

Corollary 191 *The inverse of the quadratic Fourier transform $S_{W,m}$ is given by:*

$$(S_{W,m})^{-1} = S_{W^*,m^*} \text{ with } \begin{cases} W^*(x,x') = -W(x',x) \\ m^* = n - m \mod 4. \end{cases} \tag{6.41}$$

(That is, if $W = (P, L, Q)$, then $W^* = (-Q, -L^T, -P)$.)

Proof. In view of (6.40) we have

$$(S_{W,m})^{-1} = V_Q F^{-1} (M_{L,m})^{-1} V_P$$

and a straightforward calculation, using the formulas in (6.38), shows that

$$F^{-1}(M_{L,m})^{-1} = F^{-1} M_{L^{-1},m} = M_{L^T,m} F^{-1} = M_{-L^T,n-m} F$$

and hence

$$(S_{W,m})^{-1} = V_Q M_{-L^T,n-m} F V_P$$

from which (6.41) immediately follows. ∎

It follows from this result that $\mathrm{Mp}(n)$ is a group: as we have seen, $\mathrm{Mp}(n)$ contains the identity (in fact $S_{W,m} S_{W^*,m^*} = S_{W,m}(S_{W,m})^{-1} = I$), and the inverse of a product $S = S_{W_1,m_1} \cdots S_{W_k,m_k}$ is

$$S^{-1} = S_{W_k^*,m_k^*} \cdots S_{W_1^*,m_1^*}$$

and is hence also in $\mathrm{Mp}(n)$. It also follows from the factorization formula (6.40) that the elements of $\mathrm{Mp}(n)$ are unitary operators acting on the space $L^2(\mathbb{R}^n_x)$ of square integrable functions: the operators V_{-P}, $M_{L,m}$ and F are defined on the Schwartz space $\mathcal{S}(\mathbb{R}^n_x)$ and are obviously unitary for the L^2-norm, hence so is $S_{W,m}$. Every $S \in \mathrm{Mp}(n)$ being, by definition, a product of the $S_{W,m}$ is therefore also a unitary operator on $L^2(\mathbb{R}^n_x)$.

Remark 192 *It follows from the discussion above that the metaplectic group* $\mathrm{Mp}(n)$ *is generated by the set of all operators* $M_{L,m}$, V_P *together with the Fourier transform* F. *We could in fact have defined* $\mathrm{Mp}(n)$ *as being the group generated by these "simpler" operators, but this would have led to more inconveniences than advantages, because we would then have lost the canonical relationship between free symplectic matrices and quadratic Fourier transforms.*

It follows from Lemma 181 about products of free symplectic matrices that we have the following criterion for deciding whether a product of quadratic Fourier transforms is itself a quadratic Fourier transform:

Lemma 193 *We have* $S_{W,m}S_{W',m'} = S_{W'',m''}$ *for some* $(W'', m'') \in \aleph$ *if and only if*

$$\det(P' + Q) \neq 0 \tag{6.42}$$

and in this case $W'' = (P'', L'', Q'')$ *is given by Eq. (6.19).*

(The proof of this result is somewhat technical, and will not be given here; see Leray [101] or de Gosson [55, 59].)

6.3.3 The Metaplectic Group and Wave Optics

In Chapter 3, Section 3.6 we discussed the occurrence of symplectic matrices in optics. We were actually at this stage only considering light as being made of "corpuscles", whose trajectories were the rays of geometrical optics. We were thus totally ignoring the wave-like behavior of light. It turns out that we "obtain" physical optics (i.e. the optics that takes into account the phenomena of diffraction and interference) by using $\mathrm{Sp}(n)$'s companion group $\mathrm{Mp}(n)$. In Chapter 3, Section 3.6 we discussed the corpuscular nature of light, and we showed that the motion of the light corpuscles was governed by symplectic geometry. Various experiences show that light actually also has a wave-like behavior; this leads us to postulate that there is a "wave function" Ψ, which we propose to determine. We write such a wave function in polar form

$$\Psi(x, t) = a(x, t) \exp\left(\frac{2\pi i}{\lambda}\Phi(x, t)\right)$$

where λ is the wavelength of the light, and set out to determine how the values of Ψ at two different reference lines t' and t are related. We argue as follows: as light propagates from the point x' to the point x, it will undergo both a phase change and attenuation. The phase change, expressed in radians, is simply

$$\Delta\Phi = \frac{2\pi L}{\lambda}$$

where L is the optical length of the trajectory from x' to x; using the expression (4.20) this is

$$\Delta\Phi = 2\pi n \frac{t-t'}{\lambda} + 2\pi \frac{\Delta L}{\lambda}.$$

Taking into account the expression (4.21) of the eikonal, it follows that the light will contribute along the optical path leading from (x',t') to (x,t) by the quantity

$$K \exp\left[\frac{2\pi i}{\lambda}\left(\Phi(x,t) + n(t-t') + n\frac{(x-x')^2}{2(t-t')}\right)\right]$$

where K is an attenuation factor, to be determined. We begin by noting that the net contribution at x of all these terms is obtained by integrating over x'. Assume from now on that the index of refraction n is constant in position and time (it is, for instance, equal to one in vacuum); then K does not depend on x or x', and we thus have:

$$\Psi(x,t) = K \int \exp\left[\frac{2\pi i}{\lambda}\left(n(t-t') + n\frac{(x-x')^2}{2(t-t')}\right)\right] \Psi(x',t')\, dx'.$$

To calculate K, we remark that since the total intensity of light must be the same on the t-plane as that on the initial $(t'=0)$-line, we must have

$$\int |a(x,t)|^2\, dx = \int |a(x',t')|^2\, dx'$$

and this condition leads, after some calculations, to

$$|K|^2 = \frac{n}{\lambda(t-t')}$$

and we can fix the argument of K by requiring that $\Psi(x,t) \to \Psi(x,t')$ as $t \to t'$. Using for instance the method of stationary phase, or the theory of Fresnel integrals, this finally yields the value

$$K = e^{ia(t)}\left(\frac{n}{\lambda t}\right)^{1/2} \quad , \quad \arg t = \begin{cases} 0 & \text{for } t > 0 \\ 1 & \text{for } t < 0 \end{cases}$$

where $\alpha(t)$ is some undetermined continuous function, vanishing at $t = 0$. The phase factor $e^{ia(t)}$ can be determined by the following argument. The mapping, which to every t associates the operators S_t, is the "lift" to the metaplectic group of the mapping $t \mapsto s_t$. The matrices s_t obviously satisfying the group property $s_t s_{t'} = s_{t+t'}$, it follows that we must also have $S_t S_{t'} = S_{t+t'}$, and it is not difficult to show that this is only possible if one makes the choice $a(t) = 0 \bmod 2\pi$. Neglecting the term $\exp(2\pi int/\lambda)$, we choose for "wave function"

$$\Psi(x,t) = \left(\frac{n}{i\lambda}\right)^{1/2} \int_{-\infty}^{\infty} \exp\left[2\pi in \frac{(x-x')^2}{2\lambda t}\right] \Psi_0(x')\, dx' \tag{6.43}$$

and a straightforward computation of partial derivatives shows that it satisfies the partial differential equation

$$i\lambda \frac{\partial \Psi}{\partial t} = -\frac{\lambda^2}{4\pi} \frac{\partial^2 \Psi}{\partial x^2}.$$

Formally, this is exactly Schrödinger's equation for a free particle with unit mass if one replaces λ by Planck's constant h. Formula (6.43) can be written

$$\Psi(x,t) = \int_{-\infty}^{\infty} G(x,x',t)\Psi_0(x')\, dx'$$

where the kernel

$$G(x,x',t) = \left(\frac{n}{i\lambda}\right)^{1/2} \exp\left[2\pi in \frac{(x-x')^2}{2\lambda t}\right]$$

is viewed as a "point source" of particles emanating from x' at time 0.

The theory of the metaplectic group thus allows us to associate to a family of optical matrices, a family of operators acting on the wave functions of optics. It turns out that this analogy can be used to derive Schrödinger's equation, and hence quantum mechanics, from classical mechanics, if one makes the assumption that to every material particle with mass m and velocity v is associated a wave, whose length is given by the de Broglie relation $\lambda = h/mv$, where h is Planck's constant. This will be done in the next Chapter.

6.4 The Projections Π and Π^ε

Let us now investigate more precisely the relation between $\mathrm{Mp}(n)$ and its "little brother" $\mathrm{Sp}(n)$. It will result in the construction of a projection

$\Pi : \mathrm{Mp}(n) \longrightarrow \mathrm{Sp}(n)$ which is a surjective group homomorphism with kernel $\{-I, +I\}$.

6.4.1 Construction of the Projection Π

Recall that the operators $S_{W,m}$ are associated to the free symplectic matrices s_W by the integral formula (6.28). The relation between the set $\mathrm{Sp}_0(n)$ of all free symplectic matrices, and the set $\mathrm{Mp}_0(n)$ of all quadratic Fourier transforms is two-to-one: to every s_W are associated exactly two elements $(W, m) \in \aleph$ (see notation 186), corresponding to the two possible choices (modulo 4π) of the argument of the determinant $\mathrm{Hess}_{x,x'}(W)$ of the matrix of second derivatives of W.

Notation 194 *We denote by Π_0 the mapping $\mathrm{Mp}_0(n) \longrightarrow \mathrm{Sp}_0(n)$ which to each quadratic Fourier transform $S_{W,m}$ associates the free symplectic matrix s_W.*

The mapping Π_0 has the following properties, which makes it a perfect candidate for being a "partial" projection:

Lemma 195 *The mapping $\Pi_0 : \mathrm{Mp}_0(n) \longrightarrow \mathrm{Sp}_0(n)$ satisfies*

$$\begin{cases} \Pi_0((S_{W,m})^{-1}) = (s_W)^{-1} \\ \Pi_0(S_{W,m}S_{W',m'}) = s_W s_{W'} \end{cases} \tag{6.44}$$

when $s_W s_{W'}$ is itself a free symplectic matrix.

Proof. Recall that $S_{W,m}^{-1} = S_{W^*,m^*}$ where $W^*(x,x') = -W(x',x)$ and $m^* = n - m$ (see Corollary 191). If $s_{W^*} = \Pi_0(S_{W^*,m^*})$ then

$$(x,p) = s_{W^*}(x',p') \Longleftrightarrow \begin{cases} p = \nabla_x W^*(x,x') \\ p' = -\nabla_{x'} W^*(x,x') \end{cases}$$

that is

$$(x,p) = s_{W^*}(x',p') \Longleftrightarrow \begin{cases} p = -\nabla_{x'} W(x,x') \\ p' = \nabla_x W(x,x') \end{cases}$$

and hence $(x,p) = s_{W^*}(x',p')$ is equivalent to $(x',p') = s_W(x',p')$, which proves the first formula in (6.44). The proof of the second formula (6.44) follows from Lemma 193, using Proposition 181; see de Gosson [59] (pages 86–88). ∎

The formulae (6.44) suggest that it might be possible to extend the projection Π_0 to a globally define homomorphism

$$\Pi : \mathrm{Mp}(n) \longrightarrow \mathrm{Sp}(n)$$

which is at the same time a group homomorphism:

Theorem 196 *(1) The mapping Π_0 which to $S_{W,m} \in \mathrm{Mp}(n)$ associates $s_W \in \mathrm{Sp}(n)$ can be extended into a mapping $\Pi : \mathrm{Mp}(n) \longrightarrow \mathrm{Sp}(n)$ such that*

$$\Pi(SS') = \Pi(S)\Pi(S'); \tag{6.45}$$

(2) That mapping Π is determined by the condition

$$(x,p) = \Pi(S_{W,m})(x',p') \Longleftrightarrow \begin{cases} p = \nabla_x W(x,x') \\ p' = -\nabla_{x'} W(x,x') \end{cases}$$

(3) Π is surjective (=onto) and two-to-one, hence Π is a covering mapping, and $\mathrm{Mp}(n)$ is a double cover of $\mathrm{Sp}(n)$.

Proof. We will be rather sketchy in the proof of the first part of the Proposition (the reader wanting to see the complete argument is referred to de Gosson [59]). The obvious idea, if one wants to define $\Pi(S)$ for arbitrary S is to write S as a product of quadratic Fourier transforms:

$$S = S_{W_1,m_1} \cdots S_{W_k,m_k}$$

and then to simply define the projection of S by the formula

$$\Pi(S) = s_{W_1} \cdots s_{W_k}.$$

There is however a true difficulty here, because one has to show that the right hand side does not depend on the way we have factored S (the factorization of an element of $\mathrm{Mp}(n)$ is never unique: for instance, the identity operator can be written in infinitely many ways as $S_{W,m}(S_{W,m})^{-1} = S_{W,m}S_{W^*,m^*}$). However, once this is done, formula (6.45) showing that Π is a homomorphism is straightforward. Let us show in detail the last part of the theorem, namely that Π is onto and two-to-one. Let $s = s_{W_1} \cdots s_{W_k}$ be an arbitrary element of $\mathrm{Sp}(n)$. Then, for any choice of $(W_j, m_j) \in \aleph$ $(1 \leq j \leq k)$, we have

$$\Pi(S_{W_1,m_1} \cdots S_{W_k,m_k}) = s_{W_1} \cdots s_{W_k}$$

and hence Π is onto. Let us next prove that Π is two-to-one. For this it suffices to show that $\mathrm{Ker}(\Pi) = \{\pm I\}$. Now, the inclusion $\mathrm{Ker}(\Pi) \supset \{\pm I\}$ is rather obvious. In fact, $\Pi(I) = I$ and

$$\Pi(-I) = \Pi(S_{W,m}S_{W^*,m^*}) = s_W s_{W^*} = I.$$

Let us prove the opposite inclusion $\mathrm{Ker}(\Pi) \subset \{\pm I\}$ by induction on the number k of terms in the factorization $S = S_{W_1,m_1} \cdots S_{W_k,m_k}$. We first note that if $\Pi(S) = I$, then we must have $k \geq 2$, because the identity is not a free symplectic matrix. Suppose next that

$$\Pi(S_{W,m}S_{W',m'}) = I.$$

Then, either $s_W s_{W'} = I$, or $s_W s_{W'} = -I$, depending on whether $m' = m^*$ or $m' = m^* + 2$. This establishes the result when $k = 2$. Suppose now that we have proven the implication

$$\left.\begin{aligned} S = S_{W_1,m_1} \cdots S_{W_k,m_k} \\ \Pi(S) = I \end{aligned}\right\} \Rightarrow S = \pm I,$$

and let us prove that we then have

$$\left.\begin{aligned} S' = S_{W_1,m_1} \cdots S_{W_{k+1},m_{k+1}} \\ \Pi(S') = I \end{aligned}\right\} \Rightarrow S' = \pm I.$$

Since Π is a group homomorphism, we can write

$$\Pi(S') = \Pi(S)\Pi(S_{W'_{k+1},m_{k+1}}) = s s_{W'_{k+1}}$$

and it then follows from the induction assumption that we must have either $S = S_{W^*_{k+1},m^*_{k+1}}$ or $S = -S_{W^*_{k+1},m^*_{k+1}}$. But this means that we have either

$$S' = S_{W^*_{k+1},m^*_{k+1}} S_{W_{k+1},m_{k+1}} = I$$

or

$$S' = S_{W^*_{k+1},m^*_{k+1}} S_{W^*_{k+1},m^*_{k+1}} = -I$$

which concludes the proof of the inclusion $\mathrm{Ker}(\Pi) \subset \{\pm I\}$. ■

Every symplectic matrix is the product of two free symplectic matrices, and a similar result holds for operators in $\mathrm{Mp}(n)$ replacing the phrase "free symplectic matrix" by "quadratic Fourier transform":

Proposition 197 *Every $S \in \mathrm{Mp}(n)$ is the product of two quadratic transforms $S_{W,m}$ and $S_{W',m'}$.*

Proof. Let S be an arbitrary element of $\mathrm{Mp}(n)$, $s = \Pi(S)$ its projection. Set $s = s_W s_{W'}$ and let $S_{W,m}$, $S_{W',m'}$ be two quadratic Fourier transforms with projections s_W and $s_{W'}$:

$$\Pi(S_{W,m}) = s_W \text{ and } \Pi(S_{W',m'}) = s_{W'}.$$

Then $S = S_{W,m}S_{W',m'}$ or $S = -S_{W,m}S_{W',m'} = S_{W,m+2}S_{W',m'}$. Either way S can be written as the product of two quadratic Fourier transforms. ∎

Corollary 198 *The projections of the operators $F, M_{L,m}$ and V_p are*

$$\Pi(F) = J \quad , \quad \Pi(M_{L,m}) = m_L \quad , \quad \Pi(V_p) = v_p. \tag{6.46}$$

Proof. The fact that $\Pi(F) = J$ is obvious because $F = S_{(0,I,0),0}$. To prove that $\Pi(M_{L,m}) = m_L$ it suffices to note that the equality

$$\Pi(M_{L,m}F) = \Pi(M_{L,m})\Pi(F)$$

implies that we have $\Pi(M_{L,m}) = \Pi(M_{L,m}F)J^{-1}$ and hence

$$\Pi(M_{L,m}) = \begin{pmatrix} 0 & L^{-1} \\ -L^T & 0 \end{pmatrix}\begin{pmatrix} 0 & -I \\ I & 0 \end{pmatrix} = \begin{pmatrix} L^{-1} & 0 \\ 0 & L^T \end{pmatrix}.$$

The equality $\Pi(V_P) = v_P$ is proven in a similar fashion, using for instance the equality $\Pi(V_{-P}F) = \Pi(V_{-p})J$. ∎

6.4.2 The Covering Groups $\mathrm{Mp}^{\varepsilon}(n)$

The metaplectic group $\mathrm{Mp}(n)$ endowed with the projection Π is a twofold covering group of $\mathrm{Sp}(n)$. However, a covering group can be "realized" in many different ways. Instead of choosing Π as a projection, we could as well have chosen any other mapping $\mathrm{Mp}(n) \longrightarrow \mathrm{Sp}(n)$ obtained from Π by composing it on the left with an inner automorphism of $\mathrm{Mp}(n)$, or on the right with an inner automorphism of $\mathrm{Sp}(n)$, or both. The essential point is here that the diagram

$$\begin{array}{ccc} \mathrm{Mp}(n) & \xrightarrow{F} & \mathrm{Mp}(n) \\ \Pi\downarrow & & \downarrow\Pi' \\ \mathrm{Sp}(n) & \underset{G}{\longrightarrow} & \mathrm{Sp}(n) \end{array}$$

is commutative, i.e. that we have $\Pi' \circ F = G \circ \Pi$, because for all such Π' we will have

$$\operatorname{Ker}(\Pi') = \{\pm I\}$$

and Π' will then be another honest covering mapping. We find it particularly convenient to define a new projection as follows. Set, for $\lambda > 0$:

$$M_\lambda = M_{\lambda I,0}. \tag{6.47}$$

($M_\lambda \in \operatorname{Mp}(n)$ is thus a "scaling operator" acting on functions on configuration space); we denote by m_λ $(= m_{\lambda I,0})$ its projection on $\operatorname{Sp}(n)$. Let now ε be a constant > 0 (for instance Planck's constant $\hbar$), and define

$$S^\varepsilon = M_{1/\sqrt{\varepsilon}} S M_{\sqrt{\varepsilon}} \tag{6.48}$$

for $S \in \operatorname{Mp}(n)$. The projection of S^ε on $\operatorname{Sp}(n)$ is then given by:

$$\Pi\left(S^\varepsilon\right) = s^\varepsilon = m_{1/\sqrt{\varepsilon}} s m_{\sqrt{\varepsilon}}. \tag{6.49}$$

Now, we would like, for reasons that will become clear later, to have a projection of $\operatorname{Mp}(n)$ onto $\operatorname{Sp}(n)$ that to S^ε associates, not s^ε, but rather s itself. This can be achieved by defining the new projection

$$\Pi^\varepsilon : \operatorname{Mp}(n) \longrightarrow \operatorname{Sp}(n)$$

by the formula

$$\Pi^\varepsilon\left(S^\varepsilon\right) = m_{\sqrt{\varepsilon}}\left(\Pi\left(S^\varepsilon\right)\right) m_{1/\sqrt{\varepsilon}} \tag{6.50}$$

which is of course equivalent to

$$\Pi^\varepsilon\left(S^\varepsilon\right) = \Pi(S). \tag{6.51}$$

Defining the "ε-quadratic Fourier transform" $S^\varepsilon_{W,m}$ associated with $S_{W,m}$ by

$$S^\varepsilon_{W,m} = M_{1/\sqrt{\varepsilon}} S_{W,m} M_{\sqrt{\varepsilon}} \tag{6.52}$$

we have explicitly

$$S^\varepsilon_{W,m}\psi(x) = \left(\tfrac{1}{2\pi i\varepsilon}\right)^{n/2} \Delta(W) \int e^{\frac{i}{\varepsilon}W(x,x')}\psi(x')\, d^n x'. \tag{6.53}$$

This is easily checked using the fact that W is homogeneous of degree two in (x, x'); the projection of $S^\varepsilon_{W,m}$ on $\operatorname{Sp}(n)$ is then the free matrix s_W:

$$\Pi^\varepsilon(S^\varepsilon_{W,m}) = s_W. \tag{6.54}$$

When we use the covering mapping Π^ε instead of Π, we will talk about the "metaplectic group $\operatorname{Mp}^\varepsilon(n)$". The reader should keep in mind that this is just a convenient way to say that we are using the projection Π^ε instead of Π; of course $\operatorname{Mp}(n)$ and $\operatorname{Mp}^\varepsilon(n)$ are identical as groups!

Remark 199 *It follows from formula (6.52) that if ε and ε' are two positive numbers, then we have*

$$S^{\varepsilon}_{W,m} = M_{\sqrt{\varepsilon'/\varepsilon}} S^{\varepsilon'}_{W,m} M_{\sqrt{\varepsilon/\varepsilon'}}.$$

This means, in the language of representation theory that the representations $\mathrm{Mp}^{\varepsilon}(n)$ and $\mathrm{Mp}^{\varepsilon'}(n)$ are equivalent. On the other hand it is possible to define $\mathrm{Mp}^{\varepsilon'}(n)$ also for $\varepsilon' < 0$, but one can then show that $\mathrm{Mp}^{\varepsilon}(n)$ and $\mathrm{Mp}^{\varepsilon'}(n)$ are inequivalent representations (see Folland, [46], Chapter 4, Theorem 4.5.7).

6.5 The Maslov Index on $\mathrm{Mp}(n)$

We have defined in Chapter 5 a notion of Maslov index for paths in Lagrangian manifolds. We are now going to define an integer-valued function modulo 4 on the metaplectic group, which we also call "Maslov index". Is that to say that we are using a clumsy and confusing terminology? Yes, and no. It is of course somewhat unfortunate to use the same name for two different things. However –whether one deplores it, or not– this is common usage in the literature; it comes from the fact that both indices are closely related: they are, so to say, by-products of a "master object", the Leray index , studied in last Chapter.

By definition, the Maslov index of a quadratic Fourier transform $S_{W,m}$ is the integer m modulo 4. Since every $S \in \mathrm{Mp}(n)$ is the product of operators of the type $S_{W,m}$, it is natural to ask whether it is possible to extend this definition so we can attach a "Maslov index" to an arbitrary element of $\mathrm{Mp}(n)$. We thus want to construct a $\mathbb{Z}_4$-valued function $m(\cdot)$, the "Maslov index on $\mathrm{Mp}(n)$", whose restriction to the quadratic Fourier transforms is given by $m(S_{W,m}) = m$.

We begin by making an essential observation: any continuous path $\gamma : [a,b] \longrightarrow \mathrm{Sp}(n)$ (not necessarily a loop) can be lifted, in infinitely many ways, to a continuous path in $\mathrm{Mp}(n)$. This is achieved by first assigning to each $\gamma(t) = s_t$ the two operators in $\mathrm{Mp}(n)$ with projection s_t, and then by only retaining the operators S_t such that $t \longmapsto S_t$ is continuous. If $\gamma = (s_t)$ is a loop (i.e. if $s_a = s_b$), then the "lift" $t \longmapsto S_t$ ($t \in [a,b]$) will be a loop only if γ "turns" an even number of times around the "hole" in $\mathrm{Sp}(n)$: this simply reflects the fact that $\mathrm{Mp}(n)$ is a double covering of $\mathrm{Sp}(n)$. If we require that the path $t \longmapsto S_t$ passes through a given element of $\mathrm{Mp}(n)$ at some time t_0, then the choice is unique. This "lifting procedure" is of course not specific to the situation we are considering here; in fact, it is a well-known property of homotopy theory (it is called the "path lifting property"). Moreover, if the path $t \longmapsto s_t$ is a one-parameter subgroup of $\mathrm{Sp}(n)$: $s_{t+t'} = s_t s_{t'}$, $s_0 = I$, and if we impose that $S_0 = I$, then $t \longmapsto S_t$ will also be a one-parameter subgroup of $\mathrm{Mp}(n)$: $S_{t+t'} = S_t S_{t'}$.

6.5.1 Maslov Index: a "Simple" Example

We recall that $\mathrm{Sp}(n)$ is a connected Lie group, whose first homotopy group is isomorphic to $\pi_1(S^1) = (\mathbb{Z}, +)$. Intuitively, $\mathrm{Sp}(n)$ thus has a "hole", and to each loop γ in $\mathrm{Sp}(n)$ corresponds an integer $k = k(\gamma)$ only depending on the homotopy class of γ, and counting the number of times the loop "turns around the hole".

Suppose now that $n = 1$, so (s_t) is a one-parameter subgroup of $\mathrm{Sp}(1) = S\ell(2, \mathbb{R})$; we assume that $s_0 = s_{2\pi} = I$. Up to a homotopy, we may assume that

$$s_t = \begin{pmatrix} \cos t & \sin t \\ -\sin t & \cos t \end{pmatrix}$$

which is the flow of the harmonic oscillator

$$H = \frac{1}{2}\left(p^2 + x^2\right).$$

The lift (S_t) of (s_t) passing through $I \in \mathrm{Mp}(n)$ at time $t = 0$ is explicitly given by

$$S_t\psi(x) = \int_{-\infty}^{\infty} G(x, x', t)\psi(x')dx' \tag{6.55}$$

where the kernel G is given by

$$G(x, x', t) = i^{-[t/\pi]}\sqrt{\frac{1}{2\pi\hbar|\sin t|}}\exp\left[\frac{i}{2\hbar\sin t}\left((x^2 + x'^2)\cos t - xx'\right)\right] \tag{6.56}$$

for $t \neq k\pi$ and

$$G(x, x', k\pi) = i^{-k}\delta(x - (-1)^k x') \tag{6.57}$$

(k an integer); by convention $\arg i = \pi/2$ and the brackets $[\cdot]$ denote the "integer part function". These formulae readily follow from the fact that the generating function of s_t is

$$W(x, x', t) = \frac{1}{2\sin t}\left((x^2 + x'^2)\cos t - xx'\right)$$

for $t \neq k\pi$. The factor $i^{-[t/\pi]}$ in (6.56) is there to ensure us that the mapping $t \longmapsto S_t$ is continuous, and that $\lim_{t\to 0} S_t = I$, as desired (see Dittrich and

Reuter, [39], Ch. 16, pages 196–198, for detailed calculations; it turns out that G is the Green function for Schrödinger's equation for the harmonic oscillator; we will return to this property in the next Chapter). Formula (6.56) can be rewritten as

$$G(x,x',t) = \sqrt{\frac{1}{2\pi i\hbar \sin t}} \exp\left[\frac{i}{2\hbar \sin t}\left((x^2+x'^2)\cos t - 2xx'\right)\right] \tag{6.58}$$

provided that we define

$$\arg \underset{x,x'}{\mathrm{Hess}}\left[-W(x,x',t)\right] = \arg\left(\frac{1}{\sin t}\right) = -\left[\frac{t}{\pi}\right]\pi \tag{6.59}$$

for $t \neq k\pi$. We will call the integer

$$m(S_t) = -\left[\frac{t}{\pi}\right]$$

the *Maslov index* of the operator S_t. We notice that in particular $m(S_{k\pi}) = -k$; this is consistent with the appearance of the factor i^{-k} on the right-hand side of (6.57).

In view of the group property of the S_t we have

$$m(S_tS_{t'}) = m(S_{t+t'}) = -\left[\frac{t+t'}{\pi}\right]$$

and hence

$$m(S_tS_{t'}) - m(S_t) - m(S_{t'}) = -\left[\frac{t+t'}{\pi}\right] + \left[\frac{t}{\pi}\right] + \left[\frac{t'}{\pi}\right] \tag{6.60}$$

and this is in general different from zero. To evaluate the right-hand side of this equation, we notice the following property of the integer-part function: for all real numbers a and b that are not integers, we have

$$[a+b] - [a] - [b] = \mathrm{Inert}\left(\frac{\sin(a+b)\pi}{\sin a\pi \sin b\pi}\right) \tag{6.61}$$

where the "index of inertia" $\mathrm{Inert}(\alpha)$ of a real number α is defined as being $+1$ if $\alpha < 0$, and zero otherwise. This is easily proven if one notes that for $k < a < k+1$ and $m < b < m+1$ we have

$$[a+b] - [a] - [b] = \begin{cases} 0 & \text{if } k+m < a+b < k+m+1 \\ 1 & \text{if } k+m+1 < a+b < k+m+2. \end{cases}$$

Formula (6.61) can thus be rewritten

$$m(S_t S_{t'}) = m(S_t) + m(S_{t'}) - \operatorname{Inert}\left(\frac{\sin(t+t')}{\sin t \sin t'}\right)$$

that is

$$m(S_t S_{t'}) = m(S_t) + m(S_{t'}) - \operatorname{Inert}\left(\frac{\cos t}{\sin t} + \frac{\cos t'}{\sin t'}\right).$$

Expressed in terms of the generating function W, this yields

$$m(S_t S_{t'}) = m(S_t) + m(S_{t'}) - \operatorname{Inert}\left[\frac{\partial^2 W}{\partial x'^2}(0, x', t) + \frac{\partial^2 W}{\partial x'^2}(x', 0, t')\right] \quad (6.62)$$

for t and t' non-integer multiples of π.

This formula is the key to the general definition of the Maslov index in the next Subsection.

6.5.2 Definition of the Maslov Index on $\mathrm{Mp}(n)$

We begin by stating two essential lemmas. The first of these lemmas gives us an explicit formula for calculating the Maslov index of a product of two quadratic Fourier transforms when that product is itself a quadratic Fourier transform.

Lemma 200 *Suppose that the quadratic Fourier transform $S_{W'',m''}$ is a product $S_{W,m}S_{W',m'}$ with $W = (P, L, Q)$, $W' = (P', L', Q')$; then*

$$m'' = m + m' - \operatorname{Inert}(P' + Q) \mod 4 \quad (6.63)$$

where $\operatorname{Inert}(P' + Q)$ *is the number of negative eigenvalues of $P' + Q$.*

The proof of this result is due to Leray (see [101], Ch. I, §1,2, pages 19–20). The second lemma shows that there are invariants modulo 4 attached to arbitrary products of quadratic Fourier transforms:

Lemma 201 *If $S_{W,m}S_{W',m'} = S_{W'',m''}S_{W''',m'''}$ then $P' + Q$ and $P''' + Q''$ have the same rank, modulo 4:*

$$\operatorname{rank}(P' + Q) = \operatorname{rank}(P''' + Q'') \mod 4 \quad (6.64)$$

and moreover

$$m + m' - \operatorname{Inert}(P' + Q) \equiv m'' + m''' - \operatorname{Inert}(P''' + Q'') \quad (6.65)$$

(also modulo 4).

The proof of this result was given by the author in [55] (also see de Gosson [56, 59]). It relies on the following asymptotic estimate, whose proof is long and technical (it involves repeated use of the method of stationary phase): defining Gaussians $f_\lambda \in \mathcal{S}(\mathbb{R}^n_x)$ by

$$f_\lambda(x) = \exp\left(-\lambda|x|^2\right), \quad \lambda > 0$$

one shows that

$$S_{W,m}S_{W',m'}f_\lambda(0) = C_{W,W'}i^{m+m'-n/2}\left(e^{i\frac{\pi}{4}}\right)^s \lambda^{-r/2} + O\left(\frac{1}{\lambda}\right) \tag{6.66}$$

for $\lambda \to \infty$. The integers s and r are, respectively, $\operatorname{sign}(P'+Q)$ and $\operatorname{rank}(P'+Q)$, and the factor $C_{W,W'}$ is a positive constant depending only on W and W'. If $S_{W,m}S_{W',m'} = S_{W'',m''}S_{W''',m'''}$, then (6.66) implies that we will have

$$C_{W,W'}i^{m+m'}\left(e^{i\frac{\pi}{4}}\right)^s \lambda^{-r/2} = C_{W'',W''}i^{m''+m'''}\left(e^{i\frac{\pi}{4}}\right)^{s'} \lambda^{-r'/2} + O\left(\frac{1}{\lambda}\right)$$

with $s' = \operatorname{sign}(P''' + Q'')$ and $r' = \operatorname{rank}(P''' + Q'')$; the lemma follows.

These two lemmas motivate the following definition:

Definition 202 *The Maslov index of $S = S_{W,m}S_{W',m'}$ is the integer modulo 4 defined by*

$$m(S) = m + m' - \operatorname{Inert}(P' + Q) \tag{6.67}$$

if $W = (P, L, Q)$, $W' = (P', L', Q')$.

In view of Lemma 201 the left-hand side of (6.67) is independent of the factorization of S, so that $m(S)$ is indeed well-defined. We observe that this definition extends the Maslov index calculated in the previous subsection, since formula (6.67) can be written

$$m(S) = m + m' - \operatorname{Inert}\operatorname*{Hess}_{x'}\left[-(W(0, x') + W'(x', 0))\right] \tag{6.68}$$

which is (6.62) if $S = S_t$, $S' = S_{t'}$.

Proposition 203 *The Maslov indices of the operators I and $-I$ are given by*

$$m(I) = n, \quad m(-I) = n + 2 \tag{6.69}$$

and the Maslov indices of $M_{L,m}$, V_P are:

$$m(V_P) = n, \quad m(M_{L,m}) = m + n \tag{6.70}$$

(all equalities being understood modulo 4).

Proof. Since we have $I = S_{W,m}S_{W^*,m^*}$ with $m^* = n - m$ (Corollary 191) and $W^* = (-Q, -L^T, -P)$ if $W = (P, L, Q)$, Eq. (6.69) follows at once from the definition (6.67) of the Maslov index. The Maslov indices of F and F^{-1} are 0 and n (see (6.33)), so we have

$$V_P = (V_P F)F^{-1} = S_{(P,I,0),0}S_{(0,-I,0),n}$$

and hence $m(V_P) = 0 - \operatorname{Inert} 0 = 0$. Similarly

$$M_{L,m} = (M_{L,m}F)F^{-1} = S_{(0,L,0),m}S_{(0,-I,0),n}$$

and hence

$$m(M_{L,m}) = m + n - \operatorname{Inert}(0) = m + n$$

which completes the proof. ■

6.6 The Cohomological Meaning of the Maslov Index*

We note that Eq. (6.63) implies that the difference $m(S_{W,m}S_{W',m'}) - m(S_{W,m}) - m(S_{W',m'})$ depends only on the projections s_W, $s_{W'}$ of $S_{W,m}$, $S_{W',m'}$. We are going to see that, more generally, we have for any S, S' in $\operatorname{Mp}(n)$

$$m(S) + m(S') - m(SS') = q(s, s') \tag{6.71}$$

where q is a function $\operatorname{Sp}(n) \times \operatorname{Sp}(n) \longrightarrow \mathbb{Z}_4$. The function q is a "group cocycle" (one also says "multiplier"), that is

$$q(ss', s'') + q(s, s') = q(s, s's'') + q(s', s''). \tag{6.72}$$

(This immediately follows from (6.71) together with the fact that $m((SS')S'') = m(S(S'S''))$.) We also remark that by definition of m we have

$$q(s_W, s_{W'}) = \operatorname{Inert}(P' + Q).$$

We are going to see that the cocycle q can be expressed in terms of the index of inertia of a triple of Lagrangian planes, and that

$$q(s, s') = \operatorname{Inert}(ss'\ell_p, s\ell_p, \ell_p)$$

where ℓ_p is the p-plane $0 \times \mathbb{R}^n_p$. That is,

$$m(SS') = m(S) + m(S') - \operatorname{Inert}(ss'\ell_p, s\ell_p, \ell_p). \tag{6.73}$$

6.6.1 Group Cocycles on $\mathrm{Sp}(n)$

Let σ be the signature function for triples of Lagrangian planes. Let ℓ be a fixed Lagrangian plane, and s, s' two symplectic matrices. We define, for every $\ell \in Lag(n)$, a mapping

$$\sigma_\ell : \mathrm{Sp}(n) \times \mathrm{Sp}(n) \longrightarrow \mathbb{Z}$$

by the formula

$$\sigma_\ell(s, s') = \sigma(ss'\ell, s\ell, \ell). \tag{6.74}$$

When ℓ is the "p-plane" $\ell_p = \mathbb{R}^n_p$ we will use the abbreviated notation $\sigma(s, s') = \sigma_{\ell_p}(s, s')$. Thus, by definition:

$$\sigma(s, s') = \sigma(ss'\ell_p, s\ell_p, \ell_p). \tag{6.75}$$

It turns out that σ_ℓ is a group cocycle, that is:

$$\sigma_\ell(ss', s'') + \sigma_\ell(s, s') = \sigma_\ell(s, s's'') + \sigma_\ell(s', s'') \tag{6.76}$$

for all symplectic matrices s, s', s''. This property immediately follows from the cocycle formula (5.42) and the $\mathrm{Sp}(n)$-invariance of the signature of a triple of Lagrangian planes. Also notice the obvious formulae:

$$\sigma_\ell(s, s^{-1}) = \sigma_\ell(s, I) = 0 \quad , \quad \sigma_\ell(s', s) = \sigma_\ell(s^{-1}, s'^{-1}). \tag{6.77}$$

The following result allows easy explicit calculations of the cocycle σ:

Lemma 204 *For $W = (P, L, Q)$, $W' = (P', L', Q')$, set: $s_W = \begin{pmatrix} A & B \\ C & D \end{pmatrix}$, $s_{W'} = \begin{pmatrix} A' & B' \\ C' & D' \end{pmatrix}$ and $s_W s_{W'} = \begin{pmatrix} A'' & B'' \\ C'' & D'' \end{pmatrix}$. We have*

$$\sigma(s_W, s_{W'}) = -\operatorname{sign}(B^{-1}B''(B')^{-1}) = -\operatorname{sign}(P' + Q). \tag{6.78}$$

Proof. We will use the notations $\ell_p = \mathbb{R}^n_p$ and $\ell_x = \mathbb{R}^n_x$. The second equality (6.78) follows from Proposition 175. In fact, by (6.10) we have:

$$s_{W''} = s_W s_{W'} = \begin{pmatrix} * & L^{-1}(P' + Q)L'^{-1} \\ \\ * & * \end{pmatrix}$$

and hence

$$B'' = L^{-1}(P' + Q)L'^{-1} = B(P' + Q)B'$$

so that we have $P' + Q = B^{-1}B''(B')^{-1}$. Let us first prove (6.78) in the particular case where $s_{W'}(\ell_p) = \ell_x$, that is when

$$s_{W'} = \begin{pmatrix} A' & B' \\ C' & 0 \end{pmatrix}.$$

Using successively the antisymmetry, the $\mathrm{Sp}(n)$-invariance, and again the antisymmetry of the signature, we have, since $s_{W'}(\ell_p) = \ell_x$:

$$\sigma\left(s_W, s_{W'}\right) = -\sigma\left(\ell_x, s_W^{-1}\ell_p, \ell_p\right). \tag{6.79}$$

Now

$$s_W^{-1} = \begin{pmatrix} D^T & -B^T \\ -C^T & A^T \end{pmatrix}$$

and thus

$$s_W^{-1}\begin{pmatrix} 0 \\ p \end{pmatrix} = \begin{pmatrix} -B^T p \\ A^T p \end{pmatrix}$$

so the Lagrangian plane $s_W^{-1}\ell_p$ has the equation $Ax + Bp = 0$; since B is invertible because s_W is free, this equation can be written $p = -B^{-1}Ax$. It follows, by Lemma 155 and (6.79), that we have

$$\begin{aligned}\sigma\left(s_W, s_{W'}\right) &= \mathrm{sign}(-B^{-1}A) \\ &= -\,\mathrm{sign}\left(B^{-1}(AB')(B')^{-1}\right) \\ &= -\,\mathrm{sign}\left(B^{-1}B''(B')^{-1}\right)\end{aligned}$$

which is (6.78) in the case $s_{W'}\ell_p = \ell_x$. We can reduce the general case to the former, using the fact that the symplectic group acts transitively on all pairs of transverse Lagrangian planes. In fact, since

$$s_{W'}\ell_p \cap \ell_p = \ell_x \cap \ell_p = 0$$

we can find $h \in \mathrm{Sp}(n)$ such that $(\ell_p, s_{W'}\ell_p) = h(\ell_p, \ell_x)$, that is $s_{W'}\ell_p = h\ell_x$ and $h\ell_p = \ell_p$. It follows, using again the antisymmetry and $\mathrm{Sp}(n)$-invariance of σ that:

$$\sigma\left(s_W, s_{W'}\right) = -\sigma\left(\ell_x, (s_W h)^{-1}\ell_p, \ell_x\right)$$

which is (6.79) with s_W replaced by $s_W h$. Changing $s_{W'}$ into $h^{-1}s_{W'}$ (and hence leaving $s_W s_{W'}$ unchanged) we are led back to the first case. Since $h\ell_p = \ell_p$, h must be of the type

$$h = \begin{pmatrix} L & 0 \\ P & (L^{-1})^T \end{pmatrix}, \ \det(L) \neq 0,\ P = P^T.$$

Writing again s_W in block-matrix form we have

$$sh = \begin{pmatrix} * & B(L^T)^{-1} \\ * & * \end{pmatrix}, \quad h^{-1}s_{W'} = \begin{pmatrix} * & L^{-1}B' \\ * & * \end{pmatrix}$$

and hence

$$\begin{aligned}\sigma\left(s_W, s_{W'}\right) &= -\operatorname{sign}(L^T B^{-1} B'' B'^{-1} L) \\ &= -\operatorname{sign}(B^{-1} B'' B'^{-1})\end{aligned}$$

proving (6.78) in the general case. ∎

6.6.2 The Fundamental Property of $m(\cdot)$

We begin by noting the following elementary Lemma:

Lemma 205 *(1) For any n-plane $\ell : Xx + Pp = 0$ in $\mathbb{R}^n_x \times \mathbb{R}^n_p$ we have*

$$\dim(\ell \cap \ell_p) = \operatorname{corank}(P) = n - \operatorname{rank}(P). \tag{6.80}$$

(2) For any symplectic matrix $s = \begin{pmatrix} A & B \\ C & D \end{pmatrix}$ we have

$$\operatorname{rank}(B) = n - \dim(s\ell_p \cap \ell_p). \tag{6.81}$$

Proof. (**1**) The intersection $\ell \cap \ell_p$ consists of all (x, p) which satisfy both conditions $Xx + Pp = 0$ and $x = 0$. It follows that

$$(x,p) \in \ell \cap \ell_p \iff Pp = 0$$

and hence (6.80). (**2**) Formula (6.81) follows from the equivalence

$$(x,p) \in s\ell_p \cap \ell_p \quad \Leftrightarrow \quad Bp = 0$$

which is an obvious consequence of the definition of ℓ_p. ∎

Next lemma expresses $m(S)$ in terms of the cocycle $\sigma(\cdot,\cdot)$:

Proposition 206 *The Maslov index of $S = S_{W,m}S_{W',m'}$ is given by*

$$m(S) = m + m' - \tfrac{1}{2}\left(n - \dim(s\ell_p \cap \ell_p) + \sigma(s_W, s_{W'})\right). \tag{6.82}$$

Proof. Definition (6.67) of the Maslov index can be rewritten:

$$m(S) = m + m' - \tfrac{1}{2}\left(\operatorname{rank}(P' + Q) - \operatorname{sign}(P' + Q)\right). \tag{6.83}$$

In view of the second part of Lemma 205 we have

$$\begin{aligned}\operatorname{rank}(P' + Q) &= \operatorname{rank}(B^{-1}B''B'^{-1}) \\ &= \operatorname{rank}(B'') \\ &= n - \dim(s\ell_p \cap \ell_p)\end{aligned}$$

and in view of formula (6.78) in Lemma 204

$$\operatorname{sign}(P' + Q) = -\sigma(s_W, s_{W'}).$$

Formula (6.82) follows. ∎

Proposition 206 allows us to express the Maslov index of a product of quadratic Fourier transforms in terms of the group cocycle σ. We are going to extend that result to arbitrary products of elements of the metaplectic group. Recall the "cohomological" notations from the last Chapter (see Example 139): for any pair (ℓ, ℓ') of Lagrangian planes in $\mathbb{R}_x^n \times \mathbb{R}_p^n$ we set $\dim(\ell, \ell') = \dim(\ell \cap \ell')$ and denote by $\partial \dim$ the coboundary of dim:

$$\partial \dim(\ell, \ell', \ell'') = \dim(\ell, \ell') - \dim(\ell, \ell'') + \dim(\ell', \ell'').$$

For $s, s' \in \operatorname{Sp}(n)$ we define

$$\dim(s, s') = \partial \dim(ss'\ell_p, s\ell_p, \ell_p).$$

Obviously $\dim(s, s')$ is a group cocycle on $\operatorname{Sp}(n)$, that is:

$$\dim(ss', s'') + \dim(s, s') = \dim(s, s's'') + \dim(s', s'').$$

Definition 207 *The index of inertia of a pair (s, s') of symplectic matrices is the integer*

$$\operatorname{Inert}(s, s') = \operatorname{Inert}(ss'\ell_p, s\ell_p, \ell_p)$$

where $\operatorname{Inert}(\ell, \ell', \ell'')$ is the index of inertia of the triple (ℓ, ℓ', ℓ'') of Lagrangian planes. That is:

$$\operatorname{Inert}(s, s') = \tfrac{1}{2}\left(n + \partial \dim(s, s') + \sigma(s, s')\right)$$

where $\sigma(\cdot, \cdot)$ is the group cocycle of the last subsection.

With these notations, we can rewrite formula (6.82) as

$$m(S_{W,m}S_{W',m'}) = m(S_{W,m}) + m(S_{W',m'}) - \text{Inert}(s_W, s_{W'}).$$

In fact, since $s_W \ell_p \cap \ell_p = s_{W'} \ell_p \cap \ell_p = 0$ (because s_W and $s_{W'}$ are free), we have, by definition of $\partial \dim$:

$$\begin{aligned}\partial \dim(s_W, s_{W'}) &= \dim(s_W s_{W'} \ell_p \cap s_W \ell_p) - \dim(s_W s_{W'} \ell_p \cap \ell_p) + \dim(s_W \ell_p \cap \ell_p) \\ &= \dim(s_{W'} \ell_p \cap \ell_p) - \dim(s_W s_{W'} \ell_p \cap \ell_p) + \dim(s_W \ell_p \cap \ell_p) \\ &= -\dim(s_W s_{W'} \ell_p \cap \ell_p)\end{aligned}$$

and hence

$$\text{Inert}(s_W, s_{W'}) = \tfrac{1}{2}\left(n - \dim(s\ell_p \cap \ell_p) + \sigma(s_W, s_{W'})\right)$$

which is precisely (6.82).

We have, more generally:

Theorem 208 *The Maslov index on* Mp(n) *has the following two properties, which characterizes it:* ***(1)*** *$m(S)$ remains constant when S moves continuously in* Mp(n) *in such a way that* $\dim(s\ell_p \cap \ell_p)$ *remains constant. In particular, $m(\cdot)$ is a locally constant function on the set of all quadratic Fourier transforms.* ***(2)*** *For all $S, S' \in$* Mp(n) *we have*

$$m(SS') = m(S) + m(S') - \text{Inert}(s, s') \tag{6.84}$$

where s, s' are the projections of S, S' on Sp(n).

We established this result in [55] (also see [56, 59]; the proof being again long and technical, we do not reproduce it here). It actually follows from Theorem 151 and the following essential relation between $m(\cdot)$ and the (reduced) Leray index:

Theorem 209 *The Maslov index $m(\cdot)$ and the reduced Leray index m are related by the formula*

$$m(S) = m(s_\infty \ell_{p,\infty}, \ell_{p,\infty}) \tag{6.85}$$

where s_∞, $\ell_{p,\infty}$ and $s_\infty \ell_{p,\infty}$ are defined as follows: ***(1)*** *s_∞ is the homotopy class of a path Σ in* Sp(n) *going from I to s (the projection of S);* ***(2)*** *$\ell_{p,\infty}$ is the homotopy class of any loop Λ in $Lag(n)$ through ℓ_p, and* ***(3)*** *$s_\infty \ell_{p,\infty}$ is then the homotopy class of the path $\Sigma\gamma$ in $Lag(n)$.*

The proof of that result is beyond the scope of this book; see de Gosson [54, 58, 59].

We also mention that we have reconstructed the Leray index modulo 4 from the Maslov index on $\mathrm{Mp}(n)$ in [56]. This shows that the Maslov index of this Chapter, and that defined in Chapter 3, really are (as already pointed out) by-products of the Leray index. The latter is the fundamental object in any theory involving Maslov indices.

Remark 210 *The set* $\mathrm{Sp}(n) \times \mathbb{Z}$ *equipped with the operation*

$$(s, m) * (s', m') = (ss', m + m' - \mathrm{Inert}(s, s')) \tag{6.86}$$

is a (non commutative) group $[\mathrm{Sp}(n) \times \mathbb{Z}]$ *with unit* $(I, 0)$, *and the inverse of* (s, m) *is*

$$(s, m)^{-1} = (s^{-1}, n - m). \tag{6.87}$$

Using the properties of the Maslov index that will be defined below, one can identify $\mathrm{Mp}(n)$ *with the subgroup*

$$G = \{(s, m) : m = m(\pm S), \Pi(S) = s\}$$

of $[\mathrm{Sp}(n) \times \mathbb{Z}] / 4\mathbb{Z}$. *(See de Gosson [55, 56, 59].)*

Our study of the metaplectic group would not be complete if we didn't mention the group IMp obtained by replacing the homogeneous quadratic forms $W = (P, L, Q)$ by general (non-degenerate) quadratic polynomials.

6.7 The Inhomogeneous Metaplectic Group

The metaplectic group $\mathrm{Mp}(n)$ was defined using "quadratic Fourier transforms" associated to homogeneous non-degenerate quadratic forms. If we relax the homogeneity condition in the definition of the quadratic Fourier transforms we obtain the "inhomogeneous metaplectic group." We begin by studying a related notion, the Heisenberg group, which appears in many contexts related to quantization (some additional properties of that group are given in Appendix C).

6.7.1 The Heisenberg Group

Consider the set

$$H(n) = \mathbb{R}^n_x \times \mathbb{R}^n_p \times S^1$$

equipped with the multiplicative law:

$$(z, \zeta)(z', \zeta') = \left(z + z', \zeta\zeta' e^{\frac{i}{2}\omega(z,z')}\right) \tag{6.88}$$

with $z = (x, p)$ and $z' = (x', p')$ (that law is just the exponentiated version of the definition given in Appendix C). It is straightforward to check that $H(n)$ is a group with unit $(0, 1)$, and that the inverse of (z, ζ) is given by:

$$(z, \zeta)^{-1} = (-z, \zeta^{-1}).$$

It is customary to call $H(n)$ the *Heisenberg group.* For a detailed study of that group see Folland [46], Schempp [135] or Stein [151]. It intervenes in many areas of pure and applied mathematics, and strangely enough in medical imaging (see Schempp's book [137]). We next construct an explicit unitary representation of the Heisenberg group acting on the square integrable functions on $\mathbb{R}^n_x$. We proceed as follows: to every point $(z_0, \zeta_0) = (z_0, e^{it_0})$ in $H(n)$, we associate the (obviously unitary) operator

$$T(z_0, \zeta_0) : L^2(\mathbb{R}^n_x) \to L^2(\mathbb{R}^n_x)$$

defined by

$$T(z_0, \zeta_0)f(x) = \zeta_0^{-1} e^{-\frac{i}{2\hbar}p_0 \cdot x_0} e^{\frac{i}{\hbar}p_0 \cdot x} f(x - x_0). \tag{6.89}$$

Notice that if p_0 and t_0 are zero, then $T(z_0, \zeta_0)$ is just a translation:

$$T((x_0, 0), 1)f(x) = f(x - x_0), \tag{6.90}$$

and if x_0 and t_0 are zero, it is multiplication by a complex number:

$$T(0, p_0)f(x) = e^{\frac{i}{\hbar}p_0 \cdot x} f(x). \tag{6.91}$$

The inverse of the operator $T(z_0, \zeta_0)$ is given by

$$T(z_0, \zeta_0)^{-1} = T(-z_0, \zeta_0^{-1}).$$

Proposition 211 *The mapping T which to every (z_0, ζ_0) associates the unitary operator defined by (6.89) is a true unitary representation of $H(n)$. That is, $T(0, 1)$ is the identity on $L^2(\mathbb{R}^n_x)$, and we have:*

$$T((z_0, \zeta_0)(z_1, \zeta_1)) = T(z_0, \zeta_0)T(z_1, \zeta_1). \tag{6.92}$$

Moreover, for every $s \in \mathrm{Sp}(n)$, *we have the* symplectic covariance *formula:*

$$T(sz_0, \zeta_0) = ST(z_0, \zeta_0)S^{-1} \tag{6.93}$$

where S *is any of the two operators in* $\mathrm{Mp}(n)$ *with projection* s.

Proof. Formula (6.92) is straightforward to check by a direct calculation. To prove (6.93) it is sufficient to assume that S is a quadratic Fourier transform $S_{W,m}$. Suppose indeed we have shown that

$$T(s_W z_0, \zeta_0) = S_{W,m}T(z_0, \zeta_0)S_{W,m}^{-1}. \tag{6.94}$$

Writing an arbitrary element S of $\mathrm{Mp}(n)$ as a product $S_{W,m}S_{W',m'}$, we will have

$$\begin{aligned} T(sz_0, \zeta_0) &= S_{W,m}(S_{W',m'}T(z_0, \zeta_0)S_{W',m'}^{-1})S_{W,m}^{-1} \\ &= S_{W,m}T(s_{W'}z_0, \zeta_0)S_{W,m}^{-1} \\ &= T(s_W s_{W'}z_0, \zeta_0) \\ &= S_{W,m}S_{W',m'}T(z_0, \zeta_0)S_{W,m}^{-1}S_{W',m'}^{-1} \\ &= ST(z_0, \zeta_0)S^{-1} \end{aligned}$$

that is (6.93). Let us thus prove (6.94); equivalently:

$$T(s_W z_0, \zeta_0)S_{W,m} = S_{W,m}T(s_W^{-1} z_0, \zeta_0). \tag{6.95}$$

Let us first study the term

$$g(x) = T(z_0, \zeta_0)S_{W,m}f(x).$$

By definition of a quadratic Fourier transform, we have, taking into account definition (6.89) of $T(z_0, \zeta_0)$:

$$g(x) = \left(\tfrac{1}{2\pi i\hbar}\right)^{n/2} \zeta_0^{-1}\Delta(W)e^{-\frac{1}{2\hbar}p_0\cdot x_0} \int e^{\frac{i}{\hbar}(W(x-x_0,x')+p_0\cdot x)} f(x')d^n x'.$$

In view of formula (6.13) in Proposition 179, the function

$$W_0(x, x') = W(x - x_0, x') + p_0 \cdot x \tag{6.96}$$

is a generating function of the free affine symplectomorphism $\tau(z_0)\circ s$, hence we have just shown that

$$T(z_0, \zeta_0)S_{W,m} = \zeta_0^{-1}e^{-\frac{i}{2\hbar}p_0\cdot x_0} S_{W_0,m} \tag{6.97}$$

where $S_{W_0,m}$ is the generalized quadratic Fourier transform defined by

$$S_{W_0,m}f(x) = \left(\tfrac{1}{2\pi i\hbar}\right)^{n/2} \Delta(W) \int e^{\frac{i}{\hbar}W_0(x,x')} f(x')d^n x'. \tag{6.98}$$

On the other hand, setting

$$h(x) = S_{W,m}T(s_W^{-1}z_0, \zeta_0)f(x)$$

and $z_0' = s_W^{-1}z_0$ we have

$$h(x) = \left(\tfrac{1}{2\pi i\hbar}\right)^{n/2} \zeta_0^{-1}\Delta(W) \int e^{\frac{i}{\hbar}W(x,x')} e^{-\frac{i}{2\hbar}p_0'\cdot x_0'} e^{\frac{i}{\hbar}p_0'\cdot x'} f(x'-x_0')\, d^n x'$$

that is, performing the change of variables $x' \longmapsto x' + x_0'$:

$$h(x) = \left(\tfrac{1}{2\pi i\hbar}\right)^{n/2} \zeta_0^{-1}\Delta(W) \int e^{\frac{i}{\hbar}W(x,x'+x_0')} e^{\frac{i}{2\hbar}p_0'\cdot x_0'} e^{\frac{i}{\hbar}p_0'\cdot x'} f(x')\, d^n x'.$$

To prove that $h(x) = g(x)$ as claimed it suffices to show that

$$W(x, x' + x_0') + \tfrac{1}{2}p_0' \cdot x_0' + p_0' \cdot x' = W_0(x, x') - \tfrac{1}{2}p_0 \cdot x_0$$

that is

$$W(x, x' + x_0') + \tfrac{1}{2}p_0' \cdot x_0' + p_0' \cdot x' = W(x - x_0, x') + p_0 \cdot x - \tfrac{1}{2}p_0 \cdot x_0.$$

Replacing x by $x + x_0$ this amounts to prove the identity

$$W(x + x_0, x' + x_0') + \tfrac{1}{2}p_0' \cdot x_0' + p_0' \cdot x' = W(x, x') + p_0 \cdot x + \tfrac{1}{2}p_0 \cdot x_0.$$

But the latter immediately follows from formula (6.11) in Proposition 175.

∎

6.7.2 The Group IMp

Let us show that the set IMp of all operators TS (or ST) where $S \in \mathrm{Mp}(n)$ and T is of the type (6.89) is a group. We begin by showing that IMp is closed under products. Setting $T = T(z_0, \zeta_0)$, $T' = T(z_0', \zeta_0')$, we have, using successively the metaplectic covariance formula (6.93), and the product formula (6.92):

$$\begin{aligned}(TS)(T'S') &= TT(sz_0', \zeta_0')SS' \\ &= T((z_0, \zeta_0)(sz_0', \zeta_0'))SS' \\ &= \left[T(z_0 + sz_0', \zeta_0\zeta_0' \exp\left(\tfrac{i}{2}\omega(z, z')\right))\right] SS'\end{aligned}$$

so that $(TS)(T'S') \in \mathrm{IMp}$. Now, $T(0,1)$ is the identity operator; using again metaplectic covariance, it is immediate to check that the inverse of TS is given by

$$(TS)^{-1} = T(-sz_0, \zeta_0^{-1})S^{-1}$$

so that the inverse of an element of IMp is also in IMp.

Definition 212 *The group* IMp *of all operators TS (or ST) is called the inhomogeneous metaplectic group.*

Every $S \in \mathrm{Mp}(n)$ can be written as the product of two quadratic Fourier transforms; similarly:

Proposition 213 *The inhomogeneous metaplectic group* IMp *is generated by the generalized quadratic forms $S_{W_0,m}$ associated to the (not necessarily homogeneous) non-degenerate quadratic forms W_0 by (6.98). In fact, every $U \in \mathrm{IMp}$ can be written as a product $uS_{W_0,m}S_{W,m}$ where u is a complex number with modulus one, and $S_{W,m} \in \mathrm{Mp}(n)$ (i.e. W is homogeneous).*

Proof. Every $U = TS \in \mathrm{IMp}$ can be written in the form $U = TS_{W,m}S_{W',m'}$; if $T = T(z_0, \zeta_0)$ we have, by (6.97):

$$TS_{W,m} = \zeta_0^{-1}e^{-\frac{i}{2}p_0'\cdot x_0'}S_{W_0,m}$$

where $s_W(x_0', p_0') = (x_0, p_0)$ and hence $U = uS_{W_0,m}S_{W',m'}$. Conversely, if $U = uS_{W_0,m}S_{W',m'}$, we can define (x_0', p_0') by the conditions

$$\begin{cases} s_W(x_0', p_0') = (x_0, p_0) \\ p_0 = \nabla_x W_0(x_0, x_0') \\ p_0' = -\nabla_{x'} W_0(x_0, x_0') \end{cases}$$

and then find $\zeta_0 \in S^1$ such that

$$u = \zeta_0^{-1}e^{-\frac{i}{2}p_0'\cdot x_0'}.$$

We then have $U = T(z_0, \zeta_0)S_{W,m}S_{W',m'}$, which is in IMp. ∎

Let us now study the relationship between IMp and the inhomogeneous symplectic group $\mathrm{ISp}(n)$ which was defined in Section 3.5. Recall that $\mathrm{ISp}(n)$ consists of all linear mappings of the type $\tau \circ s$ (or $s \circ \tau$) where $s \in \mathrm{Sp}(n)$

and τ is a translation in phase space. We identified $\mathrm{ISp}(n)$ with the group of all $(2n+1)\times(2n+1)$ matrices

$$\langle s, z_0\rangle \equiv \begin{pmatrix} s & z_0 \\ 0 & 1 \end{pmatrix}$$

where $s \in \mathrm{Sp}(n)$, $z_0 \in \mathbb{R}^n_x \times \mathbb{R}^n_p$ (written as a column vector) and 0 is the row vector whose entries all are zero. The product of two such matrices being

$$\begin{pmatrix} s & z_0 \\ 0 & 1 \end{pmatrix}\begin{pmatrix} s' & z_0' \\ 0 & 1 \end{pmatrix} = \begin{pmatrix} ss' & sz_0' + z_0 \\ 0 & 1 \end{pmatrix}$$

we see at once that the mapping

$$\Pi : \mathrm{IMp} \longrightarrow \mathrm{ISp}(n)$$

defined by

$$\Pi(T(z_0, \zeta_0)) = \tau(z_0)s$$

is a group homomorphism. That homomorphism is obviously surjective, so that Π is a priori a good candidate for being a covering projection. However, its kernel is

$$\mathrm{Ker}(\Pi) = \left\{T(0, \zeta_0) : \zeta_0 \in S^1\right\}$$

and can hence be identified with the whole circle group, so that Π is not a true covering mapping (IMp is in fact a projective representation of $\mathrm{ISp}(n)$; see de Gosson [63] or Folland [46] for details).

6.7.3 The Heisenberg-Weyl Operators

The consideration of the Heisenberg group leads in a natural way to the definition of the Heisenberg-Weyl operators:

Definition 214 *Let $z_0 \in \mathbb{R}^n_x \times \mathbb{R}^n_p$ be an arbitrary point of phase space. The operator $T(z_0) = T(z_0, 1)$ is the Heisenberg-Weyl operator.*

Using formula (6.89) and setting $z_0 = (x_0, p_0)$ the action of $T(z_0)$ on a function f is explicitly given by

$$T(z_0)f(x) = e^{\frac{i}{\hbar}(p_0\cdot x - \frac{1}{2}p_0\cdot x_0)}f(x - x_0) \tag{6.99}$$

from which immediately follows that Heisenberg-Weyl operators are unitary:

$$||T(z_0)f||_{L^2} = ||f||_{L^2} \quad , \quad f \in L^2(\mathbb{R}^n). \tag{6.100}$$

The operator $T(z_0)$ has a very interesting dynamical interpretation. Consider the linear function

$$H_0(z) = \omega(z, z_0) = p \cdot x_0 - p_0 \cdot x.$$

Viewing this function as a (somewhat unphysical) Hamiltonian, the corresponding Hamilton equations are $\dot{x} = x_0$ and $\dot{p} = p_0$ and the solutions are thus

$$x(t) = x(0) + x_0 t \quad , \quad p(t) = p(0) + p_0 t;$$

thus $z(t) = (x(t), p(t))$ is obtained from an initial point $z(0)$ by translating that point: $z(t) = z(0) + tz_0$; we will therefore call H_0 the "translation Hamiltonian corresponding to the vector z_0". Let us now have a look at the corresponding Schrödinger equation, that is

$$i\hbar \frac{\partial \Psi}{\partial t} = \widehat{H}_0 \Psi; \tag{6.101}$$

where $\widehat{H}_0 = \omega(\widehat{z}, z_0)$; here $\widehat{z} = (\widehat{x}, \widehat{p})$ with $\widehat{x} = x$ and $\widehat{p} = -i\hbar\nabla_x$ so that

$$\omega(\widehat{z}, z_0) = \widehat{p} \cdot x_0 - p_0 \cdot \widehat{x} = -i\hbar x_0 \cdot \nabla_x - p_0 \cdot x.$$

That equation is easy to solve using for instance the method of characteristics, and one finds that

$$\Psi(x,t) = e^{\frac{i}{\hbar}(tp_0 \cdot x - \frac{i}{2}t^2 p_0 \cdot x_0)} \Psi(x - tx_0)$$

which is immediately recognized as

$$\Psi(x,t) = T(tz_0)\Psi(x).$$

Thus, the Heisenberg-Weyl operator $T(z_0)$ appears as being simply the time-one propagator for the Schrödinger equation corresponding to the translation Hamiltonian H_0. Since the solution of (6.101) can be formally written

$$\Psi(x,t) = e^{-\frac{i}{\hbar}\widehat{H}_0 t} \Psi(x)$$

we can rewrite the Heisenberg-Weyl operator in the form

$$T(z_0) = e^{-\frac{i}{\hbar}(\widehat{p} \cdot x_0 - p_0 \cdot \widehat{x})}. \tag{6.102}$$

The operators $T(z_0)$ satisfy the following commutation and product formulae:

$$T(z_0)T(z_1) = e^{\frac{i}{\hbar}\omega(z_0, z_1)} T(z_1)T(z_0) \tag{6.103}$$

and

$$T(z_0+z_1)=e^{-\frac{i}{2\hbar}\omega(z_0,z_1)}T(z_0)T(z_1) \tag{6.104}$$

(for a proof see *e.g.* [63]).

We will come back to the Heisenberg-Weyl operators in Chapter 8, in connection with quantization. Let us for the moment deepen our understanding of the relationship between the metaplectic group Mp(n) and wave optics.

6.8 The Groups Symp(n) and Ham(n)*

The compose of two symplectomorphisms is still a symplectomorphism. Suppose in fact that f and g are symplectomorphisms defined on phase space $\mathbb{R}^n_x \times \mathbb{R}^n_p$; then by the chain rule

$$(f\circ g)'(z)=f'(g(z))g'(z)$$

so that $(f\circ g)'(z)$ is a symplectic matrix if $f'(g(z))$ and $g'(z)$ are. Since the inverse of f is also symplectic in view of the inversion formula

$$(f^{-1})'(z)=f'(z')^{-1}\big|_{z'=f^{-1}(z)}$$

it follows that the symplectomorphisms of $\mathbb{R}^n_x\times\mathbb{R}^n_p$ form a subgroup Symp(n) of the group Diff(n) of all diffeomorphisms of that space.

6.8.1 A Topological Property of Symp(n)

The group Symp(n) is closed in Diff(n) for the C^1-topology. By this we mean that if $(f_j)_j$ is a sequence of symplectomorphisms such that both $(f_j)_j$ and $(f'_j)_j$ converge locally uniformly on compact subsets in Diff(n), then the limit f of $(f_j)_j$ is a symplectomorphism. This property is a straightforward consequence of the definition of a symplectomorphism: each f_j satisfies

$$f'_j(z)^T J\, f'_j(z)=J \tag{6.105}$$

and hence $f'(z)=\lim_{j\to\infty} f'_j(z)$ is such that $f'(z)^T Jf(z)=J$, which implies that f is itself symplectomorphism.

It turns out that Symp(n) has a much stronger property: it is closed in Diff(n) even in the C^0-topology:

Proposition 215 *Let $(f_j)_j$ be a sequence in* $\operatorname{Symp}(n)$ *that converges towards a diffeomorphism f locally uniformly on compact subsets of $\mathbb{R}_x^n \times \mathbb{R}_p^n$. Then $f \in \operatorname{Symp}(n)$.*

The proof of this result is highly non-trivial, and relies on the existence of symplectic capacities; see Hofer-Zehnder [89], §2.2, pages 58–63.

Proposition 215 (which is a typical result from the area of symplectic topology) shows, in particular, that it is not possible to approximate volume-preserving diffeomorphisms by using sequences of symplectomorphisms. This is of course strongly related to our discussion of the symplectic camel property: it is just another manifestation of the fact that Hamiltonian flows are really very much more than the flow of an arbitrary incompressible vector field.

6.8.2 The Group $\operatorname{Ham}(n)$ of Hamiltonian Symplectomorphisms

In what follows, the word "Hamiltonian" will mean an arbitrary smooth function on $\mathbb{R}_z^{2n} = \mathbb{R}_x^n \times \mathbb{R}_p^n$ (or on $\mathbb{R}_{z,t}^{2n+1} = \mathbb{R}_x^n \times \mathbb{R}_p^n \times \mathbb{R}_t$). we denote by $\operatorname{Diff}(n)$ the group of all diffeomorphisms of phase space.

Let H be a Hamiltonian; the associated flow $t \longmapsto f_t$ is a path of symplectomorphisms passing through the identity at time $t = 0$. It turns out that, conversely, every such path determines a Hamiltonian function:

Proposition 216 *Let $t \mapsto f_t$ be a continuous path in* $\operatorname{Symp}(n)$*, defined in some interval $[a, b]$ containing 0, and such that f_0 is the identity operator. There exists a function $H = H(z, t)$ such that (f_t) is the time-dependent flow of the Hamilton vector field $X_H = (\nabla_p H, -\nabla_x H)$.*

Proof. Consider the time-dependent vector field X_t on $\mathbb{R}_z^{2n}$ defined, at every z, by

$$\frac{d}{dt} f_t(z) = X_t(f_t(z)).$$

Keeping the variable t fixed, the Lie derivative of the standard symplectic form Ω in the direction X_t is

$$L_{X_t}\Omega = \lim_{\Delta t \to 0} \frac{f_{t+\Delta t}^*\Omega - f_t^*\Omega}{\Delta t} = 0$$

since $f_t^*\Omega = \Omega$ for all t. This implies, using Cartan's homotopy formula

$$L_{X_t}\Omega = i_{X_t} d\Omega + d\,(i_{X_t}\Omega)$$

and recalling that $d\Omega = d(d(pdx)) = 0$, that we have $d\left(i_{X_t}\Omega\right) = 0$. Since we are working on Euclidean space $\mathbb{R}^{2n}_z$, it follows that there exists a function H_t such that

$$i_{X_t}\Omega = -dH_t.$$

Defining $H(z,t) = H_t(z)$, (f_t) is the flow determined by H, and the proposition is proven. ∎

The result above motivates the following definition:

Definition 217 *(1) A path $t \mapsto f_t$ in* Symp(n) *defined in some interval $[a,b]$ containing* 0*, and such that f_0 is the identity operator, is called a* Hamiltonian path. *(2) Any diffeomorphism f such that $f = f_{t=1}$ for some Hamiltonian path $t \mapsto f_t$ is called a* Hamiltonian symplectomorphism.

It turns out that, conversely, to every path of Hamiltonian symplectomorphisms we can associate a Hamiltonian function:

Proposition 218 *Let $t \longmapsto f_t$ be a smooth path in the group* Diff(n)*. If every f_t is a symplectomorphism, and f_0 is the identity, the family (f_t) is the time-dependent flow determined by some Hamiltonian function $H = H(z,t)$.*

Proof. Let us define a time-dependent vector field X_t by

$$\frac{d}{dt}f_t(z) = X_t(f_t(z)).$$

Viewing the variable t as fixed, the Lie derivative of the symplectic form Ω in the direction X_t is zero, that is $L_{X_t}\Omega = 0$. Since by Cartan's homotopy formula and the exactness of Ω

$$L_{X_t}\Omega = i_{X_t}d\Omega + d(i_{X_t}d\Omega) = d(i_{X_t}d\Omega)$$

we thus have $d(i_{X_t}d\Omega) = 0$, and there exists a function $H_t = H_t(z)$ such that

$$i_{X_t}d\Omega = dH_t.$$

Defining H, for each value of (z,t), by $H(z,t) = -H_t(z)$, the proposition follows. ∎

We are going to see that Hamiltonian symplectomorphisms form a connected normal subgroup Ham(n) of Symp(n). We begin by proving two lemmas.

Lemma 219 *Let $t \longmapsto f_t$ and $t \longmapsto g_t$ be two Hamiltonian paths. Then, the compose $t \longmapsto f_t \circ g_t$ is also a Hamiltonian path, and so is $t \longmapsto f_t^{-1}$. In fact, if $t \longmapsto f_t$ is determined by H and $t \longmapsto g_t$ by K, then $t \longmapsto f_t \circ g_t$ and $t \longmapsto f_t^{-1}$ are determined by the Hamiltonians $H\#K$ and $H^{\div}$ defined by*

$$\begin{cases} H\#K(z,t) = H(z,t) + K(f_t^{-1}(z),t) \\ H^{\div}(z,t) = -H(f_t(z),t) \end{cases}$$

respectively.

Proof. Denote by $X_t = J\nabla_z H$, $Y_t = J\nabla_z K$, $Z_t = J\nabla_z H\#K$, and $W_t = J\nabla_z H^{\div}$ the time-dependent Hamilton fields associated to H, K, $H\#K$, and $H^{\div}$, respectively. Using the inverse function rule together with the characterization (6.105) of symplectomorphisms, we have

$$\begin{aligned} Z_t(z) &= J\nabla_z H(z,t) + J(f_t^{-1})'(z)^T \nabla_z K(f_t^{-1}(z),t) \\ &= X_t(z) + f_t'(z)\left(f_t^{-1}(z)\right) J\nabla_z K\left(f_t^{-1}(z),t\right) \\ &= X_t(z) + f_t'(z)\left(f_t^{-1}(z)\right) Y_t\left(f_t^{-1}(z)\right). \end{aligned}$$

On the other hand, let T_t be the vector field determined by the composed path $t \longmapsto f_t \circ g_t$:

$$T_t(f_t \circ g_t(z)) = \frac{d}{dt}(f_t \circ g_t(z)).$$

By the chain rule

$$\begin{aligned} T_t(f_t \circ g_t(z)) &= \frac{df_t}{dt}(g_t(z)) + f_t'(g_t(z))\frac{dg_t}{dt}(z) \\ &= X_t(f_t(g_t(z))) + f_t'(z)\left(g_t(z)\right) Y_t\left(g_t(z)\right) \end{aligned}$$

and hence $T_t = Z_t$. A similar calculation shows that W_t is the Hamilton vector field associated with M. ■

Lemma 220 *Let $(f_t)_t$ be the flow determined by a Hamiltonian H and g a symplectomorphism. Then $(g \circ f_t \circ g^{-1})_t$ is the flow determined by the transformed Hamiltonian $H \circ g^{-1}$.*

Proof. We have, by definition of the derivative:

$$\begin{aligned} \frac{d}{dt}(gf_tg^{-1}(z)) &= \lim_{\Delta t \to 0} \frac{g \circ f_{t+\Delta t} \circ g^{-1}(z) - g \circ f_t \circ g^{-1}(z)}{\Delta t} \\ &= g \circ \lim_{\Delta t \to 0} \frac{f_{t+\Delta t} \circ g^{-1}(z) - f_t \circ g^{-1}(z)}{\Delta t} \\ &= g \circ X_H(g^{-1}(z)) \\ &= X_{H\circ g^{-1}}(z) \end{aligned}$$

where the last equality follows from the transformation law (Proposition 22) for Hamiltonian vector fields. ■

Let us now prove that Hamiltonian symplectomorphisms form a group:

Theorem 221 *The set* $\mathrm{Ham}(n)$ *of all Hamiltonian symplectomorphisms of* $\mathbb{R}^{2n}_z = \mathbb{R}^n_x \times \mathbb{R}^n_p$ *is a connected normal subgroup of* $\mathrm{Symp}(n)$.

Proof. Let f and g be two elements of $\mathrm{Ham}(n)$: $f = f_{t=1}$ and $g = g_{t=1}$ for some Hamiltonian flows (f_t) and (g_t). Since $f \circ g = (f_t \circ g_t)_{t=1}$ it follows that $f \circ g \in \mathrm{Ham}(n)$ since $t \longmapsto f_t \circ g_t$ is the flow determined by $H \# K$ (Lemma 219). Similarly, if $f \in \mathrm{Ham}(n)$, then $f^{-1} \in \mathrm{Ham}(n)$ since $f^{-1} = f_{t=-1}$ is associated to $H^{\div}$. It follows that $\mathrm{Ham}(n)$ is indeed a group. That group is connected by construction, and it is a normal subgroup of $\mathrm{Symp}(n)$ in view of Lemma 220. ■

6.9 Canonical Quantization

We have been until now implicitly using a unique quantization scheme, namely the one which associates to a Maxwell Hamiltonian

$$H = \sum_{j=1}^{n} \frac{1}{2m_j} (p_j - A_j(x,t))^2 + U(x,t) \tag{6.106}$$

the operator

$$\widehat{H} = \sum_{j=1}^{n} \frac{1}{2m_j} \left(-i\hbar \frac{\partial}{\partial x_j} - A_j(x,t) \right)^2 + U(x,t). \tag{6.107}$$

But life is not always that simple; consider for instance the square

$$L^2 = (yp_z - zp_y)^2 + (zp_x - xp_z)^2 + (xp_y - yp_x)^2 \tag{6.108}$$

of the angular momentum

$$\mathbf{L} = \mathbf{r} \times \mathbf{p} = (yp_z - zp_y)\mathbf{i} + (zp_x - xp_z)\mathbf{j} + (xp_y - yp_x)\mathbf{k}. \tag{6.109}$$

How should we then "quantize" L^2? This innocent question has led to many discussions around the "angular momentum dilemma" which already puzzled Linus Pauling in his textbook[2] about chemistry. This example —

[2]L. Pauling, General Chemistry, 3rd ed., p. 125.

and others, which we shall develop in a while — shows that we need a convincing procedure to associate to every classical function of the position and momentum variables an operator (many physicists will object that this is not a relevant question, arguing that quantum mechanics is the primordial theory, and that we should not dictate to Nature what the quantum operator should be. While it is easy to agree with this objection (and we do), it does not solve the problem, because it is in a sense equivalent to asking "given a quantum system, what is the corresponding classical system?". This amounts to privilege "dequantization" to quantization, and it is perhaps indeed more correct from a logical point of view. For instance, returning to the angular momentum; the quantum version of the latter is the vector operator

$$\widehat{\mathbf{L}} = -i\hbar\left[(y\frac{\partial}{\partial z} - z\frac{\partial}{\partial y})\mathbf{i} + (z\frac{\partial}{\partial x} - x\frac{\partial}{\partial z})\mathbf{j} + (x\frac{\partial}{\partial y} - y\frac{\partial}{\partial x})\mathbf{k}\right]$$

and its square is traditionally taken to be the scalar operator

$$\widehat{L}^2 = -\hbar^2\left[(y\frac{\partial}{\partial z} - z\frac{\partial}{\partial y})^2 + (z\frac{\partial}{\partial x} - x\frac{\partial}{\partial z})^2 + (x\frac{\partial}{\partial y} - y\frac{\partial}{\partial x})^2\right].$$

Now we ask: what is the classical observable corresponding to $\widehat{L}^2$? While it is clear that the classical observable corresponding to $\widehat{\mathbf{L}}$ is the angular momentum (6.109), the problem is much more difficult when one deals with $\widehat{L}^2$. Both aspects will be discussed, but to be on form ground, we first have to explain what we mean by "quantization". There are severable concurring definitions (see the review paper [2] by Ali and Englis). We will use here the — perhaps simplest and most general — definition, that of *canonical quantization*[3].

Let $\mathcal{S}(\mathbb{R}^n)$ be the Schwartz space of test functions, and $\mathcal{S}'(\mathbb{R}^n)$ its dual, the space of tempered distributions. (Recall that $\Psi \in \mathcal{S}(\mathbb{R}^n)$ if and only if Ψ is infinitely differentiable, and $x_k^r\partial_{x_j}^s\Psi$ is bounded for all exponents r, s and indices j, k. In particular, Ψ and all of its derivatives go to zero when $x \to \pm\infty$.)

A canonical quantization of $\mathbb{R}^n$ is a continuous linear map

$$\mathrm{Op} : \mathcal{S}'(\mathbb{R}^{2n}) \longrightarrow \mathcal{L}(\mathcal{S}(\mathbb{R}^n), \mathcal{S}'(\mathbb{R}^n))$$

associating to each $H \in \mathcal{S}'(\mathbb{R}^{2n})$ a continuous linear operator

$$\widehat{H} = \mathrm{Op}(H) : \mathcal{S}(\mathbb{R}^n) \longrightarrow \mathcal{S}'(\mathbb{R}^n)$$

[3]There is no consensus in the mathematical and physical literature about what the "best" definition of (canonical) quantization should be. We have chosen for reasons of economy to use a rather overcomplete system of axioms. See [2] for alternatives.

and such that the following properties hold:

(CQ1) Triviality: *For all integers* $r, s \geq 0$

$$\operatorname{Op}(x_j^r)\Psi = x_j^r\Psi \ , \ \operatorname{Op}(p_j^s)\Psi = (-i\hbar\partial_{x_j})^s\Psi$$

(*hence in particular* $\operatorname{Op}(1) = I_{\mathrm{d}}$);

(CQ2) Born's CCR: *The canonical commutation relations*

$$[\widehat{x}_j, \widehat{p}_k] = i\hbar\delta_{jk} \tag{6.110}$$

hold for $1 \leq j, k \leq n$;

(CQ3) Hermiticity: *When* H *is real then* $\operatorname{Op}(H)$ *is a symmetric operator defined on the dense subspace* $\mathcal{S}(\mathbb{R}^n)$ of $L^2(\mathbb{R}^n)$.

The conditions (CQ1)–(CQ3) are sometimes complemented by other conditions; for instance:

(CQ4) Symplectic covariance: *if* $s \in \operatorname{Sp}(n)$ *then*

$$\operatorname{Op}(H \circ s^{-1}) = S\operatorname{Op}(H)S^{-1} \tag{6.111}$$

where S *is any of the two operators in* $\operatorname{Mp}(n)$ *with projection* s.

(CQ5) Dirac correspondence: *If the operators* $\operatorname{Op}(H)\operatorname{Op}(K)$, $\operatorname{Op}(K)\operatorname{Op}(H)$, *and* $\operatorname{Op}(\{H, K\})$ *are defined on a common (non-empty) domain, then*

$$[\operatorname{Op}(H), \operatorname{Op}(K)] = i\hbar\operatorname{Op}(\{H, K\}). \tag{6.112}$$

We will see that there is no canonical quantizations satisfying simultaneously all Axioms (CQ1)–(CQ5). But there is one — and only one — quantization satisfying (CQ1)–(CQ4), namely Weyl quantization, and only one satisfying (CQ1)–(CQ3) and a reduced version of (CQ5), namely Born-Jordan quantization.

6.9.1 The Ordering Problem for Monomials

When studying approximations to motions near an equilibrium point one is often led to Hamiltonian functions of the type

$$H(x,p) = \frac{1}{2}(p^2 + x^2) + \lambda(p^2 + x^2)^3; \tag{6.113}$$

while this may appear as a rather unphysical Hamiltonian, there are perturbation methods which yield Hamiltonian of this type (see Crehan [31] and the references therein). While it is clear how to quantize the first term $\frac{1}{2}(p^2 + x^2)$ (it should be $\frac{1}{2}(\widehat{p}^2 + \widehat{x}^2)$ in any reasonable quantization scheme), what should the "correct" quantization of the term $\lambda(p^2 + x^2)^3$ be? Expanding the binomial we get

$$(p^2 + x^2)^3 = p^6 + 3p^4x^2 + 3p^2x^4 + x^6$$

so we are immediately confronted to a genuine ordering problem for the monomials p^4x^2 and p^2x^4. The problem of monomial ordering is an old one, harking back to Born and Jordan's work. It is still a popular area of research. Let us have a closer look at this issue.

Assume $n = 1$. What should the operator corresponding to xp be? There are in fact infinitely a priori many "natural" choices, a quite general being the interpolation $\tau\widehat{x}\widehat{p} + (1-\tau)\widehat{p}\widehat{x}$ between $\widehat{x}\widehat{p}$ and $\widehat{p}\widehat{x}$ (τ is here a real parameter which can take any value between $-\infty$ and $+\infty$). Now, the (formal) adjoint of this operator is $\tau\widehat{p}\widehat{x} + (1-\tau)\widehat{x}\widehat{p}$; physical considerations requiring formal self-adjointness leads us to the choice $\tau = \frac{1}{2}$, so in this case the rules $x \longrightarrow \widehat{x}$ and $p \longrightarrow \widehat{p}$ lead rather unambiguously to the well-known "average rule"

$$xp \longrightarrow \tfrac{1}{2}(\widehat{x}\widehat{p} + \widehat{p}\widehat{x}). \tag{6.114}$$

Consider now the general case of a monomial x^rp^s where r and s are non-negative integers. We are confronted with infinitely many ordering choices; the one initially proposed by Max Born (b.1882) and Pascual Jordan (b.1902) in their foundational papers [23, 24] is the association $x^rp^s \longrightarrow \mathrm{Op_{BJ}}(x^rp^s)$ where

$$\mathrm{Op_{BJ}}(x^rp^s) = \frac{1}{s+1}\sum_{k=0}^{s} \widehat{p}^{s-k}\widehat{x}^r\widehat{p}^k; \tag{6.115}$$

for instance if $r = s = 1$ this leads to the rule (6.114) and in the case $r = s = 2$ we get

$$\mathrm{Op_{BJ}}(x^2p^2) = \frac{1}{3}(\widehat{x}^2\widehat{p} + \widehat{x}\widehat{p}\widehat{x} + \widehat{p}\widehat{x}^2); \tag{6.116}$$

the Born and Jordan rule for $x^r p^s$ thus appears as the *equally weighted average* of all operator polynomials $\widehat{p}^{s-k}\widehat{x}^r\widehat{p}^k$. This rule is at first sight pulled out of thin air. It was actually very smartly (and even elegantly) motivated in [23, 24], as the only formal quantization rule allowing to give a rigorous meaning to the newly born Heisenberg "matrix mechanics"; we refer to de Gosson [67–69] for detailed discussions of these motivations. Without going into too many historical considerations (which the reader can find in the references above), the Born and Jordan rule was rather unnoticed by the physical community; in 1926 Hermann Weyl (*b*.1885) proposed a general quantization scheme for observables in [165]; by analogy with the Fourier inversion formula

$$a(x,p) = \left(\frac{1}{2\pi\hbar}\right)^n \int e^{\frac{i}{\hbar}(x\cdot x_0 + p\cdot p_0)} Fa(x_0,p_0)dp_0dx_0 \tag{6.117}$$

he suggested to associate with a classical observable $a(x,p)$ the operator

$$\mathrm{Op_W}(a) = \left(\frac{1}{2\pi\hbar}\right)^n \int e^{\frac{i}{\hbar}(\widehat{x}\cdot x_0 + \widehat{p}\cdot p_0)} Fa(x_0,p_0)dp_0dx_0 \tag{6.118}$$

obtained by replacing formally the variables x and p in (6.117) with the operators $\widehat{x}$ and $\widehat{p}$. In the monomial case (and when $n = 1$) this rule leads to the McCoy formula [115]

$$\mathrm{Op_W}(x^r p^s) = \frac{1}{2^s}\sum_{\ell=0}^{s}\binom{s}{\ell}\widehat{p}^{s-\ell}\widehat{x}^r\widehat{p}^\ell \tag{6.119}$$

which is different from the Born-Jordan rule (6.115) as soon as $r, s \geq 2$. For instance,

$$\mathrm{Op_{BJ}}(x^2p^2) = \frac{1}{4}(\widehat{x}^2\widehat{p} + 2\widehat{x}\widehat{p}\widehat{x} + \widehat{p}\widehat{x}^2). \tag{6.120}$$

It turns out that both the Born-Jordan rule (6.115) and the Weyl-McCoy rule (6.119) can be obtained from the more general formula

$$\mathrm{Op}_\tau(x^r p^s) = \sum_{\ell=0}^{s}\binom{s}{\ell}(1-\tau)^\ell\tau^{s-\ell}\widehat{p}^{s-\ell}\widehat{x}^r\widehat{p}^\ell \tag{6.121}$$

where τ is a parameter as above; however they are derived using two fundamentally different procedures: while (6.119) is immediately obtained by plugging in the value $\tau = \frac{1}{2}$ in (6.121), formula (6.115) needs more work. The key point lies in the observation that

$$\int_0^1 (1-\tau)^\ell\tau^{s-\ell}d\tau = \frac{(s-\ell)!\ell!}{(s+1)!}; \tag{6.122}$$

which is a consequence of the theory of the beta function, which is a well-known topic from real analysis. It immediately follows from (6.122) that we have

$$\begin{aligned}\int_0^1 \operatorname{Op}_\tau(x^r p^s)d\tau &= \sum_{\ell=0}^{s}\binom{s}{\ell}\frac{(s-\ell)!\ell!}{(s+1)!}\widehat{p}^{s-\ell}\widehat{x}^r\widehat{p}^\ell \\ &= \frac{1}{s+1}\sum_{\ell=0}^{s}\widehat{p}^{s-\ell}\widehat{x}^r\widehat{p}^\ell \\ &= \operatorname{Op}_{\mathrm{BJ}}(x^r p^s).\end{aligned}$$

Using the canonical commutation relations (6.110) one proves by induction that the following relations hold for all integers $r, s \geq 0$:

$$[\widehat{x}^r, \widehat{p}^s] = si\hbar\sum_{\ell=0}^{r-1}\widehat{x}^{r-1-\ell}\widehat{p}^{s-1}\widehat{x}^\ell, \tag{6.123}$$

$$[\widehat{x}^r, \widehat{p}^s] = ri\hbar\sum_{j=0}^{s-1}\widehat{p}^{s-1-j}\widehat{x}^{r-1}\widehat{p}^j, \tag{6.124}$$

$$[\widehat{x}^r, \widehat{p}^s] = \sum_{k=1}^{\min(r,s)}(i\hbar)^k k!\binom{r}{k}\binom{s}{k}\widehat{p}^{s-k}\widehat{x}^{r-k}, \tag{6.125}$$

$$[\widehat{x}^r, \widehat{p}^s] = -\sum_{k=1}^{\min(r,s)}(-i\hbar)^k k!\binom{r}{k}\binom{s}{k}\widehat{x}^{r-k}\widehat{p}^{s-k}. \tag{6.126}$$

Notice that the relations (6.125) and (6.126) are obtained from each other by taking adjoints and observing that we have $[\widehat{x}^r, \widehat{p}^s]^* = -[\widehat{p}^s, \widehat{x}^r]$. Using these relations it is not difficult to prove the following alternative formulas for the Weyl and Born-Jordan orderings:

Proposition 222 *We have, for all integers $r, s \geq 0$,*

$$\operatorname{Op}_{\mathrm{W}}(x^r p^s) = \sum_{\ell=0}^{\min(r,s)}(-i\hbar)^\ell\binom{s}{\ell}\binom{r}{\ell}\frac{\ell!}{2^\ell}\widehat{x}^{r-\ell}\widehat{p}^{s-\ell}; \tag{6.127}$$

$$\operatorname{Op}_{\mathrm{W}}(x^r p^s) = \sum_{\ell=0}^{\min(r,s)}(i\hbar)^\ell\binom{s}{\ell}\binom{r}{\ell}\frac{\ell!}{2^\ell}\widehat{p}^{s-\ell}\widehat{x}^{r-\ell} \tag{6.128}$$

and

$$\mathrm{Op_{BJ}}(x^r p^s) = \sum_{\ell=0}^{\min(r,s)} (-i\hbar)^\ell \binom{s}{\ell}\binom{r}{\ell}\frac{\ell!}{\ell+1}\widehat{x}^{r-\ell}\widehat{p}^{s-\ell} \tag{6.129}$$

$$\mathrm{Op_{BJ}}(x^r p^s) = \sum_{\ell=0}^{\min(r,s)} (i\hbar)^\ell \binom{s}{\ell}\binom{r}{\ell}\frac{\ell!}{\ell+1}\widehat{p}^{s-\ell}\widehat{x}^{r-\ell}. \tag{6.130}$$

See for instance de Gosson [68], Chapter 3 for a proof. Notice that (6.127) and (6.128) (resp. (6.129) and (6.130)) are deduced from each other by taking adjoints and observing that $\mathrm{Op_W}(x^r p^s)^* = \mathrm{Op_W}(x^r p^s)$ (resp. $\mathrm{Op_{BJ}}(x^r p^s)^* = \mathrm{Op_{BJ}}(x^r p^s)$).

Example 223 *It follows from (6.127) and (6.129) that we have*

$$\begin{aligned}\mathrm{Op_W}(x^2p^2) &= x^2\widehat{p}^2 - 2i\hbar\widehat{x}\widehat{p} - \tfrac{1}{2}\hbar^2\\ \mathrm{Op_{BJ}}(x^2p^2) &= x^2\widehat{p}^2 - 2i\hbar\widehat{x}\widehat{p} - \tfrac{2}{3}\hbar^2\end{aligned}$$

so that $\mathrm{Op_W}(x^2p^2) - \mathrm{Op_{BJ}}(x^2p^2) = \frac{1}{6}\hbar^2$. *This is of course not immediately obvious if we use the definitions (6.119) and (6.120).*

The following important result shows that, up to a factor, the Born-Jordan quantization of a monomial can be written as a commutator:

Proposition 224 *We have, for all integers* $r, s \geq 0$,

$$\mathrm{Op_{BJ}}(x^r p^s) = \frac{1}{i\hbar(r+1)(s+1)}[x^{r+1}, p^{s+1}]. \tag{6.131}$$

Proof. Replacing r and s with $r+1$ and $s+1$ in (6.124) we have

$$[\widehat{x}^{r+1}, \widehat{p}^{s+1}] = (r+1)i\hbar\sum_{j=0}^{s}\widehat{p}^{s-j}\widehat{x}^r\widehat{p}^j = (r+1)i\hbar(s+1)\,\mathrm{Op_{BJ}}(x^r p^s)$$

taking the definition (6.115) of $\mathrm{Op_{BJ}}(x^r p^s)$ formula (6.131) follows. ■

6.9.2 The Case of General Observables

We first set out to find a rule for quantizing the exponential functions $e^{\frac{i}{\hbar}x_0\cdot x}$ and $e^{\frac{i}{\hbar}p_0\cdot p}$. Recall that the Heisenberg-Weyl operator $T(z_0) = e^{-\frac{i}{\hbar}(\widehat{p}\cdot x_0 - p_0\cdot\widehat{x})}$ ($z_0 = (x_0, p_0)$) is explicitly defined by

$$T(z_0)f(x) = e^{\frac{i}{\hbar}(p_0\cdot x - \frac{1}{2}p_0\cdot x_0)}f(x - x_0). \tag{6.132}$$

In particular,

$$e^{\frac{i}{\hbar}p_0\cdot\widehat{x}}f(x)=e^{\frac{i}{\hbar}p_0\cdot x}f(x)\quad,\quad e^{-\frac{i}{\hbar}\widehat{p}\cdot x_0}f(x)=f(x-x_0). \tag{6.133}$$

Consider now the operator

$$M(z_0)=e^{\frac{i}{\hbar}(\widehat{x}\cdot x_0+\widehat{p}\cdot p_0)}$$

appearing in Weyl's definition (6.118). The Heisenberg-Weyl operator is related to $M(z_0)$ by the simple formula $T(-Jz_0)=M(z_0)$ hence

$$M(z_0)f(x)=e^{\frac{i}{\hbar}(x_0\cdot x+\frac{1}{2}p_0\cdot x_0)}f(x+p_0) \tag{6.134}$$

and in particular

$$e^{\frac{i}{\hbar}x_0\cdot\widehat{x}}f(x)=e^{\frac{i}{\hbar}x_0\cdot x}f(x)\quad,\quad e^{\frac{i}{\hbar}\widehat{p}\cdot p_0}f(x)=f(x+p_0). \tag{6.135}$$

We assume that $\widehat{x}_j$ and $\widehat{p}_j$ are the usual operators "multiplication by x_j" and $-i\hbar\partial_{x_j}$ (*cf.* the canonical quantization condition (CQ1) above).

The result below is essential because it is the key to the quantization of arbitrary observables.

Lemma 225 *Let* $X(x_0)=e^{\frac{i}{\hbar}x_0\cdot x}$ *and* $P(p_0)=e^{\frac{i}{\hbar}p_0\cdot p}$. *Let* Op *be an arbitrary quantization. We have*

$$\operatorname{Op}(X(x_0))=e^{\frac{i}{\hbar}x_0\cdot\widehat{x}}\text{ and }\operatorname{Op}(P(p_0))=e^{\frac{i}{\hbar}p_0\cdot\widehat{p}} \tag{6.136}$$

that is

$$\operatorname{Op}(X(x_0))\psi(x)=e^{\frac{i}{\hbar}x_0\cdot x}\psi(x),\quad \operatorname{Op}(P(p_0))\psi(x)=\psi(x+p_0). \tag{6.137}$$

Proof. It is sufficient to consider the case $n=1$. Writing $x=x_1$ and $p=p_1$ and expanding the exponential $e^{\frac{i}{\hbar}x_0x}$ in a Taylor series we have, in view of the continuity of Op,

$$\operatorname{Op}(X(x_0))\psi(x)=\sum_{k=0}^{\infty}\frac{1}{k!}\left(\frac{i}{\hbar}x_0\cdot x\right)^k\psi(x)=e^{\frac{i}{\hbar}x_0\cdot x}\psi(x)$$

and similarly

$$\operatorname{Op}(P(p_0))\psi(x)=\sum_{k=0}^{\infty}\frac{1}{k!}\left(\frac{i}{\hbar}p_0(-i\hbar\partial_x)\right)^k\psi(x)=\psi(x+p_0).$$

■

We now make the following essential observation: let H be an element of $\mathcal{S}(\mathbb{R}^{2n})$ and

$$\mathcal{H}(x_0,p_0)=\left(\tfrac{1}{2\pi\hbar}\right)^n\int H(x,p)e^{-\frac{i}{\hbar}(x_0\cdot x+p_0\cdot p)}d^npd^nx$$

its Fourier transform of H. In view of the Fourier inversion formula we have

$$H(x,p)=\left(\tfrac{1}{2\pi\hbar}\right)^n\int \mathcal{H}(x_0,p_0)e^{\frac{i}{\hbar}(x_0\cdot x+p_0\cdot p)}d^np_0d^nx_0.$$

Let Op be a canonical quantization; by continuity and linearity we can write

$$\operatorname{Op}(H)=\left(\tfrac{1}{2\pi\hbar}\right)^n\int \mathcal{H}(x_0,p_0)\operatorname{Op}(e^{\frac{i}{\hbar}(x_0\cdot(\cdot)+p_0(\cdot))})d^np_0d^nx_0. \qquad (6.138)$$

Viewing the integral as a distribution bracket, this formula extends to arbitrary $H\in\mathcal{S}'(\mathbb{R}^{2n})$ by continuity, noting that $\mathcal{S}(\mathbb{R}^{2n})$ is dense in $\mathcal{S}'(\mathbb{R}^{2n})$. Formula (6.138) is very important: it shows that a canonical quantization $H\longmapsto\operatorname{Op}(H)$ is uniquely determined by the knowledge of how the exponential

$$M(x_0,p_0)=X(x_0)\otimes P(p_0)=e^{\frac{i}{\hbar}(x_0\cdot x+p_0\cdot p)}$$

(sometimes called "Weyl's characteristic function") is quantized.

Theorem 226 *(1) Assume that the canonical quantization* Op *satisfies the symplectic covariance axiom (CQ4). Then* Op *is the Weyl quantization:* $\operatorname{Op}=\operatorname{Op}_{\mathrm{W}}$ *that is*

$$\operatorname{Op}_{\mathrm{W}}(H)\Psi(x)=\left(\tfrac{1}{2\pi\hbar}\right)^n\int \mathcal{H}(x_0,p_0)e^{\frac{i}{\hbar}(x_0\cdot\widehat{x}+p_0\cdot\widehat{p})}\Psi(x)d^np_0d^nx_0; \qquad (6.139)$$

equivalently

$$\operatorname{Op}_{\mathrm{W}}(H)\Psi(x)=\left(\tfrac{1}{2\pi\hbar}\right)^n\int \mathcal{H}_\sigma(x_0,p_0)T(x_0,p_0)\Psi(x)d^np_0d^nx_0 \qquad (6.140)$$

where $\mathcal{H}_\sigma(x_0,p_0))=\mathcal{H}_\sigma(p_0,-x_0)$ *is the symplectic Fourier transform of* H.

(2) Assume that the canonical quantization Op *satisfies the Dirac correspondence (CQ5) for all* $H(z)=T(p)+U(x)$ *and* $K(z)=T'(p)+U'(x)$. *Then* Op *is the Born-Jordan quantization* $\operatorname{Op}_{\mathrm{BJ}}$, *defined by*

$$\operatorname{Op}_{\mathrm{BJ}}(H)\Psi(x)=\left(\tfrac{1}{2\pi\hbar}\right)^n\int \mathcal{H}(x_0,p_0)\operatorname{sinc}\left(\frac{p_0\cdot x_0}{2\hbar}\right)e^{\frac{i}{\hbar}(x_0\cdot\widehat{x}+p_0\cdot\widehat{p})}\Psi(x)d^np_0d^nx_0 \qquad (6.141)$$

where $\operatorname{sinc}(u) = \sin(u)/u$; *equivalently*

$$\operatorname{Op_{BJ}}(H)\Psi(x) = \left(\tfrac{1}{2\pi\hbar}\right)^n \int \mathcal{H}_\sigma(x_0,p_0)\operatorname{sinc}\left(\frac{p_0\cdot x_0}{2\hbar}\right)T(x_0,p_0)\Psi(x)d^np_0d^nx_0. \tag{6.142}$$

Proof. (1) That formulas (6.139) and (6.140) are equivalent is clear, performing the change of variables $(x_0,p_0)\longmapsto(p_0,-x_0)$ in the integral in the left-hand side of (6.139) and noting that $e^{\frac{i}{\hbar}(x_0\cdot\widehat{x}+p_0\cdot\widehat{p})}$ then becomes $e^{\frac{i}{\hbar}(p_0\cdot\widehat{x}-x_0\cdot\widehat{p})} = T(x_0,p_0)$. Let now $s\in\operatorname{Sp}(n)$; the symplectic Fourier transform of $H\circ s^{-1}$ is, taking into account the relations $\det(s)=1$ and $\sigma(z_0,sz) = \sigma(s^{-1}z_0,z)$,

$$\begin{aligned}
\mathcal{H}^s_\sigma(x_0,p_0) &= \left(\tfrac{1}{2\pi\hbar}\right)^n \int H(s^{-1}(x,p))e^{-\frac{i}{\hbar}(p_0\cdot x-x_0\cdot p)}d^npd^nx\\
&= \left(\tfrac{1}{2\pi\hbar}\right)^n \int H(s^{-1}z)e^{-\frac{i}{\hbar}\sigma(z_0,z)}d^{2n}z\\
&= \left(\tfrac{1}{2\pi\hbar}\right)^n \det(s)\int H(z_0)e^{-\frac{i}{\hbar}\sigma(z_0,sz)}d^{2n}z\\
&= \left(\tfrac{1}{2\pi\hbar}\right)^n \det(s)\int H(z_0)e^{-\frac{i}{\hbar}\sigma(s^{-1}z_0,z)}d^{2n}z\\
&= \mathcal{H}_\sigma(s^{-1}(x_0,p_0)).
\end{aligned}$$

It follows that we have, performing a change of variables and using again $\det(s)=1$,

$$\begin{aligned}
\operatorname{Op_W}(H\circ s^{-1})\Psi(x) &= \left(\tfrac{1}{2\pi\hbar}\right)^n \int \mathcal{H}_\sigma(s^{-1}z_0)T(z_0)\Psi(x)d^{2n}z_0\\
&= \left(\tfrac{1}{2\pi\hbar}\right)^n \int \mathcal{H}_\sigma(z_0)T(sz_0)\Psi(x)d^{2n}z_0.
\end{aligned}$$

Now, recall that we have proven in our study of the Heisenberg group the symplectic covariance formula

$$T(sz_0,\zeta_0) = ST(z_0,\zeta_0)S^{-1} \tag{6.143}$$

(formula (6.93)) which immediately yields, taking $\zeta_0 = 0$,

$$T(sz_0) = ST(z_0)S^{-1}.$$

It follows that

$$\begin{aligned}
\operatorname{Op_W}(H\circ s^{-1})\Psi(x) &= \left(\tfrac{1}{2\pi\hbar}\right)^n \int \mathcal{H}_\sigma(z_0)ST(z_0)S^{-1}\Psi(x)d^{2n}z_0\\
&= S\operatorname{Op_W}(H)(S^{-1}\Psi)(x)
\end{aligned}$$

which is precisely the symplectic covariance property (6.111) in Axiom (CQ4).
(2) That formulae (6.141) and (6.142) are equivalent is clear by the same argument as in the proof of (**1**). Recalling that $X(x_0) = e^{\frac{i}{\hbar}x_0\cdot x}$ and $P(p_0) = e^{\frac{i}{\hbar}p_0\cdot p}$ we will have in particular, if the Dirac correspondence holds for all pairs of Hamiltonians $T(p) + U(x)$. Assume $p_0x_0 \neq 0$, then

$$\begin{aligned}[\mathrm{Op}(X(x_0)), \mathrm{Op}(Y(p_0))] &= i\hbar\,\mathrm{Op}(\{e^{\frac{i}{\hbar}x_0\cdot x}, e^{\frac{i}{\hbar}p_0\cdot p}\}) \\ &= \frac{1}{i\hbar}p_0x_0\,\mathrm{Op}(e^{\frac{i}{\hbar}(x_0\cdot x+p_0\cdot p)}) \\ &= \frac{1}{i\hbar}p_0x_0M(x_0,p_0)\end{aligned}$$

that is

$$\begin{aligned}\widehat{M}(x_0,p_0) &= \frac{i\hbar}{p_0x_0}\,[\mathrm{Op}(X(x_0))\,\mathrm{Op}(Y(p_0)) - \mathrm{Op}(Y(p_0))\,\mathrm{Op}(X(x_0))] \\ &= \frac{i\hbar}{p_0\cdot x_0}(e^{\frac{i}{\hbar}x_0\cdot\widehat{x}}e^{\frac{i}{\hbar}p_0\cdot\widehat{p}} - e^{\frac{i}{\hbar}p_0\cdot\widehat{p}}e^{\frac{i}{\hbar}x_0\cdot\widehat{x}}).\end{aligned}$$

In view of the Baker-Campbell-Hausdorff formula

$$e^{A+B} = e^{-\frac{1}{2}[A,B]}e^Ae^B = e^{\frac{1}{2}[A,B]}e^Be^A \tag{6.144}$$

which is valid for all operators A and B commuting with $[A,B]$ we have

$$\begin{aligned}e^{\frac{i}{\hbar}x_0\cdot\widehat{x}}e^{\frac{i}{\hbar}p_0\cdot\widehat{p}} &= e^{-\frac{1}{2i\hbar}p_0\cdot x_0}e^{\frac{i}{\hbar}(x_0\cdot\widehat{x}+p_0\cdot\widehat{p})} \\ e^{\frac{i}{\hbar}p_0\cdot\widehat{p}}e^{\frac{i}{\hbar}x_0\cdot\widehat{x}} &= e^{\frac{1}{2i\hbar}p_0\cdot x_0}e^{\frac{i}{\hbar}(x_0\cdot\widehat{x}+p_0\cdot\widehat{p})}\end{aligned}$$

and hence

$$\widehat{M}(x_0,p_0) = \frac{i\hbar}{p_0\cdot x_0}(e^{-\frac{1}{2i\hbar}p_0\cdot x_0} - e^{\frac{1}{2i\hbar}p_0\cdot x_0})e^{\frac{i}{\hbar}(x_0\cdot\widehat{x}+p_0\cdot\widehat{p})}.$$

Formula (6.141) follows. ■

One shows [65, 68] that the Weyl and Born-Jordan quantizations of Maxwell Hamiltonians coincides, and so do the quantization of quadratic Hamiltonians. The angular momentum yields an example where both quantizations differ:

Example 227 ***Back to the angular momentum.*** *Recall that the square of the angular momentum is*

$$L^2 = (yp_z - zp_y)^2 + (zp_x - xp_z)^2 + (xp_y - yp_x)^2.$$

One shows [68] using formulas (6.139) and (6.141) in Theorem 226 that

$$\widehat{L}^2 = \mathrm{Op}_{\mathrm{W}}(L^2) - \tfrac{3}{2}\hbar^2 = \mathrm{Op}_{\mathrm{BJ}}(L^2) - \hbar^2$$

hence, in particular:

$$\mathrm{Op}_{\mathrm{W}}(L^2) - \mathrm{Op}_{\mathrm{BJ}}(L^2) = \tfrac{1}{2}\hbar^2.$$

6.9.3 The Groenewold-van Hove Theorem

Let us very briefly discuss the famous "no-go", theorem of Groenewold and van Hove; for more we refer to the original papers [154] of van Hove and [77]. Gotay [72] and Gotay *et al.* [73, 74] discuss in great detail the mathematics of the Groenewold and van Hove obstruction; also see Guillemin and Sternberg [80], Chapter 1.

To explain what this is about, let us first note the following property, which is a quite straightforward consequence of Schrödinger's quantization rule: let H and K be two functions on $\mathbb{R}^n_x \times \mathbb{R}^n_p \times \mathbb{R}_t$, which are quadratic polynomials in the position and momentum coordinates. For instance, H and K could be quadratic Maxwell Hamiltonians

$$H = \sum_{j=1}^{n} \frac{1}{2m_j}\left(p_j - A_j(t)\cdot x\right)^2 + \frac{1}{2}K(t)x^2 + a(t)\cdot x \tag{6.145}$$

but we do not restrict ourselves to functions of this type: *any* quadratic polynomial in x_j, p_j will do. The Schrödinger quantization rule 1.4.3 associates to these functions H and K two Hermitian partial differential operators $\widehat{H}$ and $\widehat{K}$; for instance, if H is given by (6.145) above, then $\widehat{H}$ is the usual quantum Hamiltonian

$$H = \sum_{j=1}^{n} \frac{1}{2m_j}\left(-i\hbar\frac{\partial}{\partial x_j} - A_j(t)\cdot x\right)^2 + \frac{1}{2}K(t)x^2 + a(t)\cdot x. \tag{6.146}$$

One can now easily prove, by a direct calculation, that we have for Hamiltonians of this type the following simple relation between the Poisson bracket of H and K, and the commutator of $\widehat{H}$ and $\widehat{K}$:

$$[\widehat{H}, \widehat{K}] = ih\widehat{\{H, K\}}. \tag{6.147}$$

This formula (sometimes called the "Dirac correspondence") can, by the way, be used to prove that the mapping $\Pi : \mathrm{Mp}(n) \longrightarrow \mathrm{Sp}(n)$ constructed in Subsection 6.4.1 indeed is a covering map (see Leray, [101], Chapter I, de Gosson [63]). Now, it is immediate to verify that this formula no longer holds for functions H or K which are not quadratic polynomials; *a fortiori*, it does not hold for polynomials of degree higher than two. The Groenewold-van Hove theorem simply says that we cannot expect to be able to modify Schrödinger's quantization rule in such a way that we make the Dirac formula (6.147) hold for arbitrary functions H and K. More precisely:

Theorem 228 (Groenewold-van Hove) *Let $\mathcal{P}_k$ be the Poisson algebra of all real polynomials of degree $\leq k$ on phase space $\mathbb{R}^n_x \times \mathbb{R}^n_p$. There is no linear mapping $H \longmapsto \widehat{H}$ from $\mathcal{P}_k$ to the space of Hermitian operators on $\mathcal{S}(\mathbb{R}^n_x)$ such that*

$$p_j \longmapsto -i\hbar\frac{\partial}{\partial x_j}, \quad x_j \longmapsto x_j$$

and which at the same time satisfies the Dirac correspondence

$$[\widehat{H}, \widehat{K}] = ih\widehat{\{H, K\}} \tag{6.148}$$

for $H, K \in \mathcal{P}_k$ if $k > 2$.

The key to the proof of this famous "no-go" result lies in the following observation in the case $n = 1$. Assume that (6.148) holds; then the trivial equality

$$q^2p^2 = \frac{1}{9}\{q^3, p^3\} = \frac{1}{3}\{q^2p, p^2q\}$$

leads to the conflicting formulas

$$\operatorname{Op}(q^2p^2) = \frac{1}{9}\operatorname{Op}\{q^3, p^3\} = (\widehat{q})^2(\widehat{p})^2 - 2i\hbar\widehat{q}\widehat{p} - \frac{2}{3}\hbar^2 \tag{6.149}$$

and

$$\operatorname{Op}(q^2p^2) = \frac{1}{3}\operatorname{Op}\{q^2p, p^2q\} = (\widehat{q})^2(\widehat{p})^2 - 2i\hbar\widehat{q}\widehat{p} - \frac{1}{3}\hbar^2; \tag{6.150}$$

one concludes that there is thus no quantization satisfying Dirac's correspondence for all monomials.

Chapter 7

Schrödinger's Equation and the Metatron

Summary. *The metaplectic representation yields an algorithm allowing to calculate the solutions of Schrödinger's equation from the classical trajectories. Conversely, the classical trajectories can be recovered from the knowledge of the wave function. Both classical and quantum motion are thus deduced from the same mathematical object, the Hamiltonian flow.*

While it is true that there can be no argument leading from classical mechanics to quantum mechanics without some additional physical postulate involving Planck's constant, it is also true that if one accepts L. de Broglie's matter waves hypothesis, then Schrödinger's equation *emerges* from classical mechanics. We will see that this is an obvious consequence of the theory of the metaplectic group, not only when the Hamiltonian is quadratic, but also for general Maxwell Hamiltonians.

We will in fact see that both classical and quantum mechanics rely on the same mathematical object, the Hamiltonian flow, viewed as an abstract group. If one makes that group act on points in phase space, via its symplectic representation, one obtains Hamiltonian mechanics. If one makes it act on functions, via the metaplectic representation, one obtains quantum mechanics. It is remarkable that in both cases, we have an associated theory of motion: in the symplectic representation, that motion is governed by Hamilton's equations. In the metaplectic representation, it is governed by Bohm's equations. Since classical and quantum motion are distinct, but deduced from one another by the metaplectic representation, we will call particles obeying the Bohmian law of motion *metatrons*.

7.1 Schrödinger's Equation for the Free Particle

We begin by giving a rigorous "physical" derivation of Schrödinger's equation for a free particle in three-dimensional configuration space. This is the first step towards a complete answer to the question we posed in Chapter 1, namely whether Schrödinger could have found his equation using only arguments of pure mathematics.

7.1.1 The Free Particle's Phase

Consider a particle moving freely with velocity $\mathbf{v}$ in physical space $\mathbb{R}^3_{\mathbf{r}}$. In conformity with de Broglie's postulate, we associate with this particle a plane wave with phase

$$\Theta_{rel}(\mathbf{r},t) = \mathbf{k}\cdot\mathbf{r} - \omega(\mathbf{k})t + C. \tag{7.1}$$

Here C is an arbitrary constant, to be fixed following our needs. We are using the subscript "rel" in Θ_{rel} because the wave vector $\mathbf{k}$ and the frequency $\omega(\mathbf{k})$ are defined by the relativistic equations

$$\mathbf{k} = \frac{m\mathbf{v}}{\hbar}, \quad \omega(\mathbf{k}) = \frac{mc^2}{\hbar}\ .$$

Expressing Eq. (7.1) in terms of the momentum and energy, we get

$$\Theta_{rel}(\mathbf{r},t) = \frac{1}{\hbar}\left(\mathbf{p}\cdot\mathbf{r} - mc^2t\right) + C \tag{7.2}$$

and observing that for small velocities

$$\begin{aligned} mc^2 &= \frac{m_0c^2}{\sqrt{1-(v/c)^2}} \\ &= m_0c^2 + \frac{1}{2}m_0v^2 + O\left(\frac{v^4}{c^2}\right) \end{aligned}$$

($v = |\mathbf{v}|$) we can rewrite Eq. (7.2) as

$$\Theta_{rel}(\mathbf{r},t) = \frac{1}{\hbar}\Phi(\mathbf{r},t) - \frac{m_0c^2}{\hbar}t + O\left(\frac{v^4}{c^2}\right) \tag{7.3}$$

where Φ is the function

$$\Phi(\mathbf{r},t) = \mathbf{p}_0\cdot\mathbf{r} - \frac{p_0^2}{2m}t + C\hbar \tag{7.4}$$

($\mathbf{p}_0 = m_0\mathbf{v}$, $p_0 = |\mathbf{p}_0|$). When the velocity $\mathbf{v}$ is small, we can neglect the terms $O(v^4/c^2)$ in Eq. (7.3), so that $\Theta_{rel}(\mathbf{r},t)$ is approximated by

$$\Theta'_{rel}(\mathbf{r},t) = \frac{1}{\hbar}\Phi(\mathbf{r},t) - \frac{m_0c^2}{\hbar}t + C\hbar. \tag{7.5}$$

Now, there is no point in keeping the term $m_0c^2t/\hbar$ (its presence affects neither the phase nor the group velocities) so that we can take as definition of the phase

$$\Theta(\mathbf{r},t) = \frac{1}{\hbar}\left(\Phi(\mathbf{r},t) + C\right) \tag{7.6}$$

and fix the constant C by requiring that, at time $t = t_0$, the equation $\Theta = 0$ determines the phase plane $\mathbf{p}\cdot\mathbf{r} = \mathbf{p}_0\cdot\mathbf{r}_0$. This leads to

$$\Phi(\mathbf{r},t) = \mathbf{p}_0\cdot(\mathbf{r}-\mathbf{r}_0) - \frac{p_0^2}{2m}(t-t_0) \tag{7.7}$$

and Φ is thus simply the gain in action when the free particle proceeds from $\mathbf{r}_0$ at time t_0 to $\mathbf{r}$ at time t with velocity $\mathbf{v}_0 = \mathbf{p}_0/m$. It follows that the function Φ is a solution of the Hamilton-Jacobi Cauchy problem for the free particle Hamiltonian

$$\begin{cases} \dfrac{\partial\Phi}{\partial t} + \dfrac{1}{2m}\left(\nabla_{\mathbf{r}}\Phi\right)^2 = 0 \\ \Phi(\mathbf{r},t_0) = \mathbf{p}_0\cdot(\mathbf{r}-\mathbf{r}_0) \end{cases} \tag{7.8}$$

as can be verified by a direct calculation.

7.1.2 The Free Particle Propagator

We next make a pedestrian, but essential, observation. The phase of a matter wave is defined on the extended phase space $\mathbb{R}_{\mathbf{r}}\times\mathbb{R}_{\mathbf{p}}\times\mathbb{R}_t$. As such, it depends on the momentum vector $\mathbf{p}_0$, which can take arbitrarily large values (we are in the non-relativistic domain). For a free particle with mass m, the choice of the momentum can thus be any vector, and unless we have measured it, all the "potentialities" associated to these phases are present. This suggests that we define a "universal wave function" for the free particle by superposing all these potentialities. Since there is no reason for privileging some "origin" $(\mathbf{r}_0, \mathbf{p}_0, t_0)$ in extended phase space, we write $(\mathbf{r}', \mathbf{p}', t')$ instead of $(\mathbf{r}_0, \mathbf{p}_0, t_0)$, and set

$$\Phi_{\mathbf{p}'}(\mathbf{r},\mathbf{r}';t,t') = \mathbf{p}'\cdot(\mathbf{r}-\mathbf{r}') - \frac{p'^2}{2m}(t-t').$$

We next define the function

$$G(\mathbf{r},\mathbf{r}';t,t') = \left(\tfrac{1}{2\pi\hbar}\right)^3 \int e^{\frac{i}{\hbar}\Phi_{p'}(\mathbf{r},\mathbf{r}';t,t')}d^3p' \tag{7.9}$$

where d^3p' is shorthand for $dp'_x dp'_y dp'_z$ (the reason for which we impose the factor $(1/2\pi\hbar)^3$ will become clear in a moment). The integral in (7.9) is a Fresnel-type integral; it is convergent (but of course not absolutely convergent); we will calculate it in a moment. We first note that it immediately follows from Eq. (7.8) that G satisfies the Schrödinger equation

$$i\hbar\frac{\partial G}{\partial t} = -\frac{\hbar^2}{2m}\nabla^2_{\mathbf{r}}G \tag{7.10}$$

provided that differentiations under the integral sign are authorized. Let us calculate the limit of G as $t \to t'$. In view of the Fourier formula

$$\frac{1}{2\pi}\int_{-\infty}^{+\infty} e^{ikx}dk = \delta(x)$$

we have

$$\left(\tfrac{1}{2\pi\hbar}\right)^3 \int e^{\frac{i}{\hbar}\mathbf{p}\cdot(\mathbf{r}-\mathbf{r}')}d^3\mathbf{p} = \delta(\mathbf{r}-\mathbf{r}')$$

and hence

$$\lim_{t\to t'} G(\mathbf{r},\mathbf{r}';t,t') = \delta(\mathbf{r}-\mathbf{r}'). \tag{7.11}$$

It follows that G is a "propagator" or "Green function" for Schrödinger's equation:

Proposition 229 *Let $\Psi' \in \mathcal{S}(\mathbb{R}^3_{\mathbf{r}})$. The function*

$$\Psi(\mathbf{r},t) = \int G(\mathbf{r},\mathbf{r}';t,t')\Psi'(\mathbf{r})\,d^3r' \tag{7.12}$$

$(t \neq t')$ is the solution of Schrödinger's Cauchy problem

$$\begin{cases} i\hbar\dfrac{\partial \Psi}{\partial t} = -\dfrac{\hbar^2}{2m}\nabla^2_{\mathbf{r}}\Psi \\ \lim_{t\to t'}\Psi(\cdot,t') = \Psi'. \end{cases} \tag{7.13}$$

Proof. Assuming again that it is permitted to differentiate in x and t under the integration sign in (7.12) (it will be a posteriori justified below by calculating the explicit expression of G) the fact that Ψ is a solution of Schrödinger's equation follows from Eq. (7.10). Finally, to prove that $\lim_{t\to t'} \Psi(\cdot, t') = \Psi$, it suffices to note that we have

$$\lim_{t\to t'} \Psi(\mathbf{r}, t) = \int \delta(\mathbf{r} - \mathbf{r}')\Psi'(\mathbf{r}')\, d^3r' = \Psi'(\mathbf{r})$$

in view of Eq. (7.11). ∎

7.1.3 An Explicit Expression for G

We will make use of the following well-known result from the theory of Fresnel integrals:

Lemma 230 *Let λ be a real number, $\lambda \neq 0$. Then*

$$\frac{1}{\sqrt{2\pi}} \int_{-\infty}^{+\infty} e^{-iuv} e^{i\lambda u^2/2} du = \left(e^{i\pi/4}\right)^{sign(\lambda)} |\lambda|^{-1/2} e^{-iv^2/2\lambda} \tag{7.14}$$

where $sign(\lambda) = +1$ if $\lambda > 0$ and $sign(\lambda) = -1$ if $\lambda < 0$.

The proof of formula (7.14), which is sometimes called the "Fresnel formula" is well-known; it consists in uncompleting squares in the Gauss integral

$$\frac{1}{\sqrt{2\pi}} \int_{-\infty}^{+\infty} e^{-u^2/2} du = 1$$

and thereafter using analytic continuation. (See any book dealing with Gaussian integrals; for instance Leray [101], Guillemin-Sternberg [79, 80] or Folland [46] all contain proofs of the Fresnel formula.)

From Lemma 230 follows that:

Proposition 231 *The Green function G is given by the formula*

$$G(\mathbf{r}, \mathbf{r}'; t, t') = \left(\frac{m}{2\pi i\hbar(t-t')}\right)^{3/2} \exp\left(\frac{i}{\hbar} W_f(\mathbf{r}, \mathbf{r}'; t, t')\right) \tag{7.15}$$

where

$$\left(\frac{m}{2\pi i\hbar(t-t')}\right)^{3/2} = \left(e^{i\frac{3\pi}{4}}\right)^{sign(t-t')} \left(\frac{m}{2\pi i\hbar|t-t'|}\right)^{3/2} \tag{7.16}$$

and W_f is the free-particle generating function, i.e.

$$W_f(\mathbf{r}, \mathbf{r}'; t, t') = m\frac{(\mathbf{r} - \mathbf{r}')^2}{2(t-t')}. \tag{7.17}$$

Proof. Let $\mathbf{r} = (x, y, z)$ and $\mathbf{r}' = (x', y', z')$. We have $G = G_x \otimes G_y \otimes G_z$ where

$$G_x = \tfrac{1}{2\pi\hbar} \int_{-\infty}^{+\infty} \exp\left[\frac{i}{\hbar}\left(p_x(x-x') - \frac{p_x'^2}{2m}(t-t')\right)\right] dp_x'$$

and similar definitions for G_y and G_z. Setting $u = p_x$, $v = -(x-x')/\hbar$ and $\lambda = -(t-t')/m\hbar$ in (7.14) we get

$$\hbar\sqrt{2\pi}G_x = \left(e^{-i\frac{\pi}{4}}\right)^{sign(t-t')} \sqrt{\frac{m\hbar}{|t-t'|}} \exp\left[\frac{i}{\hbar}m\frac{(x-x')^2}{2(t-t')}\right].$$

Performing similar calculations with G_y and G_z we get Eq. (7.15). ■

Remark 232 *Formula (7.16) corresponds to the argument choices* $\arg i = \pi/2$ *and*

$$\arg(t-t') = \begin{cases} 0 & \textit{if } t-t' > 0 \\ \pi & \textit{if } t-t' < 0. \end{cases} \tag{7.18}$$

The result above can be extended without difficulty to the case of systems with an arbitrary number of particles. In fact, let

$$i\hbar\frac{\partial\Psi}{\partial t} = -\sum_{j=1}^{N} \frac{\hbar^2}{2m_j}\nabla_{\mathbf{r}_j}^2\Psi$$

be Schrödinger's equation for a system of N free particles; in mass matrix notation:

$$i\hbar\frac{\partial\Psi}{\partial t} = -\frac{\hbar^2}{2m}\nabla_x^2\Psi.$$

The corresponding Green function is then given by

$$G(x, x'; t, t') = \left(\frac{|m|}{2\pi i\hbar(t-t')}\right)^{3N/2} \exp\left(\frac{i}{\hbar}W_f(x, x'; t, t')\right) \tag{7.19}$$

where $|m| = m \cdots m_N$, and the argument of $t - t'$ is given by (7.18); W_f is, as before, the free-particle generating function:

$$W_f(x, x'; t, t') = m\frac{(x-x')^2}{2(t-t')}. \tag{7.20}$$

We next relate these constructions to the metaplectic representation.

7.1.4 The Metaplectic Representation of the Free Flow

Let $(s_{t,t'})$ be the time-dependent flow determined by the free-particle Hamiltonian

$$H = \frac{p^2}{2m}, \quad p = |\mathbf{p}|$$

on $\mathbb{R}^3_{\mathbf{r}} \times \mathbb{R}^3_{\mathbf{p}}$. The flow $(s_{t,t'})$ consists here of the symplectic 6×6 matrices

$$s_{t,t'} = \begin{pmatrix} I & \frac{t-t'}{m} I \\ 0 & I \end{pmatrix}$$

which are free for $t \neq t'$, and the associated smooth family of free generating functions is

$$W_f(\mathbf{r}, \mathbf{r}'; t, t') = m\frac{(\mathbf{r} - \mathbf{r}')^2}{2(t - t')}.$$

Let now

$$\Pi^\hbar : \overset{\hbar}{\mathrm{Mp}}(3) \longrightarrow \mathrm{Sp}(3)$$

be the covering mapping which to every quadratic Fourier transform

$$S^\hbar_{W,m}\Psi(\mathbf{r}) = \left(\frac{1}{2\pi i\hbar}\right)^{3/2} \Delta(W) \int e^{\frac{i}{\hbar}W(\mathbf{r},\mathbf{r}';t,t')}\Psi(\mathbf{r}')\, d^3r'$$

associates the free symplectic matrix s_W. We denote by $\pm S^\hbar_{t,t'}$ the two quadratic Fourier transforms with projections

$$\Pi^\hbar(\pm S^\hbar_{t,t'}) = s_{t,t'}.$$

They are given by the formula:

$$S^\hbar_{t,t'}\Psi'(\mathbf{r}, t) = \pm \left(\tfrac{m}{2\pi i\hbar(t-t')}\right)^{3/2} \int e^{\frac{i}{\hbar}W(\mathbf{r},\mathbf{r}',t,t')}\Psi'(\mathbf{r}')\, d^3r' \qquad (7.21)$$

where the argument of the factor in front of the integral is determined by (7.18); Ψ' is a function belonging to the Schwartz space $\mathcal{S}(\mathbb{R}^3_{\mathbf{r}})$. Writing

$$\Psi(\mathbf{r}, t) = S^\hbar_{t,t'}\Psi'(\mathbf{r})$$

one checks, by a direct calculation, that the function Ψ solves the free-particle Schrödinger equation

$$i\hbar\frac{\partial \Psi}{\partial t} = -\frac{\hbar^2}{2m}\nabla^2_{\mathbf{r}}\Psi.$$

Moreover, if one chooses the "+" sign in formula (7.21), then

$$\lim_{t \to t'} \Psi(\mathbf{r}, t) = \Psi'(\mathbf{r}).$$

We thus see that for the free particle the datum of $(s_{t,t'})$ is equivalent to the datum of the family of unitary operators $(S_{t,t'})$ which to each $\Psi' \in S(\mathbb{R}^3_x)$ associates the solution Ψ of Schrödinger's equation satisfying the initial condition $\Psi(\cdot, t') = \Psi'$.

7.1.5 More Quadratic Hamiltonians

Suppose that H is a Hamiltonian of the type

$$H = \frac{1}{2m}(p - Ax)^2 + \frac{1}{2}Kx^2 + a \cdot x \tag{7.22}$$

where A and K are $n \times n$ matrices, K a symmetric matrix (see Chapter 2, Section 3.4). A few examples of such Hamiltonians are:

(1) The Hamiltonian of the anisotropic n-dimensional harmonic oscillator:

$$H = \sum_{j=1}^{n} \frac{1}{2m_j}\left(p^2 + m_j^2 \omega_j^2 x_j^2\right).$$

(2) The Hamiltonian of the electron in a uniform magnetic field $\mathbf{B} = (0, 0, B_z)$ in the symmetric gauge (Chapter 2, Section 3.4.2):

$$H = \frac{\mathbf{p}^2}{2m} + \frac{e^2 B_z^2}{8mc^2}(x^2 + y^2) - \frac{eB_z}{2mc}(xp_y - yp_x).$$

(3) The Hamiltonian of the Coriolis force (Chapter 2, Example 18)

$$H = \frac{1}{2m}\left(\mathbf{p} - m\left(\mathbf{R} \times \mathbf{r}\right)\right)^2 - m\mathbf{g} \cdot \mathbf{r}$$

where $\mathbf{R} = (0,\ R\cos\phi,\ R\sin\phi)$ is the rotation vector of the Earth.

Assuming for simplicity that H is of the Maxwell type and time-independent (but this requirement is actually unessential), the flow determined by such a Hamiltonian is a one-parameter subgroup (s_t) of $\mathrm{Sp}(n)$. For

$t \neq 0$ and sufficiently small, s_t will moreover be a free symplectic matrix. Setting

$$W(t) = W(x, x'; t, 0)$$

(where $W(x, x'; t, t')$ is the generating function determined by H) we thus have $s_t = s_{W(t)}$, and to each s_t we can thus associate exactly two elements $\pm S^{\varepsilon}_{W(t),m(t)}$ of any of the metaplectic groups $\mathrm{Mp}^{\varepsilon}(n)$ (see Chapter 4, Subsection 6.4.2). Choosing for ε Planck's constant $\hbar$, we thus have a projection

$$\Pi^{\hbar} : \mathrm{Mp}^{\hbar}(n) \longrightarrow \mathrm{Sp}(n)$$

which to each quadratic Fourier transform

$$S^{\hbar}_{W,m}\Psi(x) = \left(\tfrac{1}{2\pi i\hbar}\right)^{n/2} \Delta(W) \int e^{\frac{i}{\hbar}W(x,x',t)}\Psi(x')\, d^n x'$$

where

$$\Delta(W) = i^m \sqrt{\left|\operatorname*{Hess}_{x,x'}(-W)\right|}$$

associates the free symplectic matrix s_W generated by W. Denoting by S_t a choice of $S_{W(t),m(t)}$ depending smoothly on t, for each $\Psi_0 \in \mathcal{S}(\mathbb{R}^n_x)$ the function

$$\Psi(x,t) = S_t \Psi_0(x) \tag{7.23}$$

is then a solution of the Schrödinger equation associated to H. Moreover, since $s_{W(t)} \to I$ when $t \to 0$, we can determine the argument of $\Delta(W(t))$ in such a way that $S_{W(t)} \to I$ in $\mathrm{Mp}(n)$ when $t \to 0$, and with that choice we will have

$$\lim_{t\to 0} \Psi(x, 0) = \Psi_0(x).$$

Thus, using the metaplectic representation we can solve exactly every Cauchy problem

$$i\hbar\frac{\partial \Psi}{\partial t} = \hat{H}\,\Psi, \quad \Psi(\cdot, 0) = \Psi_0$$

when the classical Hamiltonian is of the type (7.22) above. The solution is given by the formula

$$\Psi(x,t) = \left(\tfrac{1}{2\pi i\hbar}\right)^{n/2} \Delta(W(t)) \int e^{\frac{i}{\hbar}W(x,x',t)}\Psi_0(x')\, d^n x' \tag{7.24}$$

which is of course only valid when the generating function $W(t)$ is defined, that is (in general) for small t. However, this is by no way a drawback of the method, because:

(1) The times t for which the generating function is not defined are exceptional (see Proposition 178). The ambiguity in the choice of the sign of $W(t)$ when t crosses these values is eliminated using the theory of the Maslov index as developed in Chapter 5;

(2) Formula (7.23) actually allows us to solve Schrödinger's equation for *all* values of t, because we can calculate $\Psi(x,t)$ for any value of t by using the formula
$$\Psi(x,t) = \left(S_{t/N}\right)^N \Psi_0(x) \tag{7.25}$$
where one has chosen N so large that $s_{t/N}$ is a free symplectic matrix.

Remark 233 *Notice that formula (7.25) gives the exact value of the solution in a* finite *number of steps. It is thus a "Feynman" formula, but its derivation has nothing to do with any bizarre "sum of histories" argument: it is just a consequence of the metaplectic representation, together with the fact that* $S_tS_{t'} = S_{t+t'}$.

The method of resolution outlined above has been known among mathematicians working in representation theory and geometric quantization for quite a long time (see, e.g., Guillemin-Sternberg [79, 80], Folland [46], and the references therein). The most "natural" and "elegant" way to explain why the method is to use a Lie algebra argument (see for instance Guillemin-Sternberg [79, 80], Folland [46]). The method is, as far as I can tell, almost generally ignored by physicists, who rather invoke the theory of the Feynman integral (see Feynman ([44]), Feynman and Hibbs [45] (beware of misprints!), or Schulman [139] (7.25)).

We will devote the next sections of this Chapter to prove formula (7.24) in a purely analytic way, highlighting the crucial role played by the generating function, and a quantity derived from it, the *Van Vleck determinant.* This quantity, which is usually only invoked in connection with semi-classical approximations (see Brack and Bhaduri [25] or Gutzwiller [81]), is essentially the "density of trajectories" joining two points in state space. Our approach has moreover another appeal, which is of a more conceptual nature. It namely immediately makes us understand why the "metaplectic method" for solving Schrödinger's equation cannot be pushed beyond quadratic Hamiltonians, and this without invoking the classical Groenewold-van Hove "no-go" argument [154] already mentioned in Chapter 1. (See [46, 80] for a discussion of that topic, which belongs to the theory of geometric quantization.)

We will use the short-time actions constructed in Chapter 4, Section 4.5, to show that approximate solutions of Schrödinger's equation can still be obtained for small times t for *all* Hamiltonians, but we will do this in a spirit very different from the classical Feynman "path integral" approach. This will allow us to produce an algorithm for calculating the solutions of Schrödinger's equation which converges must faster than Feynman's integral formula without the use of dubious "sums over histories".

7.2 Van Vleck's Determinant

7.2.1 Trajectory Densities

Consider a system with time-dependent Maxwell Hamiltonian

$$H(x,p,t) = \sum_{j=1}^{n} \frac{1}{2m_j} \left(p_j - A_j(x,t)\right)^2 + U(x,t).$$

The associated Hamilton equations are

$$\dot{x}_j = \frac{1}{m_j} \left(p_j - A_j\right) \quad \text{and} \quad \dot{p}_j = \frac{1}{m_j} \left(p_j - A_j\right) \frac{\partial A_j}{\partial x_j} - \frac{\partial U}{\partial x_j}.$$

Differentiating the first equation with respect to t and then inserting the value of $\dot{p}_j$ given by the second equation, we see that the position coordinates x_j satisfy the following system of n coupled second order differential equations:

$$\ddot{x}_j + \frac{1}{m_j} \left(\frac{\partial U}{\partial x_j}(x,t) + \frac{\partial A_j}{\partial t}(x,t) \right) = 0$$

$(1 \leq j \leq n)$. That system, which describes the motion of the particle in configuration space, has a unique solution $x(t) = (x_1(t), ..., x_n(t))$ for each set of initial conditions $x_j(t') = x'$, $\dot{x}_j(t') = \dot{x}'$. Recall that if the time interval $|t - t'|$ is small enough, then there will exist a unique trajectory in configuration space joining two points x', x in a time $t - t'$, and both the initial and final velocities v' and v are unambiguously determined by the datum of (x', t') and (x, t). Suppose now that we vary the initial velocity v' by a small amount. We will then obtain another trajectory from x' at time t', and passing close to x at time t. Repeating the procedure a great number of times, we will obtain a whole family of trajectories emanating from x' at time t'. These trajectories may eventually intersect, but as long as t is sufficiently close to t', they will spread and form a "fan" of non-intersecting curves in configuration space. We now ask the following question:

"*By how much will we miss a given point if we vary slightly the momentum* p' *at the initial point* x'?"

Somewhat more precisely:

"*What is the relation between the deviations* Δx *from the arrival point* x *corresponding to small changes* $\Delta p'$ *of the initial momentum* p'*?*"

Let us pause and consider again the example of the tennis player already encountered in Example 96 of Chapter 4:

Example 234 ***The tennis player's come-back.*** *A tennis ball is being smashed by a player in the* x, y*-plane from a point* (x', y') *at time* t' *to a point* (x, y) *with an initial velocity vector* $\mathbf{v}' = (v'_x, v'_y)$*. The different trajectories through* (x', y') *will never cross each other outside that point. Let us now change slightly the initial velocity vector, i.e. we replace* $\mathbf{v}'$ *by* $\mathbf{v}' + \Delta\mathbf{v}'$*. Then, the position vector* $\mathbf{r}$ *will be changed into some new position vector* $\mathbf{r} + \Delta\mathbf{r}$*, the coordinate increments* Δx *and* Δy *being given by*

$$\Delta x = (t - t')v'_x, \quad \Delta y = (t - t')v'_y \ .$$

We can evaluate quantitatively the change of trajectory by introducing the Jacobian determinant of the transformation $\mathbf{r} \mapsto \mathbf{v}'$*; it is*

$$\det \frac{\partial(v'_x, v'_y)}{\partial(x, y)} = \begin{vmatrix} \frac{1}{t-t'} & 0 \\ 0 & \frac{1}{t-t'} \end{vmatrix} = \left(\frac{1}{t - t'}\right)^2 . \tag{7.26}$$

This quantity measures the rate of variation of the "number" of trajectories arriving at (x, y) *at time* t*, when one changes the initial velocity allowing to reach that point from* (x', y') *at time* t'*. We can thus view the determinant (7.26) as a measure of the "density of trajectories" arriving to* (x, y, t) *from* (x', y', t')*. Notice that this density becomes infinite when* $t \to t'$*; this can be intuitively interpreted by saying that for given small* $\Delta\mathbf{v}$*, smaller values of* $t - t'$ *lead to smaller position fluctuations* $\Delta\mathbf{r}$*: if we diminish* $t - t'$ *there will be a "greater concentration" of trajectories coming from* (x', y', t') *in the vicinity of* (x, y, t)*.*

In the case of a general Hamiltonian, one proceeds exactly in the same way, using the momenta rather than the velocities, and considering the limit, as $\Delta x \to 0$ in $\mathbb{R}^n_x$, of the determinant of the matrix

$$\frac{\Delta p'}{\Delta x} = \left(\frac{\Delta p'_i}{\Delta x_j}\right)_{1 \leq i,j \leq n} . \tag{7.27}$$

Definition 235 *The limit of the determinant of the matrix (7.27) is called (when it exists) the* van Vleck determinant, *or the* Van Vleck density of trajectories. *It is denoted by* $\rho(x, x'; t, t')$. *Thus, by definition:*

$$\rho(x, x'; t, t') = \det \frac{\partial p'}{\partial x}. \tag{7.28}$$

Notice that ρ can take negative values. The van Vleck determinant is thus not a "density" in the usual sense. This is a genuine problem, because it intervenes in the construction of the wave function via its square root $\sqrt{\rho}$. We will use the Maslov index to determine the "right" argument for ρ, and this will allow us to determine unambiguously $\sqrt{\rho}$.

It turns out that the van Vleck determinant will exist provided that $t-t'$ is sufficiently small, because it is related to the existence of a generating function for the flow, and can be expressed in terms of that function:

Proposition 236 *There exists* $\varepsilon > 0$ *such that the density of trajectories is defined for all* $0 < |t - t'| < \varepsilon$. *In fact,* $\rho(x, x'; t, t')$ *is defined whenever the symplectomorphism* $f_{t,t'}$ *defined by* $(x, p) = f_{t,t'}(x', p')$ *is free. When this is the case, the function* ρ *is given by the formula*

$$\rho(x, x'; t, t') = \operatorname*{Hess}_{x,x'}(-W). \tag{7.29}$$

Proof. The existence of ρ follows from the fact that there exists $\varepsilon > 0$ such that $f_{t,t'}$ is free provided that $0 < |t - t'| < \varepsilon$ (see Lemma 98). Suppose from now on that this condition holds. Then the Jacobian determinant

$$\det\left(\frac{\partial(x, x')}{\partial(p', x')}\right) = \det\left(\frac{\partial x}{\partial p'}\right)$$

is different from zero (see Lemma 90). Recalling that $p = \nabla_x W$ and $p' = -\nabla_{x'} W$, we have

$$p_i' = -\frac{\partial W}{\partial x_i'}(x, x'; t, t')$$

for $1 \leq i \leq n$, and thus

$$\frac{\partial p'}{\partial x} = \left(-\frac{\partial^2 W}{\partial x_i' \partial x_j}\right)_{1\leq i,j\leq n}$$

from which formula (7.29) follows. ∎

Example 237 ***The tennis player hits again!*** *A free generating function associated to the tennis player of Examples 96, 234 is*

$$W(\mathbf{r},\mathbf{r}';t,t') = m\frac{(\mathbf{r}-\mathbf{r}')^2}{2(t-t')} - \frac{mg}{2}(y+y')(t-t')$$

and the corresponding van Vleck density is thus

$$\rho(x,x';t,t') = \frac{m}{(t-t')^2}$$

which coincides with the value obtained above.

Notice that in this Example ρ does not depend on the positions. More generally, this will always be the case when the Hamiltonian flow is linear (or affine):

Corollary 238 *Let H be a Maxwell Hamiltonian that is a quadratic polynomial in the position and momentum coordinates. Then the associated van Vleck density depends on time only. More precisely, if the quadratic form*

$$W = \frac{1}{2}Px^2 - Lx\cdot x' + \frac{1}{2}Qx'^2 + \alpha\cdot x + \alpha'\cdot x' \tag{7.30}$$

(where $P = P(t,t')$, and so on) is the generating function determined by H, then:

$$\rho(t,t') = \det L(t,t'). \tag{7.31}$$

The proof of this result is obvious, since formula (7.30) immediately follows from definition (7.29) of ρ.

7.2.2 The Continuity Equation for van Vleck's Density

We now set out to prove that the van Vleck density satisfies the continuity equation

$$\frac{\partial\rho}{\partial t} + \operatorname{div}(\rho v) = 0$$

where v is the velocity expressed in terms of the initial and final points x' and x. To prove this essential property, we need a technical result, which makes explicit the differential equation satisfied by the Jacobian determinant of systems of (autonomous, or non-autonomous) differential equations.

Consider a differential system

$$\dot{x}(t) = f(x(t),t) \;\;,\;\; x = (x_1,...,x_n) \;\;,\;\; f = (f_1,...,f_n) \tag{7.32}$$

where the f_j are real-valued functions defined in some open set $U \subset \mathbb{R}^n$. We assume that each of the solutions $x_1, ..., x_n$ depends smoothly on n parameters $\alpha_1, ..., \alpha_n$. Setting $\alpha = (\alpha_1, ..., \alpha_n)$, we write the solution of the system as $x = x(\alpha, t)$. It turns out that the Jacobian determinant of the mapping $(\alpha, t) \mapsto x(\alpha, t)$ satisfies a simple differential equation.

Let us first prove the following straightforward Lemma on matrices depending on a parameter:

Lemma 239 *The determinant of any invertible matrix $M(t)$ depending smoothly on t satisfies the differential equation*

$$\frac{d}{dt}\det M(t) = \det M(t)\,\mathrm{Tr}\left(\frac{dM(t)}{dt}M^{-1}(t)\right) \tag{7.33}$$

where "Tr" means "trace of".

Proof. Replacing, if necessary, $M(t)$ by $AM(t)$ where A is a conveniently chosen constant invertible matrix, we may assume without loss of generality that $||M(t) - I|| < 1$, and define the logarithm of $M(t)$ by the convergent series

$$\mathrm{Log}\, M(t) = \sum_{j=0}^{\infty}(-1)^{j+1}(M(t) - I)^{j-1}.$$

Writing $M(t) = \exp(\mathrm{Log}\, M(t))$ we have

$$\det M(t) = \exp(\mathrm{Tr}(\mathrm{Log}\, M(t)))$$

and hence, differentiating both sides of this equality:

$$\frac{d}{dt}\det M(t) = \left(\frac{d}{dt}\mathrm{Tr}(\mathrm{Log}\, M(t))\right)\det M(t).$$

This yields formula (7.33), because

$$\begin{aligned}\frac{d}{dt}\mathrm{Tr}(\mathrm{Log}\, M(t)) &= \mathrm{Tr}\left(\frac{d}{dt}\mathrm{Log}\, M(t)\right)\\ &= \mathrm{Tr}\left(M^{-1}(t)\frac{dM(t)}{dt}\right)\\ &= \mathrm{Tr}\left(\frac{dM(t)}{dt}M^{-1}(t)\right)\end{aligned}$$

where the last equality follows from the fact that we have $\mathrm{Tr}(AB) = \mathrm{Tr}(BA)$ for all $m \times m$ matrices A, B. ■

Proposition 240 *Let $x = x(\alpha, t)$ be a solution of the differential system (7.32) and suppose that the Jacobian determinant*

$$Y(\alpha, t) = \det\left(\frac{\partial x(\alpha, t)}{\partial(\alpha, t)}\right) \tag{7.34}$$

does not vanish for (α, t) in some open subset D of the product $\mathbb{R}^n \times \mathbb{R}_t$. Then Y satisfies the scalar differential equation

$$\frac{\partial Y}{\partial t}(\alpha, t) = Y(\alpha, t)\operatorname{Tr}\left(\frac{\partial f}{\partial x}(x(\alpha, t))\right). \tag{7.35}$$

Proof. We are following Maslov-Fedoriuk [114], p. 78. These authors give as earliest reference for Eq. (7.35) a paper of Sobolev [147]. We first note that by the chain rule we have the following identity between Jacobian matrices:

$$\frac{\partial}{\partial t}\left(\frac{\partial x(\alpha, t)}{\partial(\alpha, t)}\right) = \frac{\partial f}{\partial x}(x(\alpha, t)) \cdot \frac{\partial x(\alpha, t)}{\partial(\alpha, t)}. \tag{7.36}$$

Choosing

$$M(t) = \frac{\partial x(\alpha, t)}{\partial(\alpha, t)}$$

in Lemma 239 above, we see that

$$\frac{d}{dt}Y(\alpha, t) = Y(\alpha, t)\operatorname{Tr}\left[\frac{\partial}{\partial t}\left(\frac{\partial(x(\alpha, t))}{\partial(\alpha, t)}\right)\left(\frac{\partial x(\alpha, t)}{\partial(\alpha, t)}\right)^{-1}\right]$$

which is precisely Eq. (7.35) in view of Eq. (7.36). ∎

7.2.3 The Continuity Equation

We now use Proposition 240 to prove the main result of this section:

Proposition 241 *The function $(x, t) \longmapsto \rho(x, x'; t, t')$ satisfies, for fixed values of x' and t', the equation*

$$\frac{\partial \rho}{\partial t} + \operatorname{div}(\rho v) = 0 \tag{7.37}$$

where v is the velocity vector at x of the trajectory passing through that point at time t, and starting from x' at time t'. Thus

$$v = \nabla_p H(x, p, t) \quad \textit{if} \quad (x, p) = f_{t,t'}(x', p').$$

Proof. Keeping x' and t' fixed, we set $\rho(x,t) = \rho(x,x';t,t')$, and consider Hamilton's equations

$$\begin{cases} \dot{x}(t) = \nabla_p H(x(t),p(t),t), & x(t') = x' \\ \dot{p}(t) = -\nabla_x H(x(t),p(t),t), & p(t') = p' \end{cases}$$

where p' can be varied at will. The solution $x(t)$ of the first equation is thus parametrized only by p', since x' is fixed. We may thus apply Proposition 240 with $f = \nabla_p H$, $\alpha = p'$ and

$$Y(p',t) = \det \frac{\partial x(p',t)}{\partial(p',t)} = \begin{vmatrix} \frac{\partial x}{\partial p'} & \frac{\partial x}{\partial t} \\ 0_{1\times n} & 1 \end{vmatrix}.$$

The function Y is simply the inverse of ρ, calculated at $(x(t),t)$:

$$Y(p',t) = \frac{1}{\rho(x(t),t)}$$

and Eq. (7.35) yields

$$\frac{d}{dt}\left(\frac{1}{\rho(x(t),t)}\right) = \frac{1}{\rho(x(t),t)} \operatorname{Tr}\left[\frac{\partial}{\partial x}(\nabla_p H(x(t),p(t),t))\right]$$

that is

$$\frac{d}{dt}\rho(x(t),t) + \rho(x(t),t) \operatorname{Tr}\left[\frac{\partial}{\partial x}(\nabla_p H(x(t),p(t),t))\right] = 0. \tag{7.38}$$

Now,

$$\operatorname{Tr}\left[\frac{\partial}{\partial x}(\nabla_p H(x(t),p(t),t))\right] = \nabla_x \cdot \nabla_p H(x(t),p(t),t) \tag{7.39}$$

so that Eq. (7.38) can be rewritten

$$\frac{d}{dt}\rho(x(t),t) + \rho(x(t),t)\nabla_x \cdot \nabla_p H(x(t),p(t),t) = 0.$$

On the other hand, the total derivative of $t \longmapsto \rho(x(t),t)$ is

$$\begin{aligned} \frac{d}{dt}\rho(x(t),t) &= \frac{\partial \rho}{\partial t}(x(t),t) + \nabla_x \rho(x(t),t)\dot{x}(t) \\ &= \frac{\partial \rho}{\partial t}(x(t),t) + \nabla_x \rho(x(t),t) \cdot \nabla_p H(x(t),p(t),t). \end{aligned} \tag{7.40}$$

The continuity equation (7.37) follows. ■

Remark 242 *When using (7.37) one must be careful to express the velocity components in terms of the position coordinates. For instance, if H is the free particle Hamiltonian, the continuity equation is*

$$\frac{\partial \rho}{\partial t}+\frac{\partial}{\partial x}\left(\rho \frac{p_x}{m}\right)+\frac{\partial}{\partial y}\left(\rho \frac{p_y}{m}\right)=0$$

and the density $\rho(\mathbf{r},\mathbf{r}';t,t')=m^2/(t-t')^2$ that we defined solves the latter only if we use the values

$$p_x=\frac{x-x'}{t-t'},\quad p_y=\frac{y-y'}{t-t'}.$$

7.3 The Short-Time Propagator

In Section 5.2 we constructed the Green function for Schrödinger's equation for the free particle and found that it was given by

$$G_f(\mathbf{r},\mathbf{r}';t-t')=\left(\frac{m}{2\pi i\hbar(t-t')}\right)^{3/2}\exp\left(\frac{i}{\hbar}m\frac{(\mathbf{r}-\mathbf{r}')^2}{2(t-t')}\right).$$

We would now like to extend that construction to the case of an arbitrary system with Maxwell Hamiltonian

$$H=\sum_{j=1}^{n}\frac{1}{2m_j}\left(p_j-A_j(x,t)\right)^2+U(x,t).$$

Observing that the factor in front of the exponential in the expression of G_f is just $(2\pi i\hbar)^{-3/2}\sqrt{\rho_f}$, where ρ_f is the van Vleck density for the free particle, an educated guess is that the general Green function could be the function

$$G^{sh}=\left(\tfrac{1}{2\pi i\hbar}\right)^{n/2}\sqrt{\rho}e^{\frac{i}{\hbar}W} \tag{7.41}$$

where W is the generating function determined by H, and ρ the associated van Vleck density. We will see that this guess is right only when the Hamiltonian H is quadratic, but that in all cases G^{sh} is asymptotically close to the true Green function when $t-t'\to 0$. We therefore adopt the following (unconventional) terminology:

Definition 243 *The function G^{sh} defined by formula (7.41) is called the* short-time propagator *for Schrödinger's equation.*

The function G^{sh} is often called the "semi-classical propagator" in the physical literature. We will *not* use this terminology, because it is misleading: "semi-classical" usually refers to properties that are valid "for small $\hbar$" (which of course has no absolute meaning), while we will discuss limiting properties of G^{sh} for small *times*.

7.3.1 Properties of the Short-Time Propagator

Let $(f_{t,t'})$ be the time-dependent flow of H. The value t' being fixed, let $\varepsilon > 0$ be such that $f_{t,t'}$ is a (local) free symplectomorphism for $0 < |t - t'| < \varepsilon$ (see Chapter 3, Lemma 98). As we know, the generating function $W = W(x, x'; t, t')$ determined by H is a solution of the Hamilton-Jacobi equation

$$\frac{\partial W}{\partial t} + H\left(x, \nabla_x W, t\right) = 0 \tag{7.42}$$

and the associated van Vleck density

$$\rho(x, x'; t, t') = \det \operatorname*{Hess}_{x,x'}(-W) \tag{7.43}$$

satisfies the continuity equation (7.37), which we will use in the following form:

Lemma 244 *The square root $a = \sqrt{|\rho|}$ of van Vleck's density satisfies the equation*

$$\frac{\partial a}{\partial t} + v \cdot \nabla_x a + \frac{1}{2} a \operatorname{div} v = 0. \tag{7.44}$$

Proof. Writing the continuity equation (7.37) in the form

$$\frac{\partial \rho}{\partial t} + \rho \operatorname{div} v + v \cdot \nabla_x \rho = 0$$

Eq. (7.44) follows, inserting $\rho = a^2$. ∎

We next state and prove the main result of this section:

Theorem 245 *Let $\varepsilon > 0$ be such that the generating function W determined by H exists for all t such that $0 < |t - t'| < \varepsilon$. Then: (**1**) The short-time propagator G^{sh} has the property that*

$$\lim_{t \to t'} G^{sh}(x, x'; t, t') = \delta(x - x'); \tag{7.45}$$

(2) The function $(x,t) \longmapsto G^{sh}$ satisfies the equation

$$i\hbar\frac{\partial G^{sh}}{\partial t} = (\widehat{H} - Q)G^{sh} \tag{7.46}$$

where the function Q is given by

$$Q = -\frac{\hbar^2}{2M}\frac{\nabla_x^2\sqrt{|\rho|}}{\sqrt{|\rho|}} \tag{7.47}$$

($\frac{1}{M}\nabla_x^2 \equiv M^{-1}\nabla_x \cdot \nabla_x$ (M the mass matrix)).

Proof. For notational simplicity we give the proof in the case $n = 1$. The proof for $n > 1$ is similar. (**1**) In view of Proposition 115 (Chapter 4) we have, for $t - t' \to 0$:

$$W(x,x';t,t') = M\frac{(x-x')^2}{2(t-t')} - \overline{A}(x,x';t')(x-x') + O(t-t') \tag{7.48}$$

where $\overline{A}(x,x';t')$ is the average of A in $[x',x]$ at time t':

$$\overline{A}(x,x';t') = \int_0^1 A(sx' + (1-s)x, t')ds. \tag{7.49}$$

It follows that the van Vleck determinant satisfies the estimate

$$\rho(x,x';t,t') = \frac{m}{t-t'} + O((t-t')^0) \tag{7.50}$$

when $t - t' \to 0$, and hence, for $t > t'$:

$$\sqrt{\rho}(x,x';t,t') = \left(\frac{m}{t-t'}\right)^{1/2} + O\left((t-t')^{1/2}\right). \tag{7.51}$$

Combining Eq. (7.48) and Eq. (7.50), we thus have the short-time approximation

$$G^{sh}(x,x';t,t') = \left(\frac{1}{2\pi i\hbar}\right)^{1/2}\left(\frac{m}{t-t'}\right)^{1/2}$$
$$\times \exp\left[\frac{i}{\hbar}\left(m\frac{(x-x')^2}{2(t-t')} - \overline{A}(x,x';t')(x-x')\right)\right] + O\left((t-t')^{1/2}\right)$$

for $t > t'$. Now,

$$\lim_{t\to t'+}\left(\frac{1}{2\pi i\hbar}\right)^{1/2}\left(\frac{m}{t-t'}\right)^{1/2}\exp\left[\frac{im}{\hbar}\frac{(x-x')^2}{2(t-t')}\right] = \delta(x-x')$$

and hence

$$\lim_{t\to t'+} G(x,x';t,t') = \exp\left[\frac{i}{\hbar}\overline{A}(x,x';t')(x-x')\right]\delta(x-x')$$

that is, since $(x-x')\delta(x-x')=0$,

$$\lim_{t\to t'+} G(x,x';t,t') = \delta(x-x').$$

A similar argument shows that we have

$$\lim_{t\to t'-} G^{sh}(x,x';t,t') = \delta(x-x')$$

as well, which ends the proof of the first part of the Theorem. (**2**) Setting $a=\sqrt{|\rho|}$ as in Lemma 244, we have:

$$i\hbar\frac{\partial G^{sh}}{\partial t} = \left(\frac{1}{2\pi i\hbar}\right)^{n/2} e^{\frac{i}{\hbar}W}\left(-\frac{\partial W}{\partial t}a + i\hbar\frac{\partial a}{\partial t}\right) \tag{7.52}$$

and similarly

$$\left(i\hbar\frac{\partial}{\partial x}+A\right)G^{sh} = \left(\frac{1}{2\pi i\hbar}\right)^{n/2} e^{\frac{i}{\hbar}W}\left(-\frac{\partial W}{\partial x}a + i\hbar\frac{\partial a}{\partial x} + Aa\right)$$

hence

$$\begin{aligned}(2\pi i\hbar)^{n/2}e^{-\frac{i}{\hbar}W}\widehat{H}G^{sh} &= \left(\frac{\partial W}{\partial x}\right)^2 - 2Aa\frac{\partial W}{\partial x} - \hbar^2\frac{\partial^2 a}{\partial x^2} + A^2 a \\ &\quad + Ua - 2i\hbar\frac{\partial a}{\partial x}\frac{\partial W}{\partial x} - i\hbar\frac{\partial^2 W}{\partial x^2}a + i\hbar\frac{\partial(Aa)}{\partial x} + i\hbar A\frac{\partial a}{\partial x}.\end{aligned}$$

It follows that

$$\begin{aligned}&(2\pi i\hbar)^{n/2}e^{-\frac{i}{\hbar}W}\left(i\hbar\frac{\partial G^{sh}}{\partial t} - \widehat{H}G^{sh}\right) \\ &\quad = -\frac{\partial W}{\partial t} - H\left(x,\frac{\partial W}{\partial x},t\right) + \frac{\hbar^2}{2m}\frac{\partial^2 a}{\partial x^2} + \frac{\partial a}{\partial t} + \frac{\partial(av)}{\partial x} + \frac{1}{2}a\frac{\partial v}{\partial x}.\end{aligned}$$

Taking Hamilton-Jacobi's equation (7.42) and Eq. (7.44) into account yields

$$i\hbar\frac{\partial G^{sh}}{\partial t} - \widehat{H}G^{sh} = \left(\frac{1}{2\pi i\hbar}\right)^{n/2} e^{\frac{i}{\hbar}W}\frac{\hbar^2}{2m}\frac{\partial^2 a}{\partial x^2}$$

which is Eq. (7.46). ∎

It follows from Theorem 245 that G^{sh} is the Green function of an integro-differential equation:

Corollary 246 *For every $\Psi' \in \mathcal{S}(\mathbb{R}^n_x)$ the function*

$$\Psi(x,t) = \int G^{sh}(x,x';t,t')\Psi'(x')\,d^n x' \tag{7.53}$$

satisfies the integro-differential Cauchy problem

$$i\hbar\frac{\partial\Psi}{\partial t} = (\widehat{H} - \widehat{Q})\Psi, \quad \Psi(\cdot,t') = \Psi'(\cdot) \tag{7.54}$$

where $\widehat{Q}$ is the operator $\mathcal{S}(\mathbb{R}^n_x) \longrightarrow \mathcal{S}(\mathbb{R}^n_x)$ defined by

$$\widehat{Q}\Psi(x,t) = \int Q(x,x';t,t')G^{sh}(x,x';t,t')\Psi'(x')\,d^n x' \tag{7.55}$$

where Q is the function defined by Eq. (7.47).

Proof. Differentiation under the integration sign on the right-hand side of the expression (7.53) leads to the equality

$$\begin{aligned} i\hbar\frac{\partial\Psi}{\partial t} - \widehat{H}\Psi &= \int \left(i\hbar\frac{\partial G^{sh}}{\partial t} - \widehat{H}G^{sh} \right) \Psi' d^n x' \\ &= \frac{\hbar^2}{2m}\int \frac{\nabla_x^2\sqrt{|\rho|}}{\sqrt{|\rho|}} G^{sh}\Psi' d^n x \end{aligned}$$

which is (7.54). ■

7.4 The Case of Quadratic Hamiltonians

It turns out that the short-time propagator is the exact propagator (i.e. the Green function for Schrödinger's equation) when the Hamiltonian is a second degree polynomial in the position and momentum variables.

7.4.1 Exact Green Function

Theorem 245 allows us to prove that the metaplectic representation yields the solutions to Schrödinger's equation for all quadratic Maxwell Hamiltonians. Suppose in fact that

$$H = \sum_{j=1}^{n} \frac{1}{2m_j}(p_j - A_j \cdot x)^2 + \frac{1}{2}Kx^2 + a \cdot x \tag{7.56}$$

($A_j = A_j(t)$ a time-dependent vector, K symmetric) which we write as usual in the short-hand form

$$H = \frac{1}{2m}(p - Ax)^2 + \frac{1}{2}Kx^2 + a \cdot x. \tag{7.57}$$

The corresponding quantum operator is:

$$\hat{H} = -\frac{1}{2m}\left(-i\hbar\nabla_x - A\right)^2 + \frac{1}{2}Kx^2. \tag{7.58}$$

Theorem 245 implies that:

Corollary 247 *When H is of the type (7.57), the function G^{sh} is the Green function for the associated Schrödinger equation, that is:*

$$i\hbar\frac{\partial G^{sh}}{\partial t} = \hat{H}G^{sh}, \quad \lim_{t \to t'} G^{sh} = \delta(x - x'). \tag{7.59}$$

Proof. The generating function W is itself a quadratic polynomial in the variables x_i, x'_j (with coefficients depending on t, t'), so that the matrix $W''_{x,x'}$, and hence the van Vleck density

$$\rho = \operatorname*{Hess}_{x,x'}(-W) = \det(-W''_{x,x'})$$

will depend on t and t' only. It follows that

$$\frac{\hbar^2}{2m}\frac{\nabla_x^2\sqrt{|\rho|}}{\sqrt{|\rho|}}e^{\frac{i}{\hbar}W} = 0$$

and hence Eq. (7.46) reduces to

$$i\hbar\frac{\partial G^{sh}}{\partial t} = \hat{H}G^{sh}.$$

That we have $\lim_{t\to t'} G^{sh} = \delta(x - x')$ is true, whether H is quadratic or not, in view of **(1)** in Theorem 245. ∎

7.4.2 Exact Solutions of Schrödinger's Equation

As a consequence of Corollary 247, we get the following "recipe" for solving exactly Schrödinger's equation associated to a quadratic Hamiltonian, by using only the generating function it determines. Assume in fact that this generating function is

$$W = \tfrac{1}{2}P(t,t')x^2 - L(t,t')x \cdot x' + \tfrac{1}{2}Q(t,t')x^2 + a(t,t') \cdot x + b(t,t') \cdot x'$$

(see Chapter 4, Subsection 4.2.2). The van Vleck density is here

$$\rho(x,x';t,t') = \det L(t,t')$$

and we define the Maslov index $m(t,t')$ by

$$m(t,t') = m(S_{t,t'})$$

where $t \longmapsto S_{t,t'}$ is the lift to $\mathrm{Mp}(n)$, passing through $I \in \mathrm{Mp}(n)$ at time $t = t'$ of the path $t \longmapsto s_{t,t'}$. Using the results of Section 6.5 of the last Chapter, we have

$$m(t,t') = \begin{cases} 0 \text{ if } 0 < t - t' < \varepsilon \\ 1 \text{ if } -\varepsilon < t - t' < 0 \end{cases} \tag{7.60}$$

if W is defined for $0 < |t - t'| < \varepsilon$. For arbitrary (t,t') it can be calculated by repeated use of Eq. (6.68), which yields

$$m(t,t') = m(t,t'') + m(t'',t') - n \\ + \operatorname*{Inert\,Hess}_{x'} \left[-(W(0,x';t,t'') + W(x',0;t'',t')) \right]$$

when $0 < |t - t''| < \varepsilon$ and $0 < |t'' - t'| < \varepsilon$.

Corollary 248 *The solution Ψ of Schrödinger's equation*

$$i\hbar \frac{\partial \Psi}{\partial t} = \widehat{H}\Psi, \quad \Psi(x,t') = \Psi'(x)$$

is given by the formula

$$\Psi(x,t) = \left(\tfrac{1}{2\pi i\hbar}\right)^{n/2} i^{m(t,t')} \sqrt{|\det L(t)|} \int e^{\frac{i}{\hbar}W(x,x';t,t')}\Psi'(x')d^n x'$$

where the Maslov index $m(t,t')$ is defined by Eq. (7.60).

Proof. It immediately follows from Corollaries 238 and 247; we leave it to the reader to check that the Maslov index $m(t,t')$ indeed ensures us that the mapping $t \longmapsto \Psi(x,t)$ is continuous. ∎

7.5 Solving Schrödinger's Equation: General Case

We now no longer assume that H is quadratic, but that it is a general Maxwell Hamiltonian on $\mathbb{R}^n_x \times \mathbb{R}^n_p \times \mathbb{R}_t$, which we write as usual in the form

$$H = \frac{1}{2m}(p - A)^2 + U$$

with $A = A(x,t)$, $U = U(x,t)$, and m the mass matrix. We consider the Schrödinger equation

$$i\hbar\frac{\partial\Psi}{\partial t} = \widehat{H}\Psi$$

where the operator $\widehat{H}$ is obtained from H by using the following generalized Schrödinger quantization rule: to every function

$$F(x,p,t) = \sum_{j=1}^{n} p_j A_j(x,t) \tag{7.61}$$

where $x = (x_1, ..., x_n)$, $p = (p_1, ..., p_n)$ this rule associates the partial differential operator $\widehat{F} = \widehat{F}(x, -i\hbar\nabla_x, t)$ obtained by replacing formally the coordinates p_j by $-i\hbar(\partial/\partial x_j)$ in the symmetrized products

$$\tfrac{1}{2}(p_j A_j(x,t) + A_j(x,t)p_j).$$

This rule leads to the operator

$$\widehat{H} = \sum_{j=1}^{n} \frac{1}{2m_j}\left(-i\hbar\frac{\partial}{\partial x_j} - A_j\right)^2 + U$$

which we write simply as:

$$\widehat{H} = \frac{1}{2m}(-i\hbar\nabla_x - A)^2 + U.$$

7.5.1 The Short-Time Propagator and Causality

The exact Green function is in general no longer identical with the short-time propagator, so that we cannot expect to solve the Cauchy problem for Schrödinger's equation in "closed form" as we did in Corollary 247. In fact, the term QG^{sh} appearing in the equation

$$i\hbar\frac{\partial G^{sh}}{\partial t} = (\widehat{H} - Q)G^{sh} \tag{7.62}$$

(cf. Theorem 245) only vanishes when

$$\frac{\hbar^2}{2m}\nabla_x^2\sqrt{|\rho|} = \sum_{j=1}^{n}\frac{\hbar^2}{2m_j}\frac{\partial^2\sqrt{|\rho|}}{\partial x_j^2} = 0.$$

For instance, if $n = 1$, this condition implies that the van Vleck density must have the particular form

$$\rho(x, x't, t') = (a(x'; t, t')x + b(x'; t, t'))^2.$$

Of course, there is nothing wrong *per se* with Eq. (7.62). One could, for instance, argue that when Schrödinger "derived" his equation, he made the "right" guess only for quadratic Hamiltonians, and that the "true" equation for wave functions is perhaps, after all, the integro-differential equation (7.62). Moreover, since the term Q is apparently "small" (because $\hbar^2$ is "small"), the solutions of Schrödinger's equation might be "approximations" to those of that equation. However, if we decide to take Eq. (7.62) as the equation governing quantum mechanics, then we would at the same time have to renounce to causality! This is because the "evolution operator" $(U_{t,t'})$ defined by

$$U_{t,t'}\Psi'(x) = \int G^{sh}(x, x'; t, t')\Psi'(x')\, d^n x' \tag{7.63}$$

is not in general a group; in fact we have in general:

$$U_{t,t'}U_{t',t''} \neq U_{t,t''}. \tag{7.64}$$

There is however a way to restore causality: suppose, for instance, that $t' < t$ and consider a subdivision

$$t' < t_1 < \cdots < t_{N-1} < t$$

of the interval $[t', t]$ such that each $|t_{j+1} - t_j|$ is sufficiently small. The "time-ordered products"

$$\Pi_{t,t'}(N) = U_{t,t_1}U_{t_1,t_2}\cdots U_{t_{N-1},t'}$$

will then converge, as $N \to \infty$, to a limit $F_{t,t'}$. The family of operators $(F_{t,t'})$ thus defined will satisfy the groupoid property

$$F_{t,t'}F_{t',t''} = F_{t,t''}. \tag{7.65}$$

This can be easily seen, noting that we have:

$$\begin{aligned} F_{t,t'}F_{t',t''} &= \lim_{N\to\infty} \Pi_{t,t'}(N) \lim_{N\to\infty} \Pi_{t',t''}(N) \\ &= \lim_{N\to\infty} \Pi_{t,t''}(N) \\ &= F_{t,t''}. \end{aligned}$$

It turns out that if we now define

$$\Psi(x,t) = F_{t,t'}\Psi'(x),$$

then the function Ψ will moreover satisfy Schrödinger's equation

$$i\hbar\frac{\partial\Psi}{\partial t} = \widehat{H}\,\Psi$$

with initial datum $\Psi(\cdot,t') = \Psi'$.

Of course, all this has to be put on a rigorous basis. This will be done in the forthcoming Subsections.

7.5.2 Statement of the Main Theorem

We begin by stating precisely the result we are going to prove.

Theorem 249 *Let N be an integer superior or equal to one, and set $\Delta t = (t-t')/N$. The function*

$$\Psi(x,t) = \lim_{N\to\infty} \prod_{j=0}^{N-1} U_{t-j\Delta t,t-(j+1)\Delta t}\Psi'(x) \tag{7.66}$$

is a solution of Schrödinger's equation

$$i\hbar\frac{\partial\Psi}{\partial t} = \widehat{H}\,\Psi,\ \ \Psi(x,t') = \Psi'(x).$$

Remark 250 *This theorem is essential from both a mathematical and physical point of view, because it makes clear that the generating function, that is, ultimately, the classical flow determined by H suffices to determine the wave function.*

The proof of Theorem 249 will be made in several steps. We will begin by proving an intermediary result, Proposition 251 below, which is interesting by itself. It provides us with a practical algorithm for calculating the solution Ψ which converges faster than the usual Feynman formula. Before we state that result, let us introduce the following notations: we set

$$\overline{G} = \left(\tfrac{1}{2\pi i\hbar}\right)^{n/2} \sqrt{\overline{\rho}} e^{\frac{i}{\hbar}\overline{W}}, \quad \overline{\rho} = \operatorname*{Hess}_{x,x'}(-\overline{W}).$$

Here $\overline{W}$ is the short-time approximation to the generating function W determined by H (see Chapter 4, Corollary 112), that is:

$$\overline{W}(x,x';t,t') = m\frac{(x-x')^2}{2(t-t')} - \overline{U}(x,x',t')(t-t')$$

$\overline{U}(x,x',t')$ being the average of U on $[x',x]$.

Proposition 251 *Let $\overline{F}_{t,t'} : \mathcal{S}(\mathbb{R}^n_x) \longrightarrow \mathcal{S}(\mathbb{R}^n_x)$ be defined by*

$$\overline{F}_{t,t'}\Psi'(x) = \int \overline{G}(x,x';t,t')\Psi'(x')\,d^nx'. \tag{7.67}$$

The solution of Schrödinger's equation is given by

$$\Psi(x,t) = \lim_{N\to\infty} \prod_{j=0}^{N-1} \overline{F}_{t-j\Delta t,t-(j+1)\Delta t}\Psi'(x). \tag{7.68}$$

The proof of Proposition 251 relies on the method of stationary phase, which we review below.

7.5.3 The Formula of Stationary Phase

We begin by recalling the formula of stationary phase which allows to calculate asymptotic expansions of integrals depending on a parameter (see for instance Leray [101], Ch. II, §1.3). Assume that Φ is a real non-degenerate quadratic form on $\mathbb{R}^n_x$:

$$\Phi(x) = \tfrac{1}{2}Mx\cdot x,\ M = M^T,\ \det M \neq 0.$$

and let Φ^* be the *dual form*

$$\Phi^*(p) = -\tfrac{1}{2}M^{-1}p\cdot p$$

(note that $\Phi^*(p)$ is just the critical value of the function $x \longmapsto \Phi(x)+p\cdot x$). Then, for $\alpha \in \mathcal{S}(\mathbb{R}^n_x)$ the integral

$$I(\lambda) = \int e^{\frac{i}{\lambda}\Phi(x)}\alpha(x)\,d^nx \quad (\lambda > 0)$$

has the following asymptotic expansion as $\lambda \to 0$ ("formula of stationary phase"):

$$I(\lambda) \sim (2\pi i\lambda)^{n/2}\left[\operatorname{Hess}\Phi\right]^{-1/2}\left[e^{i\lambda\Phi^*(\nabla_x)}\alpha(x)\right]_{x=0} \tag{7.69}$$

where the argument of $\operatorname{Hess}\Phi = \det M$ is defined by

$$\arg\det\operatorname{Hess}\Phi = \pi\operatorname{Inert}\Phi = \pi\operatorname{Inert}M \tag{7.70}$$

($\operatorname{Inert}M$: the number of negative eigenvalues of M), and

$$e^{i\lambda\Phi^*(\nabla_x)}\alpha(x) = \sum_{j=0}^{\infty}\frac{(-i\lambda)^j}{j!}\Phi^*\left(\nabla_x\right)^j\alpha(x).$$

We will use the following consequence of formula (7.69) in the proof of Theorem 249:

Lemma 252 *Let f be a smooth function of (x',x). The integral*

$$I(x,\Delta t) = \int \exp\left(\frac{i}{\hbar}m\frac{(x-x')^2}{2\Delta t}\right)f(x,x')d^nx'$$

has the following asymptotic expansion as $\Delta t \to 0$:

$$I(x,t) = \frac{(2\pi i\hbar\Delta t)^{n/2}}{\sqrt{|m|}}\left(f(x,x) + \left(\frac{i\hbar\Delta t}{2m}\nabla_x^2 f\right)(x,x)\right) + O\left(\Delta t^2\right) \tag{7.71}$$

with $|m| = \det m$, $\arg\Delta t = 0$ if $\Delta t > 0$ and π if $\Delta t < 0$.

Proof. Performing the change of variables $x - x' \longrightarrow x'$ we can rewrite the integral as

$$I(x,t) = \int \exp\left(\frac{i}{\hbar}m\frac{x'^2}{2t}\right)f(x,x-x')\,d^nx'.$$

Setting $\lambda = t$ and $\Phi(x') = \frac{1}{2\hbar}mx'^2$, we have

$$\operatorname{Hess}\Phi = \hbar^{-n}|m|,\quad \Phi^*(\nabla_x) = -\frac{\hbar}{2m}\nabla_x^2$$

and the estimate (7.71) follows, applying the formula of stationary phase until order 2. ∎

7.5.4 Two Lemmas – and the Proof

To make the calculations more tractable we assume that H is the Hamiltonian of a particle in $\mathbb{R}^n_x$ moving under the action of a scalar potential $U = U(x,t)$. The extension to the case where a vector potential is present is straightforward. We thus have, with our usual notations:

$$H = \sum_{j=1}^{n} \frac{p^2}{2m_j} + U(x,t) = \frac{p^2}{2m} + U(x,t). \tag{7.72}$$

We begin by proving two preparatory Lemmas.

Lemma 253 *Let W_f and ρ_f be the free particle generating function and the associated van Vleck density:*

$$W_f(x,x';t) = m\frac{(x-x')^2}{2t} \quad , \quad \rho_f(t) = |m|t^{-n}.$$

There exist smooth functions a_k $(k = 1,2,...)$ of (x,x',t') such that for $\Delta t = t - t' \to 0$:

$$\overline{G} \sim \left(\frac{1}{2\pi i\hbar}\right)^{n/2} \sqrt{\rho_f} e^{\frac{i}{\hbar}W_f} \left(1 - \frac{i}{\hbar}\overline{U}\Delta t + \sum_{k=1}^{\infty} a_k \Delta t^{2k}\right). \tag{7.73}$$

Proof. Since $\overline{W} = W_f - \overline{U}\Delta t$ we have

$$\overline{W}''_{x,x'} = (W_f)''_{x,x'} - \overline{U}''_{x,x'}\Delta t.$$

Now, $(W_f)''_{x,x'} = \frac{1}{\Delta t}m$ and hence

$$\overline{W}''_{x,x'} = \frac{1}{\Delta t}m\left(I_{n\times n} - \overline{U}''_{x,x'}\Delta t^2\right)$$

from which follows that

$$\begin{aligned}\overline{\rho}(x,x';t,t') &= \det\left(\frac{1}{\Delta t}m\right)\det\left(I_{n\times n} - \overline{U}''_{x,x'}\Delta t^2\right) \\ &= |m|(\Delta t)^{-n}\det\left(I_{n\times n} - \overline{U}''_{x,x'}\Delta t^2\right) \\ &= \rho_f(x,x',\Delta t)\det\left(I_{n\times n} - \overline{U}''_{x,x'}\Delta t^2\right).\end{aligned}$$

Expanding the determinant in the last equality, we find that

$$\det\left(I_{n\times n} - \overline{U}''_{x,x'}\Delta t^2\right) = 1 + \sum_{k=1}^{n} b_k \Delta t^{2k}$$

where $b_k = b_k(x, x', t')$. It follows, by the binomial theorem that

$$\left(1 + \sum_{k=1}^{n} b_k \Delta t^{2k}\right)^{1/2} \sim 1 + \frac{3}{8}\left(\sum_{k=1}^{\infty} b_k \Delta t^{2k}\right) + \cdots$$

so we have an asymptotic expansion

$$\sqrt{\overline{\rho}} \sim \sqrt{\rho_f}\left(1 + \sum_{k=1}^{\infty} b'_k \Delta t^{2k}\right) \tag{7.74}$$

with coefficients being smooth functions $b'_1, b'_2, \ldots$. On the other hand, $e^{\frac{i}{\hbar}\overline{W}} = e^{\frac{i}{\hbar}W_f} e^{-\frac{i}{\hbar}\overline{U}\Delta t}$ so that

$$e^{\frac{i}{\hbar}\overline{W}} = e^{\frac{i}{\hbar}W_f}\left(1 - \frac{i}{\hbar}\overline{U}\Delta t + \sum_{k=2}^{\infty} \frac{(-1)^k}{k!}\overline{U}^k \Delta t^k\right) \tag{7.75}$$

and Eq. (7.73) follows, by taking the product of the asymptotic expansions (7.74) and (7.75). ■

Lemma 254 *Let Ψ be the exact solution of Schrödinger's equation*

$$i\hbar\frac{\partial \Psi}{\partial t} = -\frac{\hbar^2}{2m}\nabla_x^2 \Psi + U\Psi, \quad \Psi(\cdot, t') = \Psi'(\cdot)$$

associated with the Hamiltonian (7.72), and

$$\overline{\Psi}(x,t) = \left(\tfrac{1}{2\pi i\hbar}\right)^{n/2} \int e^{\frac{i}{\hbar}\overline{W}(x,x';t,t')} \sqrt{\overline{\rho}}(x,x';t,t')\Psi'(x')\, d^n x'.$$

We have, for $t \to t'$,

$$\Psi(x,t) - \overline{\Psi}(x,t) = O\left(\Delta t^2\right) \tag{7.76}$$

where $\Delta t = t - t'$.

Proof. Let us first find a short-time expansion of $\Psi(x,t)$. In view of Taylor's formula we have

$$\Psi(x,t) = \Psi(x,t') + \frac{\partial \Psi}{\partial t}(x,t')\Delta t + O\left(\Delta t^2\right)$$

and hence, using Schrödinger's equation to compute the partial derivative and using the equality $\Psi(x,t') = \Psi'(x)$:

$$\Psi(x,t) = \left[1 + \frac{\Delta t}{i\hbar}\left(\frac{-\hbar^2}{2m}\nabla_x^2 + U(x,t')\right)\right]\Psi'(x) + O\left(\Delta t^2\right). \tag{7.77}$$

The next step in the proof of the Lemma consists in estimating $\overline{\Psi}(x,t)$; we will use for that purpose formula (7.71) in Lemma 252. In view of formula (7.73) in Lemma 253 we have

$$\begin{aligned}\overline{\Psi}(x,t) &= \left(\tfrac{1}{2\pi i\hbar}\right)^{n/2}\int e^{\frac{i}{\hbar}\overline{W}(x,x';t,t')}\sqrt{\overline{\rho}}(x,x';t,t')\Psi'(x')\,d^nx' \\ &= \left(\tfrac{1}{2\pi i\hbar}\right)^{n/2}\sqrt{\rho_f}\,(A(x,t)+B(x,t))\end{aligned}$$

where A and B are the integrals

$$\begin{aligned}A(x,t) &= \int e^{\frac{i}{\hbar}W_f}\left(1 - \frac{i}{\hbar}\overline{U}\Delta t\right)\Psi'(x')d^nx' \\ B(x,t) &= \int e^{\frac{i}{\hbar}W_f}\left(\sum_{k=1}^{\infty} a_k\Delta t^{2k}\right)\Psi'(x')d^nx'.\end{aligned}$$

Writing

$$A(x,t) = \int e^{\frac{i}{\hbar}W_f}\Psi'(x')\,d^nx' - \frac{i\Delta t}{\hbar}\int e^{\frac{i}{\hbar}W_f}\overline{U}\Psi'(x')\,d^nx'$$

and applying Lemma 252 to both integrals in the right-hand side, we get successively

$$\int e^{\frac{i}{\hbar}W_f}\Psi'(x')\,d^nx' = \frac{(2\pi i\hbar\Delta t)^{n/2}}{\sqrt{|m|}}\left(1 + \frac{i\hbar\Delta t}{2m}\nabla_x^2\right)\Psi'(x) + O\left(\Delta t^2\right)$$

and

$$\begin{aligned}\int e^{\frac{i}{\hbar}W_f}\overline{U}\Psi'(x')\,d^nx' &= \frac{(2\pi i\hbar\Delta t)^{n/2}}{\sqrt{|m|}}\overline{U}(x,x,t')\Psi'(x) + O\left(\Delta t^2\right) \\ &= \frac{(2\pi i\hbar\Delta t)^{n/2}}{\sqrt{|m|}}U(x,t')\Psi'(x) + O\left(\Delta t^2\right)\end{aligned}$$

so that

$$A(x,t) = \frac{(2\pi i\hbar\Delta t)^{n/2}}{\sqrt{|m|}}\left[\left(1+\frac{i\hbar}{2m}\nabla_x^2-\frac{i}{\hbar}U\right)\Delta t+O\left(\Delta t^2\right)\right]\Psi'(x).$$

We proceed similarly to estimate the term $B(x,t)$; Lemma 252 yields this time

$$\begin{aligned}B(x,t) &= \frac{(2\pi i\hbar\Delta t)^{n/2}}{\sqrt{|m|}}\sum_{k=1}^{\infty}a_k\Delta t^{2k}\int e^{\frac{i}{\hbar}W_f}\Psi'\,d^nx' \\ &= O\left(\Delta t^{n/2+2}\right)\Psi'(x)\end{aligned}$$

and hence, by definition of ρ_f:

$$\begin{aligned}\overline{\Psi}(x,t) &= \left(\frac{1}{2\pi i\hbar}\right)^{n/2}\sqrt{\rho_f}\,(A(x,t)+B(x,t)) \\ &= \left[\left(1+\frac{i\hbar}{2m}\nabla_x^2-\frac{i}{\hbar}\overline{U}\right)\Delta t+O\left(\Delta t^2\right)\right]\Psi'(x)\end{aligned}$$

that is

$$\overline{\Psi}(x,t) = \left[1+\frac{\Delta t}{i\hbar}\left(\frac{-\hbar^2}{2m}\nabla_x^2+\overline{U}\right)+O\left(\Delta t^2\right)\right]\Psi'(x). \tag{7.78}$$

Comparing the expressions (7.77) and (7.78) yields Eq. (7.76). ∎

Let us now prove Proposition 251. We denote by $(F_{t,t'})$ the evolution group determined by Schrödinger's equation. Thus, $F_{t,t'}$ is the operator which to the wave function at time t' associates the wave function at time t:

$$F_{t,t'} : \Psi(\cdot,t') \longrightarrow \Psi(\cdot,t).$$

In view of Eq. (7.76), we have, since $(F_{t,t'})^{-1} = F_{t',t}$:

$$\overline{F}_{t,t'} = F_{t,t'} + O\left(\Delta t^2\right) = F_{t,t'}(1+F_{t',t}(O\left(\Delta t^2\right)))$$

that is

$$\overline{F}_{t,t'} = F_{t,t'}(1+O(\Delta t^2)).$$

We now set, for $N \geq 1$:

$$\Delta t = \frac{t-t'}{N}.$$

We have

$$\begin{aligned}\prod_{j=0}^{N-1} \overline{F}_{t-j\Delta t,t-(j+1)\Delta t} &= \prod_{j=0}^{N-1} F_{t-j\Delta t,t-(j+1)\Delta t}(1+O\left(\Delta t^2\right)) \\ &= \prod_{j=0}^{N-1} F_{t-j\Delta t,t-(j+1)\Delta t} + NO\left(\Delta t^2\right).\end{aligned}$$

Since

$$NO\left(\Delta t^2\right) = NO\left(\left(\tfrac{t-t'}{N}\right)^2\right) = (t-t')O\left(\tfrac{t-t'}{N}\right) = O\left(\Delta t\right)$$

it follows, using the groupoid property $F_{t,t'}F_{t',t''} = F_{t,t''}$ that

$$\prod_{j=0}^{N-1} \overline{F}_{t-j\Delta t,t-(j+1)\Delta t} = F_{t,t'} + O(\Delta t)$$

and hence

$$\lim_{N\to\infty} \prod_{j=0}^{N-1} \overline{F}_{t-j\Delta t,t-(j+1)\Delta t} = F_{t,t'}$$

which proves Proposition 251.

Theorem 249 readily follows from this result: in view of the estimates

$$W = \overline{W} + O\left(\Delta t^2\right), \quad \rho = \overline{\rho} + O\left(\Delta t^2\right)$$

(the second immediately following from the first), we have

$$U_{t,t'}\Psi'(x') = \overline{F}_{t,t'}\Psi'(x') + O\left(\Delta t^2\right).$$

By the same argument as above, we thus have

$$\lim_{N\to\infty} \prod_{j=0}^{N-1} U_{t-j\Delta t,t-(j+1)\Delta t} = \lim_{N\to\infty} \prod_{j=0}^{N-1} \overline{F}_{t-j\Delta t,t-(j+1)\Delta t}$$

and hence

$$\lim_{N\to\infty} \prod_{j=0}^{N-1} U_{t-j\Delta t,t-(j+1)\Delta t} = F_{t,t'}$$

which is Eq. (7.66); the proof of Theorem 249 is thus complete.

Remark 255 *It would be interesting to compare the speed of convergence of the algorithm defined by Feynman's formula (1.42) with that of (7.68). Since the "propagator" $\overline{G}$ is a better approximation (for small times) to the true Green function G than the one appearing in Feynman's formula, it seems plausible that (7.68) leads to a higher accuracy for the same number N of steps.*

7.6 Metatrons and the Implicate Order

Recall from Chapter 1, Subsection 1.8.1, that in Bohmian mechanics the wavefunction Ψ can be viewed as a sort of "guiding field" determining a "quantum motion" described by the system of first order differential equations

$$\dot{\mathbf{r}}^{\Psi} = \frac{\hbar}{m} \operatorname{Im} \frac{\nabla_{\mathbf{r}} \Psi}{\Psi}. \tag{7.79}$$

More generally, the quantum motion of N particles with masses $m_1, ..., m_N$ and position vectors $\mathbf{r}_1$,..., $\mathbf{r}_N$ is governed by the equations

$$\dot{\mathbf{r}}_j^{\Psi} = \frac{\hbar}{m_j} \operatorname{Im} \frac{\nabla_{\mathbf{r}_j} \Psi}{\Psi} \qquad (1 \leq j \leq N)$$

which we can rewrite in compact form as

$$\dot{x}^{\Psi} = \frac{\hbar}{m} \operatorname{Im} \frac{\nabla_x \Psi}{\Psi} \tag{7.80}$$

where m is the mass matrix. As was shown in Subsection 1.8.2 of Chapter 1, these equations are equivalent to the Hamilton equations

$$\dot{x}^{\Psi} = \nabla_p (H + Q^{\Psi}), \;\; \dot{p}^{\Psi} = -\nabla_x (H + Q^{\Psi}) \tag{7.81}$$

where Q^{Ψ} is the quantum potential

$$Q^{\Psi} = -\frac{\hbar^2}{2m} \frac{\nabla_x^2 \sqrt{|\Psi|}}{\sqrt{|\Psi|}} \equiv -\frac{\hbar^2}{2} \frac{1}{\sqrt{|\Psi|}} m^{-1} \nabla_x \cdot \nabla_x \sqrt{|\Psi|}.$$

(This was actually the original form proposed by Bohm [18, 19].)

Notice that since the wave function Ψ in general depends on all the variables $x = (\mathbf{r}_1, ..., \mathbf{r}_N)$, Eqs. (7.80)–(7.81) will be systems of *coupled* differential equations.

Since Eqs. (7.79)–(7.80) depend on the wave function, and the latter is ultimately "produced" by the metaplectic representation, we propose to call the entity whose motion is governed by these equations a *metatron*. Since a particle in $n = 3N$ dimensional space is the same as N particles in ordinary physical space, a metatron is an essentially *non-local* entity.

We will, in this section, address a rather difficult question, which poses severe interpretational and epistemological problems, and which is best understood using Bohm and Hiley's notion of "implicate order", which we briefly discuss below. That question is:

What is a recorded metatron's phase space trajectory like?

We will see that the answer is: it is a perfectly *classical* trajectory! Is this to say that Bohmian trajectories are therefore not "real", that they are "surrealistic", as claimed by Englert *et al.* in [43]? *No*, not necessarily, because there is a distinction between what *is*, and what is *observed* by a *physical* measurement. The two-slit experiment is a textbook illustration: even if we cannot observe through which slit the particle went, it has followed a well-defined trajectory, which we can *retrodict* once the electron has provoked a scintillation on the screen behind the two slits.

7.6.1 Unfolding and Implicate Order

We begin by shortly discussing the "implicate order" and "enfolding–unfolding process" of Bohm (Bohm and Hiley [20], Hiley [87]). The easiest way to explain this idea is to use Bohm famous metaphor (which he reputedly thought about after having watched a popular science television program). Consider the following contraption: a hollow outer cylinder containing a concentric inner cylinder; between both cylinders one pours glycerine (which has high viscosity and therefore prevents diffusion). One then introduces a droplet of ink, or any other dye, at some suitable point. If the inner cylinder is slowly rotated, the ink droplet disappears after a while. There is, of course, nothing remarkable about that. However, if the inner cylinder is rotated in the opposite direction, the ink droplet re-appears! In the spirit of the implicate order, even if we couldn't see it, the droplet was "enfolded" in the glycerine, and was made manifest again by rotating the cylinder in the opposite direction.

Bohm's metaphor is intended to bring out the fact that if the basic process was activity, then the "track" left in, say, a bubble chamber could be explained by a similar *enfolding–unfolding* process. Thus, to quote Hiley [87], rather than seeing the track as the continuous movement of a material particle, it can be regarded as the continuity of a "quasi-stable form" evolving within the unfolding process. As we will see, this is exactly what happens when we observe the recorded phase space trajectory of a particle, because that track can be viewed as an "unfolding", in fact a visible trace left by something much more fundamental than a material particle – a *metatron*.

7.6.2 Prediction and Retrodiction

Equations (7.81) reduce to the classical Hamilton equations

$$\dot{x} = \nabla_p H, \;\; \dot{p} = -\nabla_x H$$

when $\nabla_x Q^\Psi = 0$, that is, when the quantum potential is constant. (See Holland, [90] (especially §6.1–6.3) for examples.) However, even in this case, we must not forget that a human observer cannot, in principle, use these Hamilton equations to predict the metatron's motion, because any prediction would require the datum of simultaneous initial conditions for both position and momentum, in contradiction with the uncertainty principle. (We are talking here about a real physical situation, involving a measurement apparatus, *not* about the platonic mathematical situation where we are, of course, free to assign any value that we like to the variables x and p.) This can actually also be seen without invoking Heisenberg's inequalities, by returning to the equations of motion. Let us begin by discussing in some detail the case where there is no external potential. (Classically, it is thus the free particle problem.) The equation of motion is here the first order system

$$\dot{\mathbf{r}}^\Psi = \frac{\hbar}{m} \operatorname{Im} \frac{\nabla_{\mathbf{r}}\Psi}{\Psi}$$

and at first sight all we need to solve is the datum of an initial position at some time t. However, this "obvious" guess is in general wrong. Suppose, for instance, that we have "localized" the metatron with infinite precision at a point with coordinates $\mathbf{r}'$, at some time t'. We may thus assume that the wave function is, for $t > t'$, the solution of Schrödinger's equation

$$i\hbar \frac{\partial \Psi}{\partial t} = -\frac{\hbar^2}{2m}\nabla^2_{\mathbf{r}}\Psi, \;\; \Psi(\mathbf{r}, \mathbf{t}') = \delta(\mathbf{r} - \mathbf{r}').$$

That solution is the Green function $\Psi(\mathbf{r}, \mathbf{t}) = G_f(x, x'; t, t')$ for fixed $\mathbf{r}'$ and t', that is

$$\Psi(\mathbf{r}, \mathbf{t}) = \left(\frac{m}{2\pi i\hbar(t-t')}\right)^{3/2} \exp\left[\frac{i}{\hbar} m \frac{(\mathbf{r} - \mathbf{r}')^2}{2(t-t')}\right] \tag{7.82}$$

so that the equation of quantum motion (7.79) is here

$$\dot{\mathbf{r}}^{\Psi} = \frac{\mathbf{r}^{\Psi} - \mathbf{r}'}{t - t'}.$$

One immediately sees that it is meaningless to impose an initial condition at time t' for that equation. The general solution is namely

$$\mathbf{r}^{\Psi} = \mathbf{v}'(t - t') + \mathbf{r}'$$

where $\mathbf{v}'$ is *any* constant vector and the corresponding trajectories can therefore be any straight line through the point $\mathbf{r}'$. This is of course very much in conformity with Heisenberg's uncertainty principle: since we have $\Delta\mathbf{r} = 0$ (the particle is sharply located), $\Delta\mathbf{p}$ (and hence $\mathbf{p}$) is undetermined, so that the momentum vector can have arbitrary magnitude and direction. (See the discussion and the examples in Holland [90], §8.4 and 8.6–8.8.)

If we cannot predict the metatron's motion, we can however *retrodict* it if we have performed two position measurements at different times. In fact, its trajectory will then be unambiguously determined if we can find, by a new sharp position measurement at any other time $t_0 \neq t'$, because in this case the Cauchy problem

$$\dot{\mathbf{r}}^{\Psi} = \frac{\mathbf{r}^{\Psi} - \mathbf{r}'}{t - t'}, \quad \mathbf{r}^{\Psi}(t_0) = \mathbf{r}_0$$

has the unique solution

$$\mathbf{r}^{\Psi} = \mathbf{v}_0(t - t') + \mathbf{r}', \quad \mathbf{v}_0 = \frac{\mathbf{r}_0 - \mathbf{r}'}{t_0 - t'}.$$

This shows that if we know the position of the metatron at two distinct times t' and t_0, we can *retrodict* its trajectory. We will actually find that the metatron has followed the classical trajectory from the initial to the final point. (This is essentially Schrödinger's "firefly argument" described in Subsection 1.6.1 of Chapter 1.) One should however note that we cannot use this information to *predict* its trajectory after time t_0, because we would then be led back to the same problem as in the beginning of the discussion. Suppose in fact that we make a new position measurement at some time $t_1 > t_0$, and that we find that the new position vector of the particle is $\mathbf{r}_1$. The velocity from $(\mathbf{r}_0, t_0)$ to $(\mathbf{r}_1, t_1)$ is thus $\mathbf{v}_1 = (\mathbf{r}_1 - \mathbf{r}_0)/(t_1 - t_0)$. However, there is no reason for the vectors $\mathbf{v}_0$ and $\mathbf{v}_1$ to be equal: they can have different magnitudes, or different directions, or both. Repeating this argument, we will eventually obtain a sequence of broken lines, in fact a sort of "Brownian motion" pattern, very different from an expected classical path, each "jump" corresponding to an unpredictable change of the velocity vector due to the act of observation.

This discussion suggests that we look at the problem from a slightly different point of view. Instead of trying to predict the metatron's trajectory, we rather assume that we have been able, by some direct or indirect method, to plot its trajectory in a *past* time interval $[t', t]$. That is, we assume that the metatron has left an intelligible, "explicate" trace. Such a possibility is not ruled out by Heisenberg's uncertainty principle. In fact, the possibility of performing simultaneous arbitrarily sharp position and momentum measurements which refer to the past was recognized by Heisenberg himself:

> *If the velocity of the electron is at first known and the position then exactly measured, the position for times previous to the measurement may be calculated... [but] it can never be used as an initial condition in any calculation of the future progress of the electron and thus cannot be subjected to experimental verification.* (Heisenberg [85])

Before we proceed further, let us make a mathematical interlude.

7.6.3 The Lie-Trotter Formula for Flows

We now state a well-known approximation result in the theory of dynamical systems, the "Lie-Trotter formula for flows" (see Appendix B for a proof). It will allow us, *en passant*, to give a tractable method for solving Hamilton's equations of motion in the classical case.

Proposition 256 (Lie-Trotter formula) *Let (f_t) be the flow of a vector field X defined on some open subset of $\mathbb{R}^m$. Let (k_t) be a family of functions $U \longrightarrow \mathbb{R}^m$ defined near $t = 0$ and such that the dependence of $k_t(u)$ on (u, t) is C^1. If we have*

$$f_t(u_0) = k_t(u_0) + o(t) \quad \textit{for} \quad t \to 0 \tag{7.83}$$

then the sequence of iterates $(k_{t/N})^N(u_0)$ converges, as $N \to \infty$, to $f_t(u_0)$:

$$f_t(u_0) = \lim_{N\to\infty} (k_{t/N})^N(u_0). \tag{7.84}$$

Definition 257 *A family of mappings (k_t) satisfying the conditions above is an* algorithm *for the flow (f_t).*

Here is an elementary application of Proposition 256:

Example 258 ***Lie's formula.*** *Let A and B be two square matrices, of the same dimension m. Due to the non-commutativity of matrix product, we have $e^A e^B \neq e^{A+B}$ in general. However, Sophus Lie (b.1842) proved in 1875 that*

$$e^{A+B} = \lim_{N\to\infty} \left(e^{A/N} e^{B/N}\right)^N. \tag{7.85}$$

While elementary proofs of this classical formula abound, they are all rather lengthy. However, if we take $X = A + B$ in Proposition 256, then $f_t = e^{t(A+B)}$ is the flow of X, and an algorithm is $k_t = e^{tA}e^{tB}$. Hence

$$e^{t(A+B)} = \lim_{N\to\infty} \left(e^{tA/N} e^{tB/N}\right)^N$$

and Lie's formula (7.85) follows, setting $t = 1$.

Let us now apply Proposition 256 to Hamiltonian systems. We begin with the time-independent case. That is, we assume that

$$H = \frac{p^2}{2m} + U(x) \tag{7.86}$$

where U only depends on the positions.

Corollary 259 *Assume that the potential U in the Hamiltonian (7.86) is twice continuously differentiable. An algorithm for the flow (f_t) of X_H is given by*

$$k_t \begin{pmatrix} x' \\ p' \end{pmatrix} = \begin{pmatrix} I & \frac{t}{m}I \\ 0 & I \end{pmatrix} \begin{pmatrix} x' \\ p' \end{pmatrix} + \begin{pmatrix} 0 \\ -\nabla_x U(x')t \end{pmatrix}. \tag{7.87}$$

That is,

$$f_t(z) = \lim_{N\to\infty} (k_{t/N})^N(z) \tag{7.88}$$

at every point z of phase space where f_t is defined.

Proof. Let $x = x(t)$ and $p = p(t)$ be the solutions of Hamilton's equations

$$\dot{x} = \nabla_p H, \ \ \dot{p} = -\nabla_x H$$

with initial conditions $x(0) = x_0$ and $p(0) = p_0$. Performing first order Taylor expansions at $t = 0$ of the position and momentum we get

$$\begin{cases} x = x_0 + \dot{x}(0)t + O(t^2) \\ p = p_0 + \dot{p}(0)t + O(t^2). \end{cases}$$

Since $\dot{x}(0) = p_0/m$ and $\dot{p}(0) = -\nabla_x U(x_0)$ this system can be written as:

$$\begin{cases} x = x_0 + \dfrac{p_0}{m}t + O(t^2) \\ p = p_0 - \nabla_x U(x_0)t + O(t^2) \end{cases}$$

and hence $k_t(x_0, p_0) = f_t(x_0, p_0) + O(t^2)$. The conclusion now follows from the Lie-Trotter formula since $k_t(x_0, p_0)$ depends in a C^1 fashion on (x_0, p_0, t).

■

Proposition 256 cannot be directly applied if H is a time-dependent Hamiltonian

$$H = \frac{p^2}{2m} + U(x,t)$$

because $X_H = (\nabla_p H, -\nabla_x H)$ then it no longer is a true vector field, because it depends on t. However, this difficulty is easily overcome if one uses instead of X_H the suspended vector field

$$\tilde{X}_H = (\nabla_p H, -\nabla_x H, 1)$$

defined in Chapter 2, Subsection 2.2.4. Recall that $\tilde{X}_H$ is a "true" vector field on the extended phase space $\mathbb{R}^{2n+1} = \mathbb{R}^n_x \times \mathbb{R}^n_p \times \mathbb{R}_t$; we may thus apply the Lie-Trotter formula to its flow. This yields the following extension of Corollary 259:

Corollary 260 *Let $(f_{t,t'})$ be the time-dependent flow determined by the Hamiltonian H. Consider, for every integer $N \geq 1$, a subdivision $t_0 = t' < t_1 < \cdots < t_N = t$ of the interval $[t', t]$ such that $t_{j+1} - t_j = \Delta t = (t - t')/N$. Then, if the potential U is twice continuously differentiable, we have*

$$f_{t,t'} = \lim_{N \to \infty} \left(k_{t,t-\Delta t} k_{t-\Delta t, t-2\Delta t} \cdots k_{t'+\Delta t, t'} \right) \tag{7.89}$$

where the algorithm $(k_{t,t'})$ is given by

$$k_{t,t'} \begin{pmatrix} x' \\ p' \end{pmatrix} = \begin{pmatrix} I & \frac{t-t'}{m} I \\ 0 & I \end{pmatrix} \begin{pmatrix} x' \\ p' \end{pmatrix} + \begin{pmatrix} 0 \\ -\nabla_x U(x', t')(t - t') \end{pmatrix}. \tag{7.90}$$

Proof. Let $(\tilde{f}_t)$ be the flow of $\tilde{X}_H$; we have (see Eq. (2.62) in Subsection 2.2.4):

$$\left(f_{t+t',t'}(x', p'), t' \right) = \tilde{f}_t(x', p', t + t')$$

and hence it suffices to show that the mappings $\tilde{k}_t$ defined by

$$\tilde{k}_t(x', p', t') = (k_{t+t',t'}(x', p'), t')$$

form an algorithm for the suspended flow $(\tilde{f}_t)$. By exactly the same argument as in the proof of Corollary 259 we have $k_{t,t'} - f_{t,t'} = o(t - t')$ and hence

$$(\tilde{k}_t - \tilde{f}_t)(x', p', t') = (k_{t+t',t'} - f_{t+t',t'})((x', p'), 0) = o(t)$$

which is precisely condition (7.83) in Proposition 256, replacing f_t by $\tilde{f}_t$ and k_t by $\tilde{k}_t$. The corollary follows. ∎

7.6.4 The "Unfolded" Metatron

We now set sail to prove our claim that a recorded metatron trajectory must be a classical trajectory. Although everything can be extended to the case of a general Maxwell Hamiltonian, we assume, for simplicity, that

$$H = \frac{p^2}{2m} + U(x, t). \tag{7.91}$$

Recall that we are considering the following situation: we have been able, by retrodiction, to determine the phase space trajectory $t \mapsto (x(t), p(t))$ of the metatron in some time interval $[t', t]$. Our claim is that this trajectory must then be the classical trajectory determined by the Hamilton equations of motion associated with the classical Hamilton function H. We begin by finding asymptotic equations for the quantum motion in short time intervals:

Lemma 261 *The equation of quantum motion for a particle located at x' at time t' has the asymptotic form*

$$\dot{x} = \frac{x - x'}{t - t'} - \frac{1}{2m}\nabla_x U(x', t')(t - t') + O((t - t')^2) \tag{7.92}$$

for small values of $t - t'$.

Proof. Let $G = G(x, x'; t, t')$ be the exact Green function for Schrödinger's equation:

$$i\hbar\frac{\partial G}{\partial t} = -\frac{\hbar^2}{2m}\nabla_x^2 G + UG, \quad G(t = t') = \delta(x - x').$$

The quantum motion in the time interval $[t', t]$ is determined by the equation

$$\dot{x} = \operatorname{Im}\left(\frac{\hbar}{G(x, x'; t, t')} m^{-1} \nabla_x G(x, x'; t, t')\right)$$

which we find convenient to write, with the usual abuse of notation

$$\dot{x} = \operatorname{Im}\left(\frac{\hbar}{m} \frac{\nabla_x G(x, x'; t, t')}{G(x, x'; t, t')}\right).$$

Writing G in polar form $\sqrt{\rho} e^{\frac{i}{\hbar}\Phi}$, it follows from the calculations of the previous sections that we have

$$\Phi(x, x'; t, t') = W_f(x, x', t - t') - \overline{U}(x, x', t')(t - t') + O((t - t')^2)$$

and that

$$\sqrt{\rho} \sim \sqrt{\rho_f}\left(1 + O((t - t')^2)\right)$$

where $\overline{U}$ is the average of the potential (at time t') between x' and x, and

$$W_f = m\frac{(x - x')^2}{2(t - t')}, \quad \rho_f = |m|\left(\frac{1}{t - t'}\right)^n$$

($|m| = \det m$) are, respectively, the free-particle generating function, and the corresponding Van Vleck density. We thus have

$$G = \sqrt{\rho_f} \exp\left[\frac{i}{\hbar}\left(W_f - \overline{U}(t - t')\right)\right] + O((t - t')^2)$$

and hence, since ρ_f does not depend on x, x':

$$\frac{\nabla_x G}{G} = \frac{m}{\hbar}\frac{x - x'}{t - t'} - \frac{1}{\hbar}\nabla_x \overline{U}(t - t') + O((t - t')^2).$$

It follows that we have

$$\dot{x} = \frac{x - x'}{t - t'} - \frac{1}{m}\nabla_x \overline{U}(x, x', t')(t - t') + O((t - t')^2) \tag{7.93}$$

where $\frac{1}{m}\nabla_x \equiv m^{-1}\nabla_x$. Since the velocity, at time t' is $p'/m = m^{-1}p'$, and the position is x', the equation of motion has the asymptotic solution

$$x = x' + \frac{p'}{m}(t - t') + O((t - t')^2) \tag{7.94}$$

in the interval $[t', t]$. This observation allows us to simplify (7.93): noting that in view of (7.94) we have

$$\begin{aligned}\nabla_x \overline{U}(x, x', t') &= \nabla_x \overline{U}(x', x', t') + O(x - x') \\ &= \nabla_x \overline{U}(x', x', t') + O(t - t')\end{aligned}$$

we can rewrite (7.93) as

$$\dot{x} = \frac{x - x'}{t - t'} - \frac{1}{m}\nabla_x \overline{U}(x', x', t')(t - t') + O((t - t')^2).$$

Since by definition

$$\overline{U}(x, x', t') = \int_0^1 U(sx + (1 - s)x', t')ds$$

we have

$$\nabla_x \overline{U}(x, x', t') = \int_0^1 s\nabla_x U(sx + (1 - s)x', t')ds$$

and hence, setting $x = x'$:

$$\nabla_x \overline{U}(x', x', t') = \int_0^1 s\nabla_x U(x', t')ds = \frac{1}{2}\nabla_x U(x', t').$$

It follows that (7.93) can be written

$$\dot{x} = \frac{x - x'}{t - t'} - \frac{1}{2m}\nabla_x U(x', t')(t - t') + O((t - t')^2)$$

which is precisely Eq. (7.92). ∎

We are now able to state and prove the main result of this section. Let $(f^{\Psi}_{t,t'})$ be the time-dependent flow determined by the Hamiltonian function $H^{\Psi} = H + Q^{\Psi}$ where Q^{Ψ} is Bohm's quantum potential associated to Green's function. It turns out that $(f^{\Psi}_{t,t'})$ is an algorithm for the *classical* flow:

Theorem 262 (Unfolding Theorem) *(1) Set $(x^{\Psi}, p^{\Psi}) = f^{\Psi}_{t,t'}(x', p')$. For small $t - t'$ we have the following asymptotic expressions for x^{Ψ} and p^{Ψ}:*

$$\begin{cases} x^{\Psi} = x' + \dfrac{p'}{m}(t - t') + O((t - t')^2) \\ p^{\Psi} = p' - \nabla_x U(x', t')(t - t') + O((t - t')^2). \end{cases} \tag{7.95}$$

(2) The phase space mappings $f^{\Psi}_{t,t'}$ form an algorithm for the Hamiltonian flow $(f_{t,t'})$ determined by the Hamiltonian H:

$$f_{t,t'} = \lim_{N\to\infty} \left(f^{\Psi}_{t,t-\Delta t} f^{\Psi}_{t-\Delta t,t-2\Delta t} \cdots f^{\Psi}_{t'+\Delta t,t'}\right) = f^{\Psi}_{t,t'} \tag{7.96}$$

for $\Delta t = (t-t')/N$.

Proof. (1) The first formula (7.95) is just Eq. (7.94) in the proof of Lemma 261. Let us prove that the second formula (7.95) holds as well. Differentiating both sides of Eq. (7.92) with respect to t, we get

$$\begin{aligned}\ddot{x} &= -\frac{x-x'}{(t-t')^2} + \frac{\dot{x}}{t-t'} - \frac{1}{2m}\nabla_x U(x',t') + O(t-t') \\ &= -\frac{1}{m}\nabla_x U(x',t') + O(t-t')\end{aligned}$$

that is, since $p = m\dot{x}$:

$$\dot{p} = -\nabla_x U(x',t') + O(t-t').$$

Integrating from t' to t the second formula (7.95) follows. **(2)** The phase space mappings $k_{t,t'}$ defined by

$$k_{t,t'}\begin{pmatrix} x' \\ p' \end{pmatrix} = \begin{pmatrix} I & \frac{t-t'}{m}I \\ 0 & I \end{pmatrix}\begin{pmatrix} x' \\ p' \end{pmatrix} + \begin{pmatrix} 0 \\ -\nabla_x U(x',t')(t-t') \end{pmatrix}$$

form an algorithm for the $f^{\Psi}_{t,t'}$ in view of (7.95). However, they also form an algorithm for the classical flow $(f_{t,t'})$ (see Corollary 260); formula (7.96) follows. Hence $(f^{\Psi}_{t,t'})$ is an algorithm for the classical flow $(f_{t,t'})$. ■

The Unfolding Theorem justifies our claim that a sharp phase space track left by a metatron must be a classical particle trajectory: choose points x_0, x_1,..., x_N on this track, at times $t_0 = t' < t_1 < \cdots < t_N = t$ equally spaced (this is, however, not an essential requirement, it suffices in fact that $t_{j+1} - t_j = O((t-t')/N)$). Set now

$$\Delta t = t_{j+1} - t_j = \frac{t-t'}{N}$$

and denote by p_0, p_1,..., p_N the momenta at the points (x_0, t_0),..., (x_N, t_N):

$$p_j = m\dot{x}(t_j) \quad (0 \le j \le N).$$

To each point (x_j, p_j) is associated a "particle source" $\delta(x - x_j)$ at time t_j, whose evolution in each interval $[t_j, t_{j+1}]$ is guided by the field

$$\Psi_j(x, t) = G_j(x, t) \equiv G(x, x_j; t, t_j).$$

Since the initial momentum p_j at time t_j is known, the first part of the Unfolding Theorem applies, and the motion is given by (7.95) in each interval $[t_j, t_{j+1}]$. In the limit $N \to \infty$ the trajectory from (x', p') to (x, p) is the classical trajectory in view of the second part of the theorem.

7.6.5 Quantum Zeno Effect and the Mott Paradox

The Unfolding Theorem can be generalized to the case where the initial wave function is an arbitrary localized wave packet (see de Gosson, Hiley, and Cohen [70]). This generalization allows one to study in detail and in a realistic way the quantum Zeno effect, and the related "Mott problem". The Mott problem (or paradox) (Mott [120]) illustrates in a striking manner the difficulties of understanding the nature of wave packet measurement. The story goes back to 1929, when Heisenberg and Mott reflected on the following problem: in a Wilson cloud chamber, α-particles are emitted from nuclei of radioactive atoms; due to the radial symmetry of the problem, the α-particles are represented by spherical wave functions. These particles create droplets of condensation when interacting with water molecules in the vapor, thus creating a track of condensation. Since the spherical wave function spreads isotropically in all directions, it could have been expected to randomly ionize the water molecules, leaving behind a spherical track. However, when one actually performs the experiment one always finds a single linear track. Bell has illustrated the situation by comparing the cloud chamber with a stack of photographic plates: to produce this straight trajectory, the ionized gas molecules are assumed to act as an array of potential measuring devices, leaving a record of the track of the α-particle on the stack of plates. We can thus regard the α-particle as being "continuously watched". This means, in a sense, that continuous observation "dequantizes" quantum trajectories. The frequent observations render the dynamic contribution of the quantum potential negligible. This property is, of course, essentially a consequence of the quantum Zeno effect, which has been shown to inhibit the decay of unstable quantum systems when under continuous observation. Another illustration of how continuous observations of a different kind can give rise to a quantum Zeno effect has already been given in Bohm and Hiley [20]. They considered the transition of an Auger-like particle and showed that the perturbed wave function, which is proportional to Δt for times less than $1/\Delta E$, (ΔE is the energy released in the transition) will never become large and therefore cannot make a significant contribution to the quantum potential necessary for the transition to occur. Thus again, the reason that no transition will take place is the vanishing of the quantum potential, as shown in the proof of the Unfolding Theorem where we have shown that the quantum potential plays a key role in determining the behavior of a quantum particle. This potential is not a mere "add-on" but an essential feature that is a necessary consequence of the relationship between the symplectic group structure (classical mechanics) and its double cover, the metaplectic group and its non-linear generalization. Specifically there is a deep relation between the Hamiltonian flows, $f_{tt'}$, of classical physics and the flows $f^{\psi}_{tt'}$ associated with the quantum behavior in the covering space. It is this feature that is key to understanding the relationship between a particle that exhibits only classical behavior and one that gives rise to quantum phenomena.

As is well known at the particle level, we have lost causality in quantum phenomena but it has been replaced by what we will call "wave causality", which is defined through the Schrödinger equation. However, the real part of this equation under polar decomposition appears to have restored causality at the particle level. How can this be? As Hiley puts it in [70], the answer lies in the relationship between the two types of flow, $f_{t,t'}$ and $f^{\psi}_{t,t'}$. This approach gives rise to a very different way of analyzing quantum phenomena from the conventional approaches, including the approach of Bohmian mechanics

7.6.6 The Generalized Metaplectic Representation

Recall from last Chapter that the metaplectic representation of the symplectic group $\mathrm{Sp}(n)$ associates to every symplectic matrix exactly two operators $\pm S$ in $\mathrm{Mp}(n)$. This reflects the fact that the metaplectic group is a double covering of the symplectic group. In fact, we constructed the metaplectic group in a rather abstract way, starting from "quadratic Fourier transforms" $S_{W,m}$ associated to the generating functions of free symplectic matrices; once this group of operators was identified, we defined a projection $\Pi : \mathrm{Mp}(n) \to \mathrm{Sp}(n)$ (or $\Pi^{\hbar} : \mathrm{Mp}^{\hbar}(n) \to \mathrm{Sp}(n)$) by associating to each quadratic Fourier $S_{W,m}$ transform the corresponding free symplectic matrix s_W. The point with these constructions was that, by a classical theorem from the theory of covering spaces, we were able to associate, in a canonical way, to every one-parameter group (s_t) of $\mathrm{Sp}(n)$ a unique one-parameter subgroup (S_t) of $\mathrm{Mp}^\hbar(n)$, and we proved that the function $\Psi(x,t) = S_t\psi_0(x)$ then automatically satisfies the Schrödinger equation

$$i\hbar\frac{\partial\Psi}{\partial t} = \widehat{H}\Psi, \quad \Psi(t=0) = \psi_0$$

associated with the quadratic Hamiltonian whose flow is (s_t). One can of course reverse the argument, and show that if Ψ is a solution of the equation above, then the family (S_t) of unitary operators such that $\Psi(x,t) = S_t\psi_0(x)$ belongs to $\mathrm{Mp}(n)$, and project (S_t) "down to $\mathrm{Sp}(n)$" thus recovering the classical flow (s_t).

We now shortly address the question whether this correspondence can be extended to the case of arbitrary (non-quadratic) Hamiltonians. That is:

> *Can we, by inspection of the quantum evolution group (F_t) (or, more generally, $(F_{t,t'})$), recover the classical motion, that is the Hamiltonian flow $(f_{t,t'})$?*

The answer to that question is "yes". This is because the knowledge of $(F_{t,t'})$ determines the Green function, and hence the quantum motion in view of the Unfolding Theorem; in fact the formulas (7.95)

$$f^{\Psi}_{t,t'} : \begin{cases} x = x' + \dfrac{p'}{m}(t-t') + O((t-t')^2) \\ p = p' - \nabla_x U(x',t')(t-t') + O((t-t')^2); \end{cases}$$

allow us to reconstruct the classical flow using the Lie-Trotter formula (7.96). Summarizing, we can thus associate to $(F_{t,t'})$ the family $(f_{t,t'})$ of symplectomorphisms determined by the Hamilton equations.

We will call that procedure the *generalized metaplectic representation.*

Recall from Subsection 6.8.2 that we defined the group $\mathrm{Ham}(n)$ of Hamiltonian symplectomorphisms as the subgroup of the group $\mathrm{Symp}(n)$ of all symplectomorphisms: a symplectomorphism f is in $\mathrm{Ham}(n)$ if it is the time-one map of a Hamiltonian flow. That is, $f \in \mathrm{Ham}(n)$ if there exists a Hamiltonian function (not necessarily of Maxwell type) such that $f = f_1$ if $(f_t)(= (f_{t,0}))$ is the flow determined by H. This discussion shows the following: even if we have not constructed a metaplectic representation of $\mathrm{Ham}(n)$, encompassing the standard representation of $\mathrm{Sp}(n)$ by $\mathrm{Mp}(n)$, we can easily establish a two-to-one correspondence

$$f \longleftrightarrow \pm F$$

between a subset of $\mathrm{Ham}(n)$ and a subset of the group of all unitary operators acting on functions on $\mathbb{R}^n_x$: suppose that f is the time-one map of a Hamiltonian flow (f_t) determined by a Maxwell Hamiltonian, and associate to that flow (f_t) the unitary group (F_t) such that $\Psi = F_t\psi_0$ solves the Schrödinger equation with Cauchy datum ψ_0. It suffices then to take $F = F_1$.

We do not know whether it is possible to push this sketchy construction any further, and to construct a full "metaplectic representation" of $\mathrm{Symp}(n)$ (or of $\mathrm{Ham}(n)$), in the same way we constructed the metaplectic representation of $\mathrm{Sp}(n)$. The question of determining the first homotopy group of $\mathrm{Ham}(n)$ or $\mathrm{Symp}(n)$ is a difficult problem, belonging to the area of symplectic topology, and related to the so-called Arnol'd conjecture for the "flux homomorphism" (see [116], especially Chapter 10, for a review of these questions). We however conjecture that an abstract double covering group of $\mathrm{Ham}(n)$ might be constructed by using the theory of the Maslov index sketched in Chapter 5. In fact, we have shown (see [54, 56, 58]) that such a construction is possible for the double covering $\mathrm{Sp}_2(n)$ of the symplectic group. The method can perhaps be extended to encompass the non-linear case.

7.7 Phase Space and Schrödinger's Equation

One of the recurring themes of this book has been that Newtonian mechanics is described by Hamilton's equations

$$\dot{x} = \nabla_p H, \quad \dot{p} = -\nabla_x H$$

while the hallmark of quantum mechanics is Schrödinger's equation

$$i\hbar\frac{\partial\Psi}{\partial t} = \widehat{H}\Psi.$$

There is an obvious formal dissymmetry between these equations: Hamilton's equations involve explicitly phase space variables, and time, while Schrödinger's equation only contains configuration space variables, and time. Nevertheless, as we have pointed out several times before, the solutions of these fundamental equations can be expressed using "realizations" of the same abstract one-parameter group. Why do we then have such an apparent dissymmetry between Hamilton's and Schrödinger's equations? It turns out that even if Hamilton's equations place the x and p variables on an equal footing, the way we physically derived these equations *explicitly* made use of the x representation in the form of Newton's second law. The momentum variable p was actually introduced as a mathematical commodity, just because we *wanted* to reduce Newton's second law to a system of first order differential equations, to which we could apply the powerful techniques of symplectic geometry. By doing this, we had to define a function, the Hamiltonian, which is in the simplest case

$$H = \frac{p^2}{2m} + U(x,t)$$

and thus is of a very particular mathematical type: while the potential U can be an arbitrary function, the momentum appears as a quadratic form, whose matrix is the inverse of the mass matrix m. One can say that our tradition of privileging space-time already breaks the symmetry in the "physical" phase space; even if Hamilton's equations associated to a Maxwell Hamiltonian

$$H = \sum_{j=1}^{n} \frac{1}{2m_j} (p_j - A_j(x,t))^2 + U(x,t)$$

put the position and momentum variables on equal footing, they are describing, although in disguise, physics in "x-representation". To access quantum mechanics, we lifted the flow determined by *that* particular Hamiltonian that to the (generalized) metaplectic group. It is thus not too intriguing that we obtain, at the "output" an equation privileging space-time, because we precisely expressed Newton's second law in terms of space-time!– From the very beginning our view of Nature was in a sense biased: we were using an "explicate order" corresponding to our everyday sensations and view of the world, which privileges physical space and time, *not* momentum space and time. Space-time is indeed the most immediate contact we have with our World; it is the obvious order for us, human beings: velocity is only perceived as a secondary manifestation, in the form of changes of position with time (it is for this reason that cameras taking momentum space snapshots seem not to be widely commercialized...).

Before we pursue this analysis further, let us shortly discuss the meaning of phase space in quantum mechanics. According to popular belief, there is no place for phase space in the quantum world[1]. The arguments that are usually presented to sustain this belief are however rather specious, to say the least. For instance, one invokes Heisenberg's uncertainty principle as

[1]Dirac [36] maintained already in 1930 that "there is no quantum mechanics in phase space".

"proof" of the fact that it is meaningless to talk about points in phase space in quantum mechanics. While it is true that Heisenberg's inequalities

$$\Delta p_x \Delta x \geq \tfrac{1}{2}\hbar \ , \ \Delta p_y \Delta y \geq \tfrac{1}{2}\hbar \ , \ \Delta p_z \Delta z \geq \tfrac{1}{2}\hbar \tag{7.97}$$

indicate that one cannot *measure* simultaneously with infinite precision both the position and the momentum of a particle, these relations are not, however, a *prima facie* reason for rejecting the idea of the platonic or, if you prefer, implicate *existence* of a quantum mechanical phase space. (The possible existence of phase space manifestations in quantum mechanics was already discussed in connection with the retrodicted metatron trajectory in Section 7.6.) There are other good reasons for which the uncertainty principle is not sufficient for rejecting phase space manifestations. One of these reasons is that if we did, then we would also be led to question the existence of phase space in classical mechanics, because of the symplectic camel principle which implies an uncertainty principle in classical mechanics (see Chapter 3)! Another common argument used to deny the existence of a quantum-mechanical phase space is the claim that it is impossible to define a joint probability density for the positions and momenta. However, this claim is specious, and mathematically false: it is just an illegitimate induction from one particular "no-go" example, the Wigner quasi-distribution, which can take negative values (see Hudson [91], Folland [46]). The notion of joint probability density actually *does* make perfect sense in quantum mechanics: Cohen proved in [30] that if Ψ is a normalized wave function and Φ its Fourier transform, then one can construct infinitely many phase space densities whose marginal probabilities are $|\Psi|^2$ and $|\Phi|^2$. Here is Cohen's construction (we assume that $n = 1$ for simplicity). Let $\Psi = \Psi(x, t)$ be any complex function of position and time, and choose an *arbitrary* function g of $(u, v) \in \mathbb{R}^2$ such that $g \leq 1$ and

$$\int_{-\infty}^{+\infty} g(u, v)du = \int_{-\infty}^{+\infty} g(u, v)dv = 0.$$

Define now

$$f(x, p, t) = |\Psi(x, t)\Phi(p, t)|^2(1 - g(u(x, t), v(p, t)))$$

where we have set

$$u(x, t) = \int_{-\infty}^{x} |\Psi(x', t)|^2 dx' \ , \ v(p, t) = \int_{-\infty}^{p} |\Phi(p', t)|^2 dp'.$$

The function f is a probability density: clearly $f(x,p,t) \geq 0$, and we have

$$\int_{-\infty}^{+\infty} f(x,p,t)\, dpdx = 1$$

as is easily seen by using the change of variables

$$du = |\Psi(x,t)|^2 dx, \;\; dv = |\Phi(p,t)|^2 dp.$$

A straightforward calculation moreover shows that the marginal laws are associated to the densities

$$\int_{-\infty}^{+\infty} f(x,p,t)\, dp = |\Psi(x,t)|^2$$
$$\int_{-\infty}^{+\infty} f(x,p,t)\, dx = |\Phi(p,t)|^2$$

confirming our claim.

In practice one uses the Wigner quasi-distribution $W\Psi$ to study the phase space picture of quantum mechanics. It is defined, in dimension n, for $\Psi \in L^2(\mathbb{R}^n)$ by

$$W\Psi(x,p) = \left(\frac{1}{2\pi\hbar}\right)^n \int e^{-\frac{i}{\hbar}p\cdot y}\Psi(x+\tfrac{1}{2}y)\Psi^*(x-\tfrac{1}{2}y)d^n y.$$

One immediately checks that $W\Psi$ always is a real function: $(W\Psi)^* = W\Psi$. Its interest comes from the fact that it has the correct marginals: if $\Psi \in L^1(\mathbb{R}^n) \cap L^2(\mathbb{R}^n)$ then

$$\int W\Psi(x,p)d^n p \;\; = \;\; |\Psi(x)|^2$$
$$\int W\Psi(x,p)d^n x \;\; = \;\; |\Phi(p)|^2$$

and hence

$$\iint W\Psi(x,p)d^n p d^n x = ||\Psi||_{L^2}^2.$$

In particular

$$\iint W\Psi(x,p)d^n p d^n x = 1$$

if Ψ is normalized. It should however be emphasized that the Wigner quasi-distribution is not a probability distribution in the usual sense, since it can take negative values:

Example 263 *Let $\Psi \in L^2(\mathbb{R}^n)$ be an odd function: $\Psi(-x) = \Psi(x)$. We have*

$$W\Psi(0,0) = \left(\frac{1}{2\pi\hbar}\right)^n \int \Psi(\tfrac{1}{2}y)\Psi^*(-\tfrac{1}{2}y)d^n y = -\left(\frac{1}{\pi\hbar}\right)^n ||\Psi||^2_{L^2}.$$

The Wigner distribution was introduced by Eugene Wigner (b.1902) in his celebrated paper [167]; it is a fundamental topic in phase space quantum mechanics; see [63] for a detailed treatment.

We finally mention that recent work of Hiley and his collaborators seems to lead to a very promising approach to the question of quantum mechanical phase space by using the notion of "shadow phase space", which is physically most easily understood in terms of the Bohmian implicate order. Mathematically, the theory consists in rewriting Bohm's equations of motion in operator form. We refer to Brown and Hiley [26] for a detailed description of this theory, which seems very promising.

7.7.1 Mixed Representations in Quantum Mechanics

The question that we address is whether it is possible, or if it even makes sense, to express Newton's second law in an alternative way, leading to another type of (equivalent) Hamilton's equations, whose flow could then be "realized" as some group of operators satisfying another type of Schrödinger's equation involving this time the p variable (and time). The answer is that, yes, all this is possible, and moreover mathematically very straightforward!

In this, and the previous Chapter, we have been resolutely adopting the "x-representation" formalism:

(1) We have described the metaplectic group $\mathrm{Mp}(n)$ by using a particular choice of generating functions $W = W(x, x')$ defined on (twice) the configuration space. This allowed us to define the quadratic Fourier transforms $S_{W,m}$ that generate $\mathrm{Mp}(n)$;

(2) We used that "x-representation" of the metaplectic group to derive Schrödinger's equation, which is a partial differential equation involving solely the space and time variables x and t, and not the momentum variable p.

There is however nothing particularly compelling with that choice, which was finally suggested (or rather *dictated*) to us by the fact that Newton's second law is most easily expressed in terms of our everyday space-time variables. This formulation of classical mechanics is only one "explicate order" in which Newton's second law can be stated: Newtonian mechanics, in its Hamiltonian formulation, is covariant under symplectic transformations. Mathematically speaking, there is thus no "preferred frame": we are free to express the laws of mechanics in any coordinate system deduced from the usual x, p frame by a symplectomorphism (more precisely, it doesn't matter which "symplectic basis" we use to calculate coordinates). Transforming Hamilton's equations using symplectic changes of variables amounts to express Newton's second law (and the Maxwell principle) in other representations, and the obtained flows are all conjugate groups.

Recall (Chapter 1, Subsection 1.1.2) that we had written Newton's second law in the form $\dot{x} = v$, $\dot{p} = F$ where F is a "force field" and the momentum p a fundamental quantity which is conserved under some interactions (we are working here in general dimension n). Making the assumption that there exists a potential function U such that $F = -\nabla_x U$, Newton's second law becomes

$$\dot{x} = \frac{p}{m} \ , \ \dot{p} = -\nabla_x U \tag{7.98}$$

and these equations are precisely Hamilton's equations for the function

$$H = \frac{p^2}{2m} + U.$$

Let us now make the simple change of variables $(x^*, p^*) = J(x, p)$, that is

$$p^* = -x, \ \ x^* = p$$

and define a function U^* of p^* by $U^*(p) = U(x)$. Equations (7.98) are in these variables

$$p^* = -\frac{x^*}{m}, \ \ \dot{x}^* = \nabla_{p^*} U^* \tag{7.99}$$

and they are also Hamilton equations, not for H however, but for the new Hamiltonian function

$$H^* = \frac{1}{2m} x^{*2} + U^*(p). \tag{7.100}$$

One immediately checks that the flow (f_t^*) determined by H^* is related to the flow (f_t) determined by H by the formula $f_t^* = J f_t J^{-1}$, so that (f_t) and (f_t^*) are just two *conjugate groups* of symplectomorphisms. Let us now denote by (F_t) the group of unitary operators obtained by lifting (f_t) to the metaplectic representation, and by (F_t^*) that obtained by lifting (f_t^*). We ask: what is the relation between these two groups? The answer is that we have $F_t^* = F F_t F^{-1}$ where F is the (quantum) Fourier transform.

Let us check this on the free particle in one dimension. The flows (s_t) and (s_t^*) of the Hamiltonians $H = p^2/2m$ and $H^* = x^2/2m$ consist of the symplectic matrices

$$s_t = \begin{pmatrix} 1 & \frac{t}{m} \\ 0 & 1 \end{pmatrix}, \quad s_t^* = \begin{pmatrix} 1 & 0 \\ -\frac{t}{m} & 1 \end{pmatrix}.$$

The lift (S_t) of (s_t) to $\mathrm{Mp}(n)$, is defined by $\Psi(x,t) = S_t\psi_0(x)$ where

$$\Psi(x,t) = \left(\tfrac{m}{2\pi i\hbar(t-t')}\right)^{1/2} \int e^{\frac{i}{\hbar}W(x,x',t)}\psi_0(x')\,dx'$$

with $W = m(x-x')^2/2t$; the function Ψ is, as we know, the solution of Schrödinger's equation

$$i\hbar\frac{\partial\Psi}{\partial t} = -\frac{\hbar^2}{2m}\frac{\partial^2\Psi}{\partial x^2}. \tag{7.101}$$

If we instead lift the conjugate flow (s_t^*) to $\mathrm{Mp}(n)$, we obtain a group (S_t^*) acting this time on functions of p. In fact, the function Φ defined by $\Phi(p,t) = S_t^*\phi_0(p)$ satisfies the equation

$$i\hbar\frac{\partial\Phi}{\partial t} = \frac{p^2}{2m}\Phi \tag{7.102}$$

which is just Schrödinger's equation in the p-representation (notice that equations (7.101) and (7.102) are deduced from one another by a Fourier transformation interchanging x and p).

The situation above is actually quite general, and only reflects the properties of the group $\mathrm{Ham}(n)$ of Hamiltonian symplectomorphisms, briefly discussed in Subsection 6.8.2 of Chapter 6: Hamilton equations are transformed into other Hamilton equations by a symplectic change of variables. This fact leads to the following general property, which we state and prove only in the linear case:

Proposition 264 *Let H be a quadratic function of the coordinates (x,p), and s a symplectic matrix. Then:* **(1)** *the flows (s_t) and (s_t^*) of H and $H^* = H\circ s^{-1}$ satisfy the relation $s_t^* = s\circ s_t\circ s^{-1}$, and:* **(2)** *The metaplectic representations (S_t) and (S_t^*) of these flows are related by the formula $S_t^* = SS_tS^{-1}$.*

Proof. Part **(1)** is just Lemma 220 in Chapter 6; part **(2)** follows, noting that the projection on $\mathrm{Sp}(n)$ of the group (S_t^*) is

$$\Pi^\hbar(S_t^*) = \Pi^\hbar(S)\Pi^\hbar(S_t)\Pi^\hbar(S^{-1}) = ss_ts^{-1}$$

where $\Pi^\hbar$ is the projection $\mathrm{Mp}^\hbar(n) \longrightarrow \mathrm{Sp}(n)$ (see Subsection 6.4.2 of Chapter 6). ■

This result shows the following: even if Newton's law together with the Maxwell principle leads to Hamiltonians of the type

$$H = \sum_{j=1}^{n} \frac{1}{2m_j}(p_j - A_j(x,t))^2 + U(x,t)$$

(the "Maxwell Hamiltonians"), these Hamiltonians belong, so to say, to the "x-representation". Proposition 264 shows that any Hamiltonian function $H \circ s^{-1}$ where $s \in \mathrm{Sp}(n)$ is in fact exactly as good for classical mechanics! (More generally, this applies to every compose $H \circ f^{-1}$, where f is an arbitrary symplectomorphism.) What about Schrödinger's equation? It is no more difficult, using **(2)** in Proposition 264 to rewrite it in another representation:

Theorem 265 *Let s be a symplectic matrix and $S \in \mathrm{Mp}(n)$ have projection s. (1) If Ψ is a solution of Schrödinger's equation*

$$i\hbar\frac{\partial\Psi}{\partial t} = \widehat{H}\Psi$$

then $\Phi = S\Psi$ is a solution of

$$i\hbar\frac{\partial\Phi}{\partial t} = \widehat{K}\Phi \tag{7.103}$$

where $\widehat{K} = S^{-1}\widehat{H}S$. (2) When H is quadratic, then $\widehat{K}$ is just the operator

$$\hat{K} = \widehat{H \circ s^{-1}} \tag{7.104}$$

obtained by applying Schrödinger's quantization rule to the quadratic Hamiltonian $H \circ s^{-1}$.

Proof. We have $\Psi = S^{-1}\Phi$ hence

$$i\hbar\frac{\partial\Phi}{\partial t} = S^{-1}\widehat{H}\Psi = S^{-1}\widehat{H}S\Phi$$

which proves **(1)**. Let us sketch the proof of **(2)**. One first notes that if H is quadratic in the position and momentum variables, then so is $H \circ s^{-1}$. Applying the Schrödinger quantization rule to $H \circ s^{-1}$, formula (7.104) readily follows, using the properties of metaplectic operators. ■

Proposition 265 is a first step towards the branch of mathematics known as geometric quantization (see the discussion in Chapter 1, Subsection 1.4.3). The extension of Proposition 265 to the case where H is an arbitrary function on phase space, and s is replaced by an arbitrary symplectomorphism, requires more sophisticated mathematical techniques, such as the Weyl calculus of pseudo-differential operators as constructed in Leray [101] (also see Dubin *et al.* [40], Folland [46], and the references therein). For instance, the quantization of the Hamiltonian H^* of (7.100) leads to the operator

$$\widehat{H^*} = \frac{1}{2m}x^{*2} + \hat{U}^*$$

where $\hat{U}^*$ is an operator acting on functions in the p^* variables, and defined by

$$\hat{U}^*\Phi(p^*) = \left(\tfrac{1}{2\pi i\hbar}\right)^{n/2}\int e^{\frac{i}{\hbar}p^*x^*}FU^*(x^*)\Phi(x^*)d^nx^*.$$

The reader interested in further developments of these questions is referred to [40, 79, 101, 169] and the references therein.

The discussion of the topics above was intended to convince the reader that the choice of the x-representation, both in classical and quantum mechanics has nothing of an absolute necessity; it is only a convenient choice, whose origin lies in the fact that we are culturally accustomed to privilege space-time.

7.7.2 Complementarity and the Implicate Order

The physicist's answer to these questions is that this is a manifestation of the principle of complementarity, following which one can "see with the x-eye", or with "the p-eye", but not with both simultaneously. This principle is actually a consequence of the non-commutativity of operator products (already present at the level of Schrödinger's quantization rule discussed in Chapter 1). From the Bohmian perspective, complementarity is just a manifestation of the implicate order: as Hiley notes in [87], there is no room for non-commutativity in the classical world, and yet the classical world does contain lots of non-commutativity, associated with activity or process (try for instance taking a cup from a cupboard before opening the door!). In [87] Hiley presents a very subtle analysis of phase space in quantum mechanics, based on Bohm's implicate order (see Subsection 7.6.1). To understand his argument, we begin by quoting the following metaphor from Hiley's paper.

Suppose that we have a collection of spheres and cubes, with two different colors, say red and blue. We assume that we cannot, at this point, view these properties directly, but need an "operator" S to determine the shape, and an operator C to determine the color. Now, let us try to collect together a set of red spheres using these operators. First we measure the color using C, and collect the reds together in one group, separating them from the blues. We then take the red set and use the operator S to which of these red objects are spheres. We have thus collected a set of objects that were red according to the first measurement and spheres according the second measurement. We might be tempted to conclude that we now have a collection of red spheres. However, this is only true if the operators C and S commute! If $SC \neq CS$ then we find, by inspection, that half of our spheres have changed color and are now blue!

The point with that metaphor is that in the quantum worlds we cannot display every property in one picture. In terms of the implicate order, color would be one explicate order and shape another, but at the deepest level the process has neither color or shape: the implicate order is, following this viewpoint, a structure of relationships, described by an "algebra of process" (Hiley [86]). This is exactly what happens in Hamiltonian mechanics, replacing "shape" by "domain" and "color" by "representation": the flow (f_t) is an abstract group, a process without domain and without representation. It is only when we decide to realize that group, to make it operate on *something* (points in phase space, or functions) that we are confronted to choices of "explicate order".

Hiley's argument in fact not only encompasses the case of quantum observables, but extends to phase space properties related to the symplectic camel. Imagine an egg (ellipsoid) with capacity πR^2 in phase space, and a symplectic cylinder with radius R (and hence also capacity πR^2). We can squeeze the egg inside the cylinder using a symplectomorphism, and then rotate the set squeezed egg+cylinder: the egg will remain inside the cylinder. If, however, we perform these operations in reverse order, that is, if we first rotate the egg and the cylinder, we might very well be unable to squeeze the rotated egg inside the rotated cylinder. This is because the compose of both operations is not in general a symplectomorphism, and thus does not preserve the capacities. In fact, as we have seen in Chapter 3 (formula (3.18) in Proposition 50),

$$\mathrm{Sp}(n) \cap O(2n, \mathbb{R}) = U(n)$$

so that there are phase space rotations which are not symplectic, and hence do not preserve the symplectic capacity of the egg.

In conclusion, we can say that the Hamiltonian flow, viewed as an abstract group (f_t) contains all the information, both classical and quantum: it is thus an "implicate order". One explicate order is obtained by realizing (f_t) as a group of symplectomorphisms acting on phase space. However, when we realize (f_t) as a subgroup of the metaplectic group $\mathrm{Mp}(n)$ the corresponding explicate orders do not live in phase space, because of the mathematical fact that $\mathrm{Mp}(n)$ consists of operators acting on Hilbert spaces of functions defined on a copy of $\mathbb{R}^n$. According to whether one writes the elements of $\mathrm{Mp}(n)$ in the x-representation or the p-representation, that Hilbert space is $L^2(\mathbb{R}^n_x)$ or $L^2(\mathbb{R}^n_p)$ (one might of course as well consider intermediate representations, by mixing position and momentum variables, cf. Theorem 265).

Appendix A

Symplectic Linear Algebra

Lagrangian Planes

A Lagrangian plane is an n-dimensional linear subspace of the phase space $\mathbb{R}^n_x \times \mathbb{R}^n_p$ on which the symplectic form Ω vanishes identically. Here is an interesting interpretation in terms of "real" and "imaginary" subspaces of phase space. Identifying $\mathbb{R}^n_x \times \mathbb{R}^n_p$ with the direct sum $\mathbb{R}^n_x \oplus (i\mathbb{R}^n_p)$ we can write $z = (x, p) = x + ip$, thus identifying phase space points with n-tuples $z \in \mathbb{C}^{2n}_z$. Equipping $\mathbb{C}^{2n}_z$ with the natural complex scalar product

$$\langle z, z' \rangle_{\mathbb{C}} = x \cdot x' - p \cdot p' + i(p \cdot x' - p' \cdot x)$$

we see that

$$\Omega(z, z') = \operatorname{Im}\langle z, z' \rangle_{\mathbb{C}}.$$

A Lagrangian plane ℓ can thus be viewed as a "real" subspace in the sense that

$$z, z' \in \ell \times \ell \Longleftrightarrow \operatorname{Im}\langle z, z' \rangle_{\mathbb{C}} = 0.$$

If ℓ is a Lagrangian plane so is its image $s\ell$ by any linear symplectic transformation $s \in \operatorname{Sp}(n)$. This property is an immediate consequence of the relation $\Omega(sz, sz') = \Omega(z, z')$ (definition of $\operatorname{Sp}(n)$). The set of all Lagrangian planes in $\mathbb{R}^n_x \times \mathbb{R}^n_p$ is denoted by $\operatorname{Lag}(n)$ and called the *Lagrangian Grassmannian*. As a manifold, it is both connected and compact, and its fundamental group is the integer group $(\mathbb{Z}, +)$. The action

$$\operatorname{Sp}(n) \times \operatorname{Lag}(n) \longrightarrow \operatorname{Lag}(n)$$

is transitive: for every pair (ℓ, ℓ') there exists $s \in \operatorname{Sp}(n)$ such that $\ell = s\ell'$. Moreover $\operatorname{Sp}(n)$ also acts transitively on all pairs of transversal Lagrangian planes (two Lagrangian planes ℓ and ℓ' are transverse if $\ell \cap \ell' = 0$).

The Equation of a Lagrangian Plane

Any n-dimensional subspace of phase space can be represented by an equation of the type

$$Xx + Pp = 0$$

where X and P are two $n \times n$ matrices such that $\operatorname{rank}(X, P) = n$ (here X, P is viewed as an $n \times 2n$ matrix). The following result gives a necessary and sufficient condition for a subspace to be Lagrangian:

Proposition 266 *Let ℓ be an n-dimensional subspace of $\mathbb{R}^n_x \times \mathbb{R}^n_p$. That subspace can be represented by the equation*

$$Xx + Pp = 0 \tag{A.1}$$

with $\operatorname{rank}(X, P) = n$ *and ℓ is Lagrangian if and only if PX^T is symmetric: $PX^T = XP^T$ (or, equivalently, if PX^T is symmetric).*

Proof. The condition $\operatorname{rank}(X, P) = n$ implies that we can rearrange the lines and the rows of X so that, say $\operatorname{rank}(P) = n$. WE leave it to the reader that such a rearrangement corresponds to writing the equation $Xx + Pp = 0$ in a new coordinate system obtained by applying a linear symplectic transformation to the (x, p) coordinates, and it is thus sufficient to prove the proposition in this case. Let now (x, p) and (x', p') be two points of ℓ:

$$Xx + Pp = 0 \ , \ Xx' + Pp' = 0.$$

Since rank $(P) = n$ we have $\det P \neq 0$ and we may thus solve these equations and p and p':

$$p = -P^{-1}Xx, \ p' = -P^{-1}Xx'$$

so that

$$\begin{aligned} \Omega(x, p; x', p') &= P^{-1}Xx \cdot x' - P^{-1}Xx' \cdot x \\ &= (P^{-1}X - X^T(P^{-1})^T)x \cdot x'. \end{aligned}$$

Now, ℓ is by definition a Lagrangian plane if and only if

$$(P^{-1}X - X^T(P^{-1})^T)x \cdot x = 0$$

for all x, x', that is if and only if

$$P^{-1}X = X^T(P^{-1})^T$$

which is equivalent to PX^T being symmetric. To show that this condition is equivalent to the symmetry of PX^T it suffices to note that since

$$J = \begin{pmatrix} 0 & I \\ -I & 0 \end{pmatrix}$$

is symplectic, $J\ell$ will also be a Lagrangian plane, and that an equation for $J\ell$ is

$$Px - Xp = 0$$

so that $(-X)P^T$ (and hence XP^T) is symmetric. ■

Remark 267 *The argument above has given us a geometrical proof of the following property from linear algebra: let X and Y be two square matrices of same dimension such that (X, Y) has maximal rank. Then X^TY is symmetric if and only if XY^T is.*

The symplectic group $\mathrm{Sp}(n)$ acts transitively on the set of Lagrangian planes. It turns out that so does the unitary group

$$U(n) = \mathrm{Sp}(n) \cap O(2n).$$

Since every $u \in U(n)$ is of the type

$$u = \begin{pmatrix} A & -B \\ B & A \end{pmatrix}$$

with $A^TB = B^TA$, $AB^T = BA^T$, and $AA^T + BB^T = I$ it follows that a Lagrangian plane can always be represented by an equation

$$\ell : Xx + Pp = 0 \ \text{ with } \begin{cases} PX^T = XP^T \\ XX^T + PP^T = I. \end{cases} \tag{A.2}$$

Appendix B

The Lie-Trotter Formula for Flows

The simplest form of the Lie-Trotter product formula is obtained by considering exponentials of matrices. Due to the non-commutativity of matrix product, the naively "expected" formula $\exp(A+B) = \exp(A)\exp(B)$ does not hold in general. One can in fact the "Campbell–Hausdorff formula", which says that

$$\exp(A)\exp(B) = \exp\left(A + \sum_{k=2}^{\infty} M_k(A,B)\right)$$

where each term $M_k(A,B)$ is a linear combination of k-fold commutators of the matrices A and B (see Mneimé and Testard [118] for explicit calculations). For instance, when A and B commute with $[A,B]$, one gets the well-known formula

$$\exp(A)\exp(B) = \exp(A+B)\exp\left(\frac{1}{2}[A;B]\right).$$

However, there is a more useful way to see things, at least from a computational standpoint. In fact, one can show that for large values of the integer N the estimate

$$\left[\exp\left(\frac{A+B}{N}\right)\right]^N = \exp\left(\frac{A}{N}\right)\exp\left(\frac{B}{N}\right) + O\left(\frac{1}{N^2}\right)$$

holds. From this, we immediately get the classical formula

$$\exp(A+B) = \lim_{N\to\infty}\left[\exp\left(\frac{A}{N}\right)\exp\left(\frac{B}{N}\right)\right]^N \tag{B.1}$$

which was proven by Sophus Lie (b.1842) in 1875.

B.1 The Lie-Trotter Formula for Vector Fields

It turns out that formulas of the type (B.1) can easily be derived from a well-known general result on approximations of flows associated to a vector field. We begin with a classical lemma from the theory of first-order differential equations:

Lemma 268 *Let X be a vector field satisfying the local Lipschitz condition*

$$\left\|X(u) - X(u')\right\| \le a\left\|u-u'\right\| \tag{B.2}$$

for all z, z' in an open set U. The flow (f_t) of X then satisfies the estimate

$$\left\|f_t(u) - f_t(u')\right\| \le e^{a|t|}\left\|u-u'\right\| \text{ for } z, z' \in U. \tag{B.3}$$

Proof. From the equality

$$f_t(u) = u + \int_0^t X(f_s(u))ds$$

follows that $h(t) = \|f_t(u) - f_t(u')\|$ satisfies the estimate:

$$\begin{aligned} h(t) &\le \left\|u-u'\right\| + \left\|\int_0^t \left[X(f_s(u)) - X(f_s(u'))\right] ds\right\| \\ &\le \left\|u-u'\right\| + a\int_0^t h(s)ds. \end{aligned}$$

The inequality (B.3) follows, using Gronwall's inequality for differential systems (see any elementary treatise on differential equations). ∎

Proposition 269 *Let X be a vector field on some open subset U of $\mathbb{R}^m$, and (f_t) the flow of X. Let (k_t) be a family of functions $k_t : U \to \mathbb{R}^m$ defined for t in some open interval containing 0 such that the dependence of $k_t(u)$ on $(u,t)$is C^1. Assume that for every $u_o \in U$ we have*

$$k_0(u_0) = u_0 \text{ and } X(U_0) = \frac{\partial}{\partial t}k_t(u_0)\mid_{t=0}. \tag{B.4}$$

Then the following estimate holds

$$f_t(u_0) = k_t(u_0) + o(t) \text{ for } t \to 0 \tag{B.5}$$

and the sequence of iterates converges towards $f_t(u_0)$*:*

$$f_t(u_0) = \lim_{N\to\infty}(k_{t/N})^N(u_0). \tag{B.6}$$

Proof. The proof of Proposition 269 is based on the use of telescoping sums. We are following here the presentation given in Abraham *et al.* [1] (See also Nelson [125] for more on these topics.) First, it is clear that (B.4) implies (B.5). In fact, by the definition of (f_t) we have

$$\left.\frac{\partial}{\partial t}(f_t(u_0) - k_t(u_0))\right|_{t=0} = 0$$

and hence, applying Taylor's formula at $t = 0$

$$f_t(u_0) - k_t(u_0) = O(t^2) \tag{B.7}$$

since $f_0(u_0) - k_0(u_0) = 0$. Let us now show that for $u_0 \in U$ all the iterates are defined and stay in U if t is small enough. Setting

$$u_j = (k_{t/N})^j(u_0)$$

we note that since k_0 is the identity, we have

$$k_t(u_0) = u_0 + O(t^2) \text{ for } t \to 0.$$

It follows from this estimate that if u is in some neighborhood U of u_0, then $k_t(u)$ will also be in U if t is chosen sufficiently small. Suppose now that u_j is defined for $1 \leq j \leq N-1$, and that we have, for these values of j,

$$u_j - u_0 = O(t).$$

Writing the difference $u_N - u_0$ as the "telescoping sum"

$$u_N - u_0 = \sum_{j=1}^{N} u_j - u_{j-1}$$

and noting that we have

$$u_j - u_{j-1} = k_{t/N}(u_{j-1}) - u_{j-1} = O(t/N)$$

for $1 \leq j \leq N$, we get

$$u_N - u_0 = (k_{t/N})^N(u_0) - u_0 = NO(t/N) = O(t).$$

This estimate being independent of the integer N it follows, by iteration, that (B.7) will hold for all j and that the u_j will stay in U if time t is chosen small enough. Let us finally prove formula (B.6). It is sufficient to show that

$$f_t(u_0) - (k_{t/N})^N(u_0) = N_o(t/N) \text{ for } N \to \infty$$

because $N_o(t/N) \to 0$ as $N \to \infty$. In view of the obvious equality

$$f_t(u_0) - (k_{t/N})^N(u_0) = (f_{t/N})^N(u_0) - (k_{t/N})^N(u_0)$$

we can write

$$f_t(z_0) - (k_{t/N})^N(z_0) = \sum_{j=1}^{N}(f_{t/N})^{N-j}f_{t/N}(z_{j-1}) - (f_{t/N})^{N-j}k_{t/N}(z_{j-1}).$$

Let us estimate each term of this sum. In view of Lemma 268 there exists $a > 0$ such that

$$\|(f_{t/N})^{N-j}f_{t/N}(z_{j-1}) - (f_{t/N})^{N-j}k_{t/N}(z_{j-1})\| \leq e^{a|t|}\|f_{t/N}(z_{j-1}) - k_{t/N}(z_{j-1})\|.$$

In fact (B.7) is immediately obtained by applying $N - j$ times to each term.

$$(f_{t/N})^{N-j}f_{t/N}(z_{j-1}) - (f_{t/N})^{N-j}k_{t/N}(z_{j-1}).$$

Using (B.5) and the inequality above, we have for $1 \leq j \leq N$,

$$\left\|(f_{t/N})^{N-j}f_{t/N}(u_{j-1}) - (f_{t/N})^{N-j}g_{t/N}(u_{j-1})\right\| = e^{a|t|}o(t/N)$$

and, adding together these N estimates, we finally get

$$\left\|f_{t/N}(u) - (k_{t/N})^N(u_0)\right\| \leq Ne^{a|t|}o(t/N) \text{ for } N \to \infty$$

which was to be proven. ∎

Definition 270 *The family of mappings* (k_t) *defined in Proposition 269 is called an algorithm for the flow* (f_t).

Corollary 271 *Let* X *and* Y *be two vector fields on* $\mathbb{R}^m$, *with respective flows* (f_t) *and* (g_t). *The flow* (h_t) *of* $X + Y$ *is then given by*

$$h_t(u) = \lim_{N \to \in}(f_{t/N} \circ g_{t/N})^N(u).$$

Proof. It suffices to choose $k_t = f_t \circ g_t$ in Proposition 269. ∎

Appendix C

The Heisenberg Groups

Let $S^1 = \{u \in \mathbb{C} : |u| = 1\}$ be the unit circle. The standard Heisenberg group is the set

$$H(n) = \mathbb{R}^n_x \times \mathbb{R}^n_p \times S^1$$

equipped with the group law:

$$(z, \zeta)(z', \zeta') = \left(z + z', \zeta\zeta' \exp \frac{i}{2}\omega(z, z') \right).$$

Here are two variants of $H(n)$ often encountered in the literature, and also called "Heisenberg groups". They may be viewed as isomorphic copies of the universal covering of $H(n)$. We begin with the "polarized Heisenberg group" $H(n)$; it is the multiplicative group of all upper triangular $(2n+2) \times (2n+2)$ matrices of the type

$$M(z, t) = \begin{pmatrix} 1 & p_1 & \dots & p_n & t \\ 0 & 1 & \dots & 0 & x_1 \\ \vdots & \vdots & \vdots & \vdots & \vdots \\ 0 & 0 & \dots & 1 & x_n \\ 0 & 0 & \dots & 0 & 1 \end{pmatrix}.$$

It is economical to write the matrices $M(z, t)$ as

$$M(z, t) = \begin{pmatrix} 1 & p^T & t \\ 0 & 1 & x \\ 0 & 0 & 1 \end{pmatrix} \tag{C.1}$$

if $z = (x, p)$. The determinant of $M(z, t)$ is 1, and its inverse is

$$M(z,t) = \begin{pmatrix} 1 & -p^T & -t + p \cdot x \\ 0 & 1 & -x \\ 0 & 0 & 1 \end{pmatrix}. \tag{C.2}$$

The polarized Heisenberg group is thus essentially the extended phase space equipped with the multiplicative law

$$(z,t)(z',t') = (z + z', t + t' + p \cdot x'); \tag{C.3}$$

the unit is $(0, 0, 0)$ and the inverse of (x, p, t) is given by

$$(x,p,t)^{-1} = (-x, -p, -t + p \cdot x). \tag{C.4}$$

We notice that $H_{pol}(n)$ contains the group of phase space translations, identified with the matrices

$$\begin{pmatrix} 1 & p^T & 0 \\ 0 & 1 & x \\ 0 & 0 & 1 \end{pmatrix}.$$

Consider now the bijection $\mathbb{R}^{2n+1} \longrightarrow \mathbb{R}^{2n+1}$ defined by

$$F(z,t) = (z, t - \tfrac{1}{2} p \cdot x) \text{ if } z = (x,p). \tag{C.5}$$

We can use that bijection to transport the algebraic structure of $H_{pol}(n)$; the new group thus obtained is called the isotropic Heisenberg group $H_{iso}(n)$; it is by construction isomorphic to $H_{pol}(n)$. One easily checks that the group law of $H_{iso}(n)$ is given by

$$(z,t)(z',t') = (z + z', t + t' + \tfrac{1}{2}\Omega(z, z')) \tag{C.6}$$

where Ω is the symplectic form. The unit element of that group is $(0, 0)$ and the inverse of (z, t) is $(-z, -t)$.

Both Heisenberg groups are simply connected and connected Lie groups (as manifolds they are just $\mathbb{R}^{2n+1}$). To determine their Lie algebra, we first note that the matrices (C.2) can be written as

$$M(z,t) = I + m(z,t) \tag{C.7}$$

where I is the $(2n + 1)$-identity matrix and

$$m(z,t) = \begin{pmatrix} 0 & p^T & t \\ 0 & 0 & x \\ 0 & 0 & 0 \end{pmatrix}. \tag{C.8}$$

We have

$$m(z,t)m(z',t') = m(0,0,p\cdot x') \tag{C.9}$$

and also

$$m(z,t)^k = 0 \text{ for every integer } k > 2. \tag{C.10}$$

It follows that

$$e^{m(z,t)} = I + m(z,t) + \tfrac{1}{2}(0,0,p,x)$$

and hence

$$e^{m(z,t)} = M(z, t + \tfrac{1}{2}p\cdot x). \tag{C.11}$$

Taking (C.5) into account we have thus shown that the ring $h(n)$ of all matrices $M(z,t)$ is the Lie algebra of the Heisenberg group $H_{iso}(n)$. It is called the "Heisenberg Lie algebra". Note that in view of (C.9) the Lie bracket is given by

$$[m(z,t), m(z',t')] = m(0,\Omega(z,z')). \tag{C.12}$$

As noticed above, both $H_{pol}(n)$ and $H_{iso}(n)$ are simply connected; they are actually isomorphic copies of the universal covering group of the standard Heisenberg group $H(n)$. Obviously the natural projection

$$\pi : H_{iso}(n) \longrightarrow H(n)$$

is a covering mapping.

Appendix D

The Bundle of s-Densities

In what follows s is a positive real number.

Definition 272 *An s-density on a vector space L (real or complex) is a mapping $\rho : L^n \longrightarrow \mathbb{K}$ ($\mathbb{K} = \mathbb{R}$ or $\mathbb{C}$) such that*

$$\rho(A\xi_1, ..., A\xi_n) = |\det(A)|^s \rho(\xi_1, ..., \xi_n)$$

for all $A \in GL(n, \mathbb{K})$ and $(\xi_1, ..., \xi_n) \neq 0$. An s-density on a manifold V is the datum for every $z \in V$ of an s-density $\rho(z)$ on the tangent space T_zV at z, depending smoothly on the point z.

It turns out that all densities of same degree s on L are proportional (and hence proportional to $|\det|^s$). Let us show this in detail. We first remark that if a density vanishes at one point $(\xi_1, ..., \xi_n) \neq 0$, then it is identically zero. In fact, for every n-tuple $(\xi_1^{'}, ..., \xi_n^{'}) \neq 0$ we can find $A \in GL(n, \mathbb{K})$ such that $(\xi_1^{'}, ..., \xi_n^{'}) = (A\xi_1, ..., A\xi_n)$ and we then have $\rho(\xi_1^{'}, ..., \xi_n^{'}) = 0$ if and only if $\rho(\xi_1, ..., \xi_n) = 0$. Since two functions are trivially proportional if one of them is zero, it is thus no restriction to assume that neither ρ nor ρ' vanishes at any point. Consequently the range of both ρ and ρ' is $\mathbb{K}^*$ (the field $\mathbb{K}$ with 0 deleted). Thus, for every n-tuple $(z_1, ..., z_n) \neq 0$ there exists another n-tuple $(\xi_1^{'}, ..., \xi_n^{'}) \neq 0$ such that

$$\rho(\xi_1, ..., \xi_n) = \rho(\xi_1^{'}, ..., \xi_n^{'}).$$

Choosing again $A \in GL(n, \mathbb{K})$ such that $(\xi_1^{'}, ..., \xi_n^{'}) = (A\xi_1, ..., A\xi_n)$ we get by definition of a density:

$$(A\xi_1, ..., A\xi_n) = |\det(A)|^s \, \rho(\xi_1, ..., \xi_n). \tag{D.1}$$

The matrix A may of course depend on the vectors $\xi_1, ..., \xi_n$ but its determinant does not, for if we change $(\xi_1, ..., \xi_n)$ into, say, $(\eta_1, ..., \eta_n)$, then we can find $B \in GL(n, \mathbb{K})$ such that $(\xi_1, ..., \xi_n) = (B\eta_1, ..., B\eta_n)$ and we will thus have

$$\begin{aligned}\rho(\eta_1, ..., \eta_n) &= \left|\det(B^{-1})\right|^s \rho(\xi_1, ..., \xi_n) \\ &= \left|\det(B^{-1})\right|^s \left|\det(A)\right|^s \rho'(B\eta_1, ..., B\eta_n) \\ &= \left|\det(A)\right|^s \rho'(\eta_1, ..., \eta_n).\end{aligned}$$

Setting $\lambda = \left|\det(A)\right|^s$, we have thus shown that $\rho = \lambda\rho'$, which proves our claim.

The one-dimensional vector space of all densities of degree s on L is classically denoted by $|\Omega|^s(L)$. It turns out that the definition of a density carries over to manifolds. Suppose in fact that V is an n-dimensional manifold; to every $z \in V$ we can associate the one-dimensional space $|\Omega|^s(T_zV)$. We denote by $|\Omega|^s(V)$ the disjoint union of all the spaces $|\Omega|^s(T_zV)$ when z ranges over V:

$$|\Omega|^s(V) = \cup_{z\in V}\{z\} \times |\Omega|^s(T_zV).$$

Defining a projection

$$\pi : |\Omega|^s(V) \longrightarrow V, \quad (z, \rho(z)) \longmapsto z$$

one checks that $(|\Omega|^s(V), \pi)$ is a line-bundle, which is common practice to call the *s-density bundle*. An s-density on V is then, by definition, a section ρ of $(|\Omega|^s(V), \pi)$, *i.e.* a continuous mapping $\rho : V \longrightarrow |\Omega|^s(V)$ such that $\pi \circ \rho$ is the identity on V.

It is not very difficult to show, using coordinates, that both definitions of s-densities are equivalent; we will not do it here.

Definition 273 *An s-density $\rho = (\rho_\alpha)$ on a manifold V is the datum of an atlas $(U_\alpha, f_\alpha)_\alpha$ of V together with the datum, for each α, of a complex-valued smooth function ρ_α on U_α such that the following compatibility conditions hold: for all indices α, β such that $U_{\alpha\beta} = U_\alpha \cap U_\beta \neq \emptyset$ we have*

$$\rho_\alpha(z) = |Df_{\alpha\beta}(z)|^s(f_{\alpha\beta})^*\rho_\beta(z), \quad z \in U_{\alpha\beta}; \tag{D.2}$$

here $Df_{\alpha\beta}$ is the Jacobian of the transition function

$$f_{\alpha\beta} = f_\alpha \circ f_\beta^{-1} : f(U_{\alpha\beta}) \longrightarrow f(U_{\alpha\beta}).$$

Let $(V_\alpha, g_\alpha)_\alpha$ be a refinement of the atlas $(U_\alpha, f_\alpha)_\alpha$. The restrictions of the ρ_α to the sets V_α also define a density which we will identify with the density ρ. Notice that if $s = 0$ then the conditions (D.2) become

$$\rho_\alpha(z) = (f_{\alpha\beta})^* \rho_\beta(z), \quad z \in U_{\alpha\beta}; \tag{D.3}$$

which are the formulas for the change of the local representation of a function on V.

If $s = 1$, we will simply say that ρ is a density on V; if $s = 1/2$ then we say that ρ is a *half-density*. Densities of a given order in a manifold V clearly form a vector space: if $\rho = (\rho_\alpha, U_\alpha, f_\alpha)_\alpha, \rho' = (\rho'_\alpha, U'_\alpha, f'_\alpha)_\alpha$ are two s-densities, their sum $\rho'' = \rho + \rho'$ is the m-density $(\rho''_\alpha, U''_\alpha, f''_\alpha)_\alpha$ where $(U''_\alpha, f''_\alpha)_\alpha$ is an atlas such that $U_\alpha = U_\alpha \cap U'_\alpha$ and $\rho''_\alpha = \rho_\alpha + \rho'_\alpha$. The product $\lambda\rho$ for $\lambda \in \mathbb{C}$ is defined in an obvious way: it is the density $(\lambda\rho_\alpha, U_\alpha, f_\alpha)$.

One can pull-back and push-forward densities: if f is a diffeomorphism of a manifold V' onto V, the pull-back $f^*\rho$ of an s-density ρ on V by f is defined by

$$f^*\rho(z')(\zeta) = \rho(f(z'))(f'(z')\zeta)$$

where $f'(z')\zeta = f'(z')\zeta_1, ..., f'(z')\zeta_n; f'(z')$ is the Jacobian matrix of f at z'. The push-forward of a density ρ on V' by f is defined by $f_*\rho = (f^{-1})^*\rho$. Both the pull-back $f^*\rho$ and push-forward $f_*\rho$ are themselves of same degree as ρ. It suffices to prove this assertion for the pull-back: for a matrix A we have

$$\begin{aligned} f^*\rho(z)(A\zeta) &= \rho(f(z'))(f'(z')A\zeta) \\ &= \rho(f'(z')) \left[(f'(z')Af'(z')^{-1})f'(z')A\zeta \right] \\ &= \left| \det(f(z')Af'(z')^{-1}) \right|^s \rho(f'(z'))(f'(z')\zeta) \\ &= \left| \det(A) \right|^s \rho(f'(z'))(f'(z')\zeta) \end{aligned}$$

hence $f^*\rho$ is an s-density as claimed.

Appendix E

The Lagrangian Grassmannian

E.1 $\mathrm{Lag}(n)$ as a Set of Matrices

We are going to show that the Lagrangian Grassmannian $\mathrm{Lag}(n)$ can be identified with the set of all symmetric unitary $n \times n$ matrices. One proceeds as follows: for an arbitrary n-plane ℓ (not necessarily Lagrangian), let P_ℓ be the orthogonal projection on ℓ: noting that the n-plane $J\ell$ is orthogonal to ℓ, it is characterized by the conditions

$$P_\ell^2 = P_\ell, \quad \mathrm{Ker}(P_\ell) = J\ell, \quad (P_\ell)^T = P_\ell. \tag{E.1}$$

Set now

$$w(\ell) = (2P_\ell - I)C,$$

where C is the "conjugation matrix" defined by:

$$C = \begin{pmatrix} -I & 0 \\ 0 & I \end{pmatrix}$$

and suppose that ℓ is a Lagrangian plane. Since the subgroup $U(n)$ of $\mathrm{Sp}(n)$ acts transitively on $\mathrm{Lag}(n)$, there exist two $n \times n$ matrices A and B such that

$$\begin{cases} A^TB = B^TA \\ AB^T = BA^T \end{cases}, \quad \begin{cases} A^TA + B^TB = I \\ AA^T + BB^T = I \end{cases}$$

and $\ell = u(0 \times \mathbb{R}_p^n)$ where

$$u = \begin{pmatrix} A & -B \\ B & A \end{pmatrix}.$$

The vectors orthogonal to ℓ are of the type (Ax, Bx); one checks that the projection operator on ℓ is

$$P_l = \begin{pmatrix} BB^T & -AB^T \\ -BA^T & AA^T \end{pmatrix}$$

and hence

$$w(\ell) = \begin{pmatrix} AA^T - BB^T & -2AB^T \\ 2BA^T & AA^T - BB^T \end{pmatrix}$$

that is

$$w(\ell) = \begin{pmatrix} A & -B \\ B & A \end{pmatrix} \begin{pmatrix} A^T & -B^T \\ B^T & A^T \end{pmatrix}.$$

Recalling that the mapping

$$A + iB \longmapsto \begin{pmatrix} A & -B \\ B & A \end{pmatrix}$$

is an isomorphism $U(n, \mathbb{C}) \longrightarrow U(n)$, it is clear that $w(\ell)$ is the image by that isomorphism of the *symmetric* matrix

$$w(\ell) = (A + iB)(A + iB)^T.$$

The subgroup $U(n)$ of $\mathrm{Sp}(n)$ acts transitively on $\mathrm{Lag}(n)$ and hence so does the unitary group $U(n, \mathbb{C})$; we will denote this action $(R, \ell) \longmapsto R\ell$. Notice that we have

$$\ell : Ax + Bp = 0 \Longleftrightarrow \ell = (A + iB)(0 \times \mathbb{R}^n_p).$$

Let $W(n, \mathbb{C})$ be the set of all symmetric unitary matrices in $U(n, \mathbb{C})$:

$$W(n, \mathbb{C}) = U(n, \mathbb{C}) \cap Sym(n).$$

We claim that $\mathrm{Lag}(n)$ can be identified with $W(n, \mathbb{C})$. We begin by proving the following Lemma, which shows that one can take square roots in $W(n, \mathbb{C})$.

Lemma 274 *For every $w \in W(n, \mathbb{C})$. there exists $u \in W(n, \mathbb{C})$. such that $w = u^2$.*

Proof. Writing $w = A + iB$, the condition $ww^* = I$ implies that $AB = BA$. It follows that the symmetric matrices A and B can be diagonalized simultaneously, i.e. that there exists $R \in O(n)$ such that $G = RAR^T$ and

$H = RBR^T$ are diagonal. Let g_j and h_j ($1 \leq j \leq n$) be the eigenvalues of G and H, respectively. Since $A^2 + B^2 = I$ we have $g_j^2 + h_j^2 = 1$ for every j. Choose now real numbers x_j, y_j such that

$$x_j^2 + y_j^2 = g_j,\ 2x_j y_j = h_j \quad (1 \leq j \leq n)$$

and let X and Y be the diagonal matrices whose entries are these numbers x_j, y_j. Then $(X + iY)^2 = G + iH$, and $u = R^T(X + iY)R$ satisfies $u^2 = w$.

■

Proposition 275 *The mapping*

$$w : \mathrm{Lag}(n) \longrightarrow W(n, \mathbb{C}),\ \ \ell \longmapsto \mathbb{W}(\ell),$$

which to every Lagrangian plane $\ell = U(0 \times \mathbb{R}_p^n)$, $U \in U(n, \mathbb{C})$, *associates the symmetric unitary matrix*

$$W(\ell) = UU^T$$

is a homeomorphism $w : \mathrm{Lag}(n) \approx W(n, \mathbb{C})$. *Moreover, for any* $R \in U(n, \mathbb{C})$, *we have*

$$W(R\ell) = RW(\ell)R^T. \tag{E.2}$$

Proof. The mapping $\ell \longmapsto W(\ell)$ is injective, because the mapping $\ell \longmapsto P(\ell)$ is. Since w is continuous and Lag(n) is compact, it suffices to show that is also surjective. But this is an immediate consequence of the Lemma above. Let us finally show that (E.2) holds. By definition of the mapping w, we have $W(\ell) = UU^T$if $\ell = U(0 \times \mathbb{R}_p^n)$ so that

$$W(R\ell) = RU(RU^T)$$

which is precisely (E.2). ■

Example 276 ***The case*** $n = 1$ ***revisited.*** Lag(1) *can be identified with* $S^1 / \{\pm 1\}$; *this amounts identifying a line through the origin in the plane with its polar angle* α *modulo* π. *The mapping* $w(\cdot)$ *is then simply the bijection* $\alpha \longmapsto e^{2i\alpha}$ *of* Lag(1) *onto* S^1.

Proposition 275 above can of course be restated without difficulty in terms of the image $W(n)$ of $W(n, \mathbb{C})$ in $U(n)$. Since we have

$$\begin{pmatrix} A & -B \\ B & A \end{pmatrix} \in W(n) \Longleftrightarrow \begin{cases} A^T = A\ ,\ B = B^T \\ AB = BA \\ A^2 + B^2 = I \end{cases},$$

the mapping $\ell \longmapsto w(\ell)$ induced by $\ell \longmapsto W(\ell)$ is given by

$$w(\ell) = \begin{pmatrix} A & -B \\ B & A \end{pmatrix} \begin{pmatrix} A^T & -B^T \\ B^T & A^T \end{pmatrix}. \tag{E.3}$$

It can be used to measure the dimension of the intersection of two Lagrangian planes, and to characterize transversality:

Proposition 277 *Let (ℓ, ℓ') be a pair of Lagrangian planes. We have*

$$\operatorname{rank}(w(\ell) - w(\ell')) = 2(n - \dim(\ell \cap \ell')). \tag{E.4}$$

In particular,

$$\ell \cap \ell' = 0 \Longleftrightarrow \det(w(\ell) - w(\ell')) \neq 0. \tag{E.5}$$

Proof. Since the action of $\operatorname{Sp}(n)$ on $\operatorname{Lag}(n)$ is transitive, it suffices to consider the case $\ell' = 0 \times \mathbb{R}^n_p$, in which case (E.4) is

$$\operatorname{rank}(w(\ell) - I_{2n\times 2n}) = 2(n - \dim(\ell \cap (0 \times \mathbb{R}^n_p))).$$

We have, by (E.3), noting that $A^TA + B^TB = I_{n\times n}$:

$$\begin{aligned} w(\ell) - I &= -2\begin{pmatrix} B^TB & A^TB \\ -B^TA & B^TB \end{pmatrix} \\ &= -2\begin{pmatrix} B^T & 0 \\ 0 & B^T \end{pmatrix}\begin{pmatrix} B & A \\ -A & B \end{pmatrix} \end{aligned}$$

and hence, since $\operatorname{rank}(A, B) = n$:

$$\operatorname{rank}(w(\ell) - I) = 2\operatorname{rank}(B)$$

which proves (E.5) in view of the second part of Lemma 205. ∎

E.1.1 The Universal Coverings $U_\infty(n, \mathbb{C})$ and $W_\infty(n, \mathbb{C})$

Recall (Chapter 3, Subsection 3.3.3) that $\operatorname{Sp}(n)$ is homeomorphic to the product of $U(n)$ and of the simply connected space $\mathbb{R}^{n(2n+1)}$. It follows that $\operatorname{Sp}(n)$ and $U(n)$ have isomorphic fundamental groups, and hence $\pi_1(\operatorname{Sp}(n)) = (\mathbb{Z}, +)$, because

$$\pi_1(U(n, \mathbb{C})) = (\mathbb{Z}, +).$$

Consider the set

$$U_\infty(n, \mathbb{C}) = \left\{(R, \theta) : R \in U(n, \mathbb{C}), \det R = e^{i\theta}\right\}$$

equipped with the multiplicative law

$$(R, \theta)(R', \theta') = (RR', \theta + \theta').$$

It is a connected Lie group. The projection

$$\pi : U_\infty(n, \mathbb{C}) \longrightarrow U(n, \mathbb{C}), \ \ (R, \theta) \longmapsto R$$

is a surjective group homomorphism, with kernel $\operatorname{Ker}(\pi) = (2\pi\mathbb{Z}, +)$ hence $U_\infty(n, \mathbb{C})$ is the universal covering of $U(n, \mathbb{C})$. A similar argument shows that

$$W_\infty(n, \mathbb{C}) = \left\{(R, \theta) : R \in W(n, \mathbb{C}), \det R = e^{i\theta}\right\}$$

is the universal covering of $W(n, \mathbb{C})$; since we have identified $W(n, \mathbb{C})$ with Lag(n) it follows that $W_\infty(n, \mathbb{C})$ can be identified with the universal covering $\mathrm{Lag}_\infty(n)$ of the Grassmannian Lagrangian:

$$\mathrm{Lag}_\infty(n) = W_\infty(n, \mathbb{C}).$$

Bibliography

[1] Abraham, R., Marsden, J.E. and Ratiu, T. *Manifolds, Tensor Analysis, and Applications*, Applied Mathematical Sciences **75** (Springer, 1988).

[2] Ali, T.S. and Engliš, M. Quantization methods: a guide for physicists and analysts. Rev. Math. Phys. **17**(04) (2005), 391–490.

[3] Anandan, J. Geometric angles in quantum and classical physics. Phys. Lett. A **129**(4) (1988), 201–207.

[4] Arnold, V.I. *Mathematical Methods of Classical Mechanics*, 2nd edition, Graduate Texts in Mathematics (Springer-Verlag, 1989).

[5] Arnold, V.I. *A characteristic class entering in quantization conditions*, Funkt. Anal. i. Priloz. **1**(1), 1–14 (in Russian) (1967); Funct. Anal. Appl. **1**, 1–14 (English translation) (1967).

[6] Arnold, V.I. Sturm theory and symplectic geometry. Funct. Anal. Appl. **19** (1985), 251–259.

[7] Auffray, J.-P. *Einstein et Poincaré* (Éditions le Pommier, 1999).

[8] Bargmann, V. On unitary ray representations of continuous groups. Ann. Math. **59** (1954), 1–46.

[9] Bell, J. *Speakable and Unspeakable in Quantum Mechanics* (Cambridge University Press, 1993).

[10] Berndl, K., Dürr, D., Goldstein, S. Peruzzi, G. and Zanghi, N. On the global existence of Bohmian mechanics. Comm. Math. Phys. **173** (1995), 647–673.

[11] Berry, M.V. Quantal phase factors accompanying adiabatic changes. Proc. Roy. Soc. London A **392** (1984), 45–57.

[12] BERRY, M.V. Classical adiabatic angles and quantal adiabatic phase. J. Phys. A: Math. Gen. **18** (1985), 15–27.

[13] BINZ, E. AND SCHEMPP, W. *Quantum hologram and relativistic hodogram: Magnetic resonance tomography and gravitational wavelet detection*, in: *Proceedings of the Second International Conference on Geometry, Integrability and Quantization*, June 7–15, 2000, Varna, Bulgaria, I.M. Mladenov and G.L. Naber, Editors (Coral Press, Sofia, 2001), 110–150.

[14] BLATTNER, R.J. *Book review* in Bull. Amer. Math. Soc. **9**(3) (1983).

[15] BLATTNER, R.J. *Quantization and representation theory*, in: *Harmonic Analysis and Homogeneous Spaces*, E.T. Taam, Editor, Proc. Sym. Pure Math. **26** (AMS, 1973), 147–165.

[16] BLATTNER, R.J. *Pairing of half-form spaces*, in: *Géométrie Symplectique et Physique Mathématique*, Colloq. Internat. CNRS **237**, Paris (1974).

[17] BOHM, D. *Quantum Theory* (Prentice Hall, 1951).

[18] BOHM, D. A suggested interpretation of the quantum theory in terms of "hidden" variables: Part I. Phys. Rev. **85** (1952), 166–179.

[19] BOHM, D. A suggested interpretation of the quantum theory in terms of "hidden" variables: Part II. Phys. Rev. **85** (1952), 180–193.

[20] BOHM, D. AND HILEY, B. *The Undivided Universe* (Routledge, 1993).

[21] BOLOTIN, S.V. Libration motion of natural dynamical systems. Moscow University Bulletin **3** (1978).

[22] BOOSS-BAVNBEK B. AND FURUTANI K. The Maslov index: a functional analytical definition and the spectral flow formula. Tokyo J. Math. **21**(1) (1998), 1–34.

[23] BORN, M. AND JORDAN, J. Zur Quantenmechanik. Z. Physik **34** (1925), 858–888.

[24] BORN, M., JORDAN, HEISENBERG, W., Zur Quantenmechanik II. *Z.* Physik **35** (1925), 557–615.

[25] BRACK, M. AND BHADURI R.K. *Semiclassical Physics* (Addison-Wesley, 1997).

[26] Brown, M.R. and Hiley, B.J. Schrödinger revisited: the role of Dirac's 'standard' ket in the algebraic approach (preprint, 2001).

[27] Buslaev V.C. Quantization and the W.K.B method. Trudy Mat. Inst. Steklov **110** (1978), 5–28 [in Russian].

[28] Cappell, S.E., Lee R. and Miller, E.Y. On the Maslov index. Comm. Pure Appl. Math. **47** (1994), 121–186.

[29] Castellani, L. Quantizaton rules and Dirac's correspondence. Il Nuovo Cimento, **48**(3) (1978), 359–368.

[30] Cohen, L. Joint quantum probabilities and the uncertainty principle. Ann. New York Acad. Sc. **480** (1986), 283.

[31] Crehan, P. The parametrisation of quantisation rules equivalent to opertor orderings, and the effect of different rules on the physical spectrum. J. Phys. A: Math. Gen. **22** (1988), 811–822.

[32] Crumeyrolle, A. *Orthogonal and Symplectic Clifford Algebras* (Kluwer Acad. Publ., 1990).

[33] Dacorogna, B and Moser, J. On a partial differential equation involving the Jacobian determinant. Ann. Inst. Henri Poincaré, Analyse Non Linéaire **7** (1990), 1–26.

[34] Dazord, P. Invariants homotopiques attachés aux fibrés symplectiques. Ann. Inst. Fourier, Grenoble **29**(2) (1979), 25–78.

[35] Demazure, M. *Classe de Maslov II*, Exposé numéro 10, Séminaire sur le fibré cotangent, Orsay (1975-76).

[36] Dirac, P.A.M. *The Principles of Quantum Mechanics*, 4th edition (Oxford Science Publications, 1999).

[37] Dirac, P.A.M. Quantised singularities in the electromagnetic field. Proc. Roy. Soc. **A133** (1931), 60–72.

[38] Dirac, P.A.M. The theory of magnetic poles. Phys. Rev. **74** (1948), 817–830.

[39] Dittrich, W. and Reuter, M. *Classical and Quantum Dynamics*, 2nd corrected and enlarged edition (Springer, 1996).

[40] DUBIN, D.A., HENNINGS, M.A. AND SMITH, T.B. *Mathematical Aspects of Weyl Quantization and Phase* (World Scientific, 2000).

[41] EINSTEIN, A. *Zum Quantensatz von Sommerfeld und Epstein*, Verhandlungen der Deutschen Phys. Ges., nr. 9/10 (1917).

[42] EKELAND, I. AND HOFER, H. Symplectic topology and Hamiltonian dynamics, I and II, Math. Zeit. **200**, 355–378 and **203** (1990), 553–567.

[43] ENGLERT, B.-G., SCULLY, M.O., SÜSSMANN, G. AND WALTHER, H. Surrealistic Bohm trajectories. Z. Naturforsch. **47**a (1992), 1175–1186.

[44] FEYNMAN, R.P. Space-time approach to non-relativitic quantum mechanics. Rev. Mod. Phys. **76** (1948), 367–387.

[45] FEYNMAN, R.P. AND HIBBS, A.R. *Quantum Mechanics and Path Integrals* (McGraw-Hill, 1965).

[46] FOLLAND, G.B. *Harmonic Analysis in Phase Space*, Annals of Mathematics Studies (Princeton University Press, 1989).

[47] FÖLSING, A. *Albert Einstein, A Biography* (Viking (Penguin Group), 1997).

[48] FRANKEL, T. *The Geometry of Physics, An Introduction* (Cambridge University Press, 1997).

[49] GALLISSOT, F. Les formes extérieures en mécanique classique. *Ann. Inst. Fourier* **4** (1952), 145–297.

[50] GIBBS, J.W. *Elementary Principles in Statistical Mechanics* (Dover Publications, 1960).

[51] GODBILLON, C. *Éléments de Toplogie Algébrique* (Hermann, 1971).

[52] GOLDSTEIN, H. *Classical Mechanics* (Addison-Wesley, 1950; 2nd edition, 1980).

[53] DE GOSSON, M. La définition de l'indice de Maslov sans hypothèse de transversalité. C.R. Acad. Sci., Paris, **309**, Série I (1990) 279–281.

[54] DE GOSSON, M. La relation entre Sp_∞, revêtement universel du groupe symplectique Sp *et* $Sp \times \mathbb{Z}$. C.R. Acad. Sci., Paris **310**, Série I (1990), 245–248.

[55] DE GOSSON, M. Maslov Indices on $Mp(n)$. Ann. Inst. Fourier, Grenoble **40**(3) (1990), 537–55.

[56] DE GOSSON, M. Cocycles de Demazure-Kashiwara et Géométrie Métaplectique. J. Geom. Phys. **9** (1992), 255–280.

[57] DE GOSSON, M. On half-form quantization of Lagrangian manifolds and quantum mechanics in phase space. Bull. Sci. Math. **121** (1997), 301–322.

[58] DE GOSSON, M. The structure of q-symplectic geometry. J. Math. Pures Appl. **71** (1992), 429–453.

[59] DE GOSSON, M. *Maslov Classes, Metaplectic Representation and Lagrangian Quantization*, Research Notes in Mathematics **95** (Wiley-VCH, 1997).

[60] DE GOSSON, M. On the classical and quantum evolution of Lagrangian half-forms in phase space. Ann. Inst. H. Poincaré **70**(6) (1999), 547–73.

[61] DE GOSSON, M. Lagrangian path intersections and the Leray index, in: *Aarhus Geometry and Topology Conference*, Contemp. Math. **258** (Amer. Math. Soc., 2000), 177–184.

[62] DE GOSSON, M. *The Cohomological Meaning of Maslov's Lagrangian Path Intersection Index*, Proceedings of the Conference in the Honor of Jean Leray, Karlskrona 1999, M. de Gosson, Editor (Kluwer Acad. Publ., 2001).

[63] DE GOSSON, M. *Symplectic Geometry and Quantum Mechanics.* Birkhäuser, Basel, series "Operator Theory: Advances and Applications" (subseries: "Advances in Partial Differential Equations"), Vol. 166 (2006).

[64] DE GOSSON, M. The symplectic camel and the uncertainty principle: the tip of an iceberg? *Found. Phys.* **39** (2009), 194–214.

[65] DE GOSSON, M. Symplectic covariance properties for Shubin and Born-Jordan pseudo-differential operators. Trans. Amer. Math. Soc. **365**(6) (2013), 3287–3307.

[66] DE GOSSON, M., Quantum blobs. Found. Phys. **43**(4) (2013), 440–457.

[67] de Gosson, M. Born-Jordan quantization and the equivalence of the Schrödinger and Heisenberg pictures. Found. Phys. **44**(10) (2014), 1096–1106.

[68] de Gosson, M. *Introduction to Born-Jordan Quantization* (Springer-Verlag, 2016).

[69] de Gosson, M. From Weyl to Born-Jordan quantization: the Schrödinger representation revisited. Phys. Rep. **623** (2016), 1–58.

[70] de Gosson, M., Hiley, B. and Cohen, E. Observing quantum trajectories: from Mott's problem to quantum Zeno effect and back. To appear in Ann. Phys. (2016), DOI: 10.1016/j.aop.2016.08.003

[71] de Gosson, M. and Luef, F. Symplectic capacities and the geometry of uncertainty: the irruption of symplectic topology in classical and quantum mechanics. Phys. Rep. **484**(5) (2009), 131–179.

[72] Gotay, M.J. Functorial geometric quantization and van Hove's theorem. Internat. J. Phys. **19** (1980), 139–161.

[73] Gotay, M.J. On the Groenewold–Van Hove problem for $\mathbb{R}^{2n}$. J. Math. Phys. **40**(4) (1999), 2107–2116.

[74] Gotay, M.J., Grundling, H.B. and Tuynman, G.M. Obstruction results in quantization theory. J. Nonlinear Sci. **6** (1996), 469–498.

[75] Gotay, M.J. and Isenberg, G.A. The symplectization of science. Gazette des Mathématiciens **54** (1992), 59–79.

[76] Gribbin, J. *In Search of Schrödinger's Cat, Quantum Physics and Reality* (Corgi Books, 1984).

[77] Groenewold, H.J. On the principles of elementary quantum mechanics. Physics **12** (1946), 405–460

[78] Gromov M., Pseudoholomorphic curves in symplectic manifolds. Invent. Math. **82** (1985), 307–347.

[79] Guillemin V. and Sternberg S. *Geometric Asymptotics*, Math. Surveys Monographs **14** (Amer. Math. Soc., 1978).

[80] Guillemin V., and Sternberg S. *Symplectic Techniques in Physics* (Cambridge University Press, 1984).

[81] Gutzwiller, M.C. *Chaos in Classical and Quantum Mechanics*, Interdisciplinary Applied Mathematics (Springer-Verlag, 1990).

[82] Hamilton, W.R. *Mathematical Papers* **2** (Cambridge University Press, 1940).

[83] Hamilton, W.R. On a general method of expressing the paths of light, and of the planets, by the coefficients of a characteristic function. Dublin University Review and Quarterly Magazine **I** (1833), 795–826.

[84] Hannay, J.H. Angle variable holonomy in adiabatic excursion of an integrable Hamiltonian. J. Phys. A: Math. Gen. **18** (1985), 221–230.

[85] Heisenberg, W. *The Physical Principles of the Quantum Theory* (Chicago University Press, 1930); reprinted by Dover, New York, 1949.

[86] Hiley, B.J. *The algebera of process*, in: *Consciousness at the Crossroads of Cognitive Science and Philosophy*, Maribor, Aug. 1994 (1995), 52–67.

[87] Hiley, B.J. *Non-commutative geometry, the Bohm interpretation and the mind-mater relationship*, in *Proc. CASYS'2000*, Liège, Belgium, Aug. 7–12, 2000.

[88] Hiley, B.J. and Peat, F.D. In *Quantum Implications: Essays in honour of David Bohm* (Routledge & Kegan Paul, 1987).

[89] Hofer, H. and Zehnder, E. *Symplectic Invariants and Hamiltonian Dynamics*, Birkhäuser Advanced Texts (Birkhäuser Verlag, 1994).

[90] Holland, P.R. *The Quantum Theory of Motion: An Account of the de Broglie-Bohm Causal Interpretation of Quantum Mechanics* (Cambridge University Press, 1993).

[91] Hudson, R.L. When is the Wigner quasi-probability density nonnegative? Rep. Math. Phys. **6** (1974), 249–252.

[92] Jammer, M. *The Conceptual Development of Quantum Mechanics,* Inst. Series in Pure and Appl. Physics (McGraw-Hill, 1966).

[93] Jauch, J.M. *Foundations of Quantum Mechanics*, Addison-Wesley Series in Advanced Physics (Addison-Wesley, 1968).

[94] KAUDERER, M. *Symplectic Matrices: First Order Systems and Special Relativity* (World Scientific, 1994).

[95] KELLER, J.B. Corrected Bohr-Sommerfeld quantum conditions for nonseparable systems. Ann. Physics **4** (1958), 180–188.

[96] KNUDSEN, J.M. AND HJORTH, P.G. *Elements of Newtonian Mechanics, Including Nonlinear Dynamics*, 2nd revised and enlarged edition (Springer, 1996).

[97] KOSTANT, B. *On the definition of quantization*, in: *Géométrie Symplectique et Physique Mathématique*, Colloq. Internat. CNRS **237**, Paris (1974).

[98] KUGA, M. *Galois' Dream* (Birkhäuser, 1993) [Japanese version *Garoa no yume*, publ. Nippon Hyoron Sha Co. (1968)].

[99] LAGRANGE, J.-L. *Mécanique analytique* (Facsimilé de la troisième édition) (Librairie Albert Blanchard, Paris, 1965).

[100] LANG, S. *Differential and Riemannian Manifolds*, Graduate Texts in Mathematics **160** (Springer, 1996).

[101] LERAY, J. *Lagrangian Analysis and Quantum Mechanics, A Mathematical Structure Related to Asymptotic Expansions and the Maslov Index* (the MIT Press, 1981); translated from *Analyse Lagrangienne* RCP 25, Strasbourg Collège de France (1976–1977).

[102] LERAY, J. *The meaning of Maslov's asymptotic method the need of Planck's constant in mathematics*, Bull. of the Amer. Math. Soc., Symposium on the Mathematical Heritage of Henri Poincaré (1980).

[103] LERAY, J. *Complément à la théorie d' Arnold de l'indice de Maslov*, Convegno di geometrica simplettica et fisica matematica, Instituto di Alta Matematica, Roma (1973).

[104] LIBERMANN, P. AND MARLE, C.-M. *Symplectic Geometry and Analytical Mechanics* (D. Reidel Publishing Company, 1987).

[105] LION, G. AND VERGNE, M. *The Weil Representation, Maslov Index and Theta Series*, Progress in Mathematics **6** (Birkhäuser, 1980).

[106] LITTLEJOHN, R.G. The semiclassical evolution of wave packets. Phys. Rep. **138** (1986), 193–291.

[107] MACKEY, G.W. *The Mathematical Foundations of Quantum Mechanics* (Benjamin, Inc., 1963).

[108] MACKEY, G.W. *Unitary Group representations* (The Benjamin/ Cummings Publ. Co., 1978).

[109] MACKEY, G.W. *The Relationship Between Classical and Quantum Mechanics in Contemporary Mathematics* **214** (Amer. Math. Soc., 1988).

[110] MAKRI, N. AND MILLER, W.H. Correct short time propagator for Feynman path integration by power series expansion in Δt. Chem. Phys. Lett. **151** (1988), 1–8.

[111] MAKRI, N. AND MILLER, W.H. Exponential power series expansion for the quantum time evolution operator. J. Chem. Phys. **90** (1989), 904–11.

[112] MANIN, YU. *Mathematics and Physics*, Progress in Physics, **3** (Birkhäuser, 1981).

[113] MASLOV, V.P. *Théorie des Perturbations et Méthodes Asymptotiques* (Dunod, 1972); translated from Russian [original Russian edition 1965].

[114] MASLOV, V.P. AND FEDORIUK, M.V. *Semi-Classical Approximations in Quantum Mechanics* (Reidel, 1981).

[115] MCCOY, N.H. On the function in quantum mechanics which corresponds to a given function in classical mechanics. Proc. Natl. Acad. Sci. U.S.A. **18**(11) (1932), 674–676.

[116] MCDUFF, D. AND SALAMON, D. *Symplectic Topology* (Oxford Science Publications, 1998).

[117] MESSIAH, A. *Quantum Mechanics*, **I**, **II** (North-Holland Publ. Co., 1991), translated from the French; original title: *Mécanique Quantique* (Dunod, 1961).

[118] MNEIMÉ, R. AND TESTARD, T. *Introduction à la Théorie des Groupes de Lie Classiques*, Collection Méthodes (Hermann, 1986).

[119] MOSER, J. On the volume element of a manifold. Trans. Amer. Math. Soc., **120** (Amer. Math. Soc., 1965), 286–294.

[120] Mott, N.F. *The Wave Mechanics of α-Ray Tracks.* Proc. Royal Soc. of London A: Mathematical, Physical and Engineering Sciences. Vol. 126. No. 800. The Royal Society (1929), 79–84.

[121] Moyal, J.E. Quantum mechanics as a statistical theory. Proc. Camb. Phil. Soc. **45** (1947), 99–124.

[122] Naber, G.L. *Topology, Geometry, and Gauge fields,* Texts in Applied Mathematics **25** (Springer, 1997).

[123] Nakahara, M. *Geometry, Topology and Physics*, Graduate Students Series in Physics (IOP Publ., 1995).

[124] Nazaikiinskii, V., Schulze, B.-W. and Sternin, B. Quantization methods in differential equations, preprint, Potsdam (2000).

[125] Nelson, E. *Topics in Dynamics I*: Flows, Mathematical Notes (Princeton University Press, 1969).

[126] Omnès, R. *The Interpretation of Quantum Mechanics*, Princeton Series in Physics (Princeton University Press, 1994).

[127] Pais, A. *Niels Bohr's Times, in Physics, Philosophy, and Polity* (Oxford University Press, 1991).

[128] Park, D. *Classical Dynamics and Its Quantum Analogues*, 2th Edition (Springer-Verlag, 1990); 1st Edition: Lecture Notes in Physics, **110** (Springer-Verlag, 1979).

[129] Penrose, R. *The Emperor's New Mind* (Oxford University Press, 1989).

[130] Philippidis, C., Dewdney, C. and Hiley, B.J. Quantum interference and the quantum potential. Nuovo Cimento **52**(1) (1979), 15–28.

[131] Piron, C. *Mécanique quantique, Bases et applications* (Presses polytechniques et universitaires romandes, Lausanne, 1990).

[132] Ponomarev, L. I. *The Quantum Dice* (IOP Publishing, 1993).

[133] Rabinowitz, P. Periodic solutions of Hamiltonian systems. Comm. Pure Appl. Math. **31** (1978), 157–184.

[134] de Rham, G. *Variétés Différentiables* (Hermann, Paris, 1960).

[135] SCHEMPP, W. *Harmonic Analysis on the Heisenberg Nilpotent Lie Group*, Pitman Research Notes in Mathematics **147** (Longman Scientifical and Technical, 1986).

[136] SCHEMPP, W. *Die Kepplerschen Strategien der geometrischen Spinorquantisierung* (Wiley, 1999).

[137] SCHEMPP, W. *Magnetic Resonance Imaging: Mathematical Foundations and Applications* (Wiley, 1997).

[138] SCHRÖDINGER, E. *Quantisierung als Eigenwertproblem*, Ann. der Physik (1926), 1st communication: **79**, 489–527, 2d communication **80**, 437–490, 3d communication **81**, 109–139.

[139] SCHULMAN, L.S. *Techniques and Applications of Path Integrals* (Wiley, 1981).

[140] SCHULMAN, L.S. A path integral for spin. Phys. Rev. **176**(5) (1961), 1558.

[141] SCHWARTZ, L. *Généralisation des espaces L^p*, Publ. Inst. Statist. Univ. Paris **6** (1957), 241–250.

[142] SEGAL, I.E. Foundations of the theory of dynamical systems of infinitely many degrees of freedom (I). Mat. Fys. Medd., Danske Vid. Selsk., **31**(12) (1959), 1–39.

[143] SEIFERT, H. Periodische bewegungen mechanischer systeme. Math. Zeit. **51** (1948), 197–216.

[144] SHALE, D. Linear symmetries of free boson fields. Trans. Amer. Math. Soc. **103** (1962), 149–167.

[145] SIBURG, K.F. Symplectic capacities in two dimensions. Manuscripta Math. **78** (1993), 149–163.

[146] SNIATYCKI, J. *Geometric Quantization and Quantum Mechanics*, Appl. Math. Sciences **30** (Springer-Verlag, 1980).

[147] SOBOLEV, S.L. *Some Applications of Functional Analysis in Mathematical Physics*, Leningrad State University, Leningrad (1959) [in Russian].

[148] Souriau, J.-M. *Structure des Systèmes Dynamiques* (Dunod, 1970); English translation by C.H. Cushman-de-Vries: *Structure of Dynamical Systems* (Birkhäuser, 1997).

[149] Souriau, J.-M. *Construction explicite de l'indice de Maslov*, Group Theoretical Methods in Physics, Lecture Notes in Physics, **50** (Springer-Verlag, 1975), 17–148.

[150] Souriau, J.-M. Indice de Maslov des variétés lagrangiennes orientables. C. R. Acad. Sci., Paris, Série A, **276** (1973), 1025–1026.

[151] Stein, E.M. *Harmonic Analysis: Real Variable Methods, Orthogonality, and Oscillatory Integrals* (Princeton University Press, 1973).

[152] Trèves, F. *Introduction to Pseudo-differential and Fourier Integral Operators* (two Volumes), University Series in Mathematics (Plenum Press, 1980).

[153] Tuynman, G. *What is prequantization, and what is geometric quantization?*, in *Proceedings, Seminar 1989-1990, Mathematical Structures in Field Theory*, 1–28. CWI Syllabus **39**. CWI, Amsterdam (1996).

[154] van Hove, L. *Sur le Problème des Relations entre les Transformations Unitaires de la Mécaniqu Quantique et les Transformations Canoniques de la Mécanique Classique*, Mém. Acad. Roy. Belg. **26** (1951), 610.

[155] van Vleck, J.H. Quantum principles and line spectra. Bull. Natl. Res. Council **10**(54) (1926), 1–316.

[156] Viterbo, C. Symplectic topology as the geometry of generating functions. Math. Ann. **292** (1992), 685–710.

[157] Voros, A. Asymptotic h-expansions of stationary quantum states. Ann. Inst. H. Poincaré, Sect. A **26** (1977), 343–403.

[158] Voros, A. An algebra of pseudo-differential operators and the asymptotics of quantum mechanics. J. Funct. Anal. **29** (1978), 104–132.

[159] Waerden, B.L. van der. *Sources of Quantum Mechanics* (North-Holland, 1967). 0

[160] Wallach, N. *Lie Groups: History, Frontiers and Applications*, **5**: Symplectic Geometry and Fourier Analysis (Math Sci Press, Brookline, MA, 1977).

[161] WEIERSTRASS, K. *Mahematische Werke*, Berlin (1858), Band I: 233–246, Band II:19–44, Nachtrag: 139–148.

[162] WEIL, A. Sur certains groupes d'opérateurs unitaires. Acta Math. **111** (1964), 143–211; also in Collected Papers, Vol. III, Springer-Verlag, Heidelberg (1980), 1–69.

[163] WEINSTEIN, A. Periodic orbits for convex Hamiltonian systems. Ann. Math. **108** (1978), 507–518.

[164] WESTENHOLZ, C. VON. *Differential Forms in Mathematical Physics*, Studies in Mathematics and its Applications **3** (North-Holland Publ. Co., 1978).

[165] WEYL, H. Quantenmechanik und Gruppentheorie. Z. Physik **46** (1927), 1–46.

[166] WHEELER, A. AND ZUREK, H.Z. (Editors). *Quantum Theory and Measurement*, Princeton Series in Physics (Princeton University Press, 1983).

[167] WIGNER, E.P. On the quantum correction for thermodynamic equilibrium. Phys. Rev. **40**(5) (1932), 749–759.

[168] WIGNER, E.P. The unreasonable effectiveness of mathematics in the natural sciences. Commun. Pure Appl. Math. **13** (1960), 1–14.

[169] WOODHOUSE, N.M.J. *Geometric Quantization*, 2nd edition (Oxford Science Publications, 1991).

Index

action, 12, 120, 149, 152
additivity of the Maslov index, 181
affine symplectomorphisms, 242
algorithm, 333
angular momentum, 68
angular momentum dilemma, 281
antisymplectomorphism, 109
averages (lemmas on), 162

Berry's phase, 41, 194, 233
Bohm, 15, 16, 29
Bohm's theory, 30
Bohmian mechanics, 28
Bohr, 15
Bohr-Sommerfeld condition, 179
Born, 15
Brahe, 2

canonical quantization, 282
canonical transformation, 84
capacity, 115
    (linear), 117
    (symplectic), 114
cart with a spring, 65, 68
catalogue, 227
causality, 9, 58
caustic, 168, 195
celestial mechanics, 3, 5, 7
Chapman-Kolmogorov's law, 9, 58
classical mechanics, 2
coboundary, 199
    operator, 199
cochain, 198
cocycle, 199
cohomological notations, 198
complementarity (principle of), 351
completely integrable, 70
configuration space, 47
conjugate momentum, 52
constant of the motion, 67
constants of the motion (in involution), 69
continuity equation, 75, 310
continuity equation (for van Vleck's density), 308
Copenhagen interpretation, 33
Copernicus, 2
Coriolis, 40
    force, 39, 53
    force (potential), 41
cyclic order, 201

Dacorogna-Moser theorem, 108
de Broglie, 15, 16
    wavelength, 16
de Rham form, 50, 213, 216
de Rham form (on a Lagrangian manifold), 220
density, 216
density (distributional), 76
density of trajectories, 305
diamagnetic term, 94
Dirac, 15
    correspondence, 283, 292

string, 42

Earth (turning), 40
EBK condition, 179
eikonal, 147
Einstein, 5, 14, 179
electromagnetic field (charged particles in), 48
electron in a uniform magnetic field, 93
electron in an electromagnetic field, 39
electron in uniform magnetic field, 165
ellipsoid, 117
energy, 68
levels, 33
shell, 120
enfolding-unfolding, 330
Euclidean group, 61
Euler, 5
Euler-Lagrange equations, 101
exact Lagrangian manifold, 170, 183
explicate order, 344
extended phase space, 38, 57
extended state space, 38

Fermat's principle, 101
Feynman's path integral, 26, 304, 322
firefly argument, 23, 332
flow, 9
force, 5
fields, 38
Foucault pendulum, 40
Fourier transform, 246
free particle, 151, 296
free particle (phase), 296
free particle in a non-trivial gauge, 154
free symplectomorphism, 142
Fresnel formula, 299

Galilean covariance, 59, 62
Galilean covariance of Newton's second law, 62
Galilean group, 60
Galilean invariance, 7
Galilean transformations, 60
Galilei, 2, 59
gauge, 8, 40
gauge (symmetric), 94
gauge and generating functions
gauge, 153
gauge transformation, 8, 41, 153
generating function, 145
generating function for J, 145
generating function in optics, 147
generators of the symplectic group, 243
geometric phase shifts, 41
geometric quantization, 19
geometrical optics
ray optics, 100
Gibbs, 73
Groenewold-van Hove theorem, 292
Gromov, 25
Gromov width, 114
Gromov's non-squeezing theorem, 24
Gromov's theorem, 109
ground level energy, 125
group cocycle (on $\mathrm{Sp}(n)$), 265
group velocity, 16

Hamilton, 5
Hamilton field, 38
Hamilton field (suspended), 38
Hamilton vector field, 57
Hamilton's equations, 6, 50
Hamilton-Jacobi equation, 13, 149, 156, 172
Hamiltonian (several particles), 54
Hamiltonian flow, 57

Hamiltonian function, 50
Hamiltonian vector field, 8
Hannay angles, 41
Hardy, 20
harmonic oscillator, 70, 151, 156, 160, 179
harmonic oscillator (energy levels), 125
harmonic oscillator (tired), 71
Heisenberg, 15, 333
Heisenberg group, 271
Heisenberg inequalities, 22
Heisenberg-Weyl operator, 275
heliocentric frame, 59
Helmholtz's theorem, 138
Hessian, 146
Hiley, 34, 35, 330, 351
Hodge star operator, 44
homotopy formula, 86

imaging problem, 101
implicate order, 347, 351
index of refraction, 100
inertial frame, 3, 59
inhomogeneous symplectic group, 98, 242
integer part, 199
integer part (antisymmetric), 200
integrable systems, 8, 66
interpretations (of QM), 32
intersection cochain, 199
invariant torus, 70
involution (constants of the motion in), 69
ISp(n), 98

Jacobi identity, 67
Jordan, 15

Keller, 179
Keller-Maslov quantization, 179
Kepler, 2
Keplerian strategy, 2
Kirchhoff, 3
Kuga, 175

Lagrange, 5, 7
Lagrange form, 37, 44
Lagrange form (n dimensions), 48
Lagrangian Grassmannian, 167
Lagrangian manifold, 12, 167
Lagrangian manifolds and symplectomorphisms, 167
Lagrangian plane, 167
Larmor frequency, 94
Leibniz, 2
lens power, 104
lenses, 102
Leray index, 198, 199, 259
Leray index (definition), 203
Leray index ($n = 1$), 200
Leray index (reduced), 212
level set, 70, 120
Lie, 334
Lie algebra of Sp(n), 88
Lie's formula, 334
Lie-Trotter formula for flows, 333
Liouville integrable, 70, 171
Liouville's condition, 73
Liouville's equation, 72–74
Liouville's theorem, 107, 108
loops in Lagrangian manifolds, 169
Lorentz force, 39

Mach, 4
Mach's principle, 4
marginal probabilities, 74
Maslov index, 179
Maslov index (of a quadratic form), 247
Maslov index (of the identity), 263
Maslov index for loops, 179

Maslov index on Mp(n), 259
mass, 5, 6
mass matrix, 47
matrix J, 10
matter waves, 15
Maxwell, 7
Maxwell Hamiltonian (quadratic), 91
Maxwell Hamiltonians, 54, 64, 350
Maxwell's principle, 40, 48
Maxwell's reciprocity law, 49
metalinear group, 249
metaplectic group, 248
metaplectic representation, 237
metaplectic representation (generalized), 342
metaplectic representation (of the free flow), 301
metatron, 295
minimum capacity principle, 127
momentum, 5
momentum vector, 6
Morse index, 196
Moser, 108
Mott's paradox, 340
Mp(n), 248

N-particle systems, 47
neutron stars, 95
Newton, 2
Newton field, 38
Newton's laws, 3
Newton's second law, 3, 38, 58
Newtonian mechanics, 2
non-commutativity, 351
non-squeezing property, 25
non-symplectic cylinders, 111
non-symplectic diffeomorphism, 111

one-parameter group, 9
ontology, 34
optical axis, 101
optical Hamiltonian, 102
optical Lagrangian, 101
optical length, 101
optical matrix
    ray transfer matrix, 103
optical medium, 100
orbit, 9
ordering problem, 284
orthogonal subgroup of Sp(n), 87

page (of a catalogue), 227
Pantheon, 40
paramagnetic term, 94
paraxial approximation, 100
paraxial optics, 102
paraxial rays, 103
Paris, 40
Penrose, 25
period of a loop, 169
periodic orbits, 121
periodic orbits on a symplectic cylinder, 122
phase, 172
phase (of a free particle), 296
phase of a Lagrangian manifold, 176
phase of an exact Lagrangian manifold, 172
phase of the circle, 174
phase space (extended), 57
phase velocity, 16
Planck's constant, 16
Planck's law, 127
Poincaré, 5
Poincaré-Cartan form, 51
Poincaré-Cartan form (fundamental property of), 138
Poisson brackets, 66, 171
position vector, 6
primitive ontology, 35

Principia, 2
principle of complete knowledge, 72
principle of density in phase, 73
propagator (free particle), 297
propagator (short-time), 312
pseudo-vector, 38, 50

quadratic Fourier transforms, 247
quadratic Maxwell Hamiltonians, 91
quadratic polynomials and symplectic Lie algebra, 88
quantization, 281
quantization rule, 319
quantized Lagrangian manifold, 184
quantum blobs, 127
quantum cell, 127
quantum mechanics, 14
quantum mechanics (statistical interpretation), 20
quantum mechanics in phase space, 23
quantum motion, 29, 192, 336
quantum potential, 338
Quantum Zeno effect, 340

ray-transfer matrix, 103
reduced action, 120
reduced distance, 104
reduced Leray index, 212
retrodiction, 332
rotation vector, 39

s-density, 216
scalar potential, 40
Schempp, 2
Schrödinger, 15
Schrödinger's argument, 236
Schrödinger's quantization rule, 18
Science and Hypothesis, 5
semi-classical mechanics, 127, 194
semi-classical propagator, 313
semi-classical wave functions, 231
shadow phase space, 347
short-time action, 165
  action, 158
short-time action (single particle), 162
short-time propagator, 312
signature (of a triple of lines), 211
signature of a triple of Lagrangian planes, 206
signature of a triple of lines, 201
skew-product, 10
Snell's law, 103
Sp(n), 86, 89
space rotations, 60
space translations, 60
speed of light, 101
standard symplectic form, 10, 81
stationary phase (formula of), 322
stationary state, 125
statistical interpretation of quantum mechanics, 20
subgroup O(n) of Sp(n), 87
subgroup U(n) of Sp(n), 87
sum over paths, 27
suspended Hamilton field, 38, 57
suspended Hamiltonian flow, 57
symmetric gauge, 302
symplectic area, 114
symplectic camel, 24, 106, 345
symplectic capacity, 114
symplectic capacity (linear), 117
symplectic covariance, 272, 283
symplectic cylinder, 109
symplectic form, 9, 81
symplectic geometry, 9
symplectic gradient, 11
symplectic group, 82, 86
symplectic matrix, 80
symplectic product, 10
symplectic radius, 114

symplectic spectrum, 117
symplectic topology, 343
symplectization of science, 9
symplectomorphism, 11, 28, 70, 83
symplectomorphism (free), 142

tennis player, 145, 306, 308
time translations, 60
time-dependent flow, 9, 57
trajectory density, 305
transition probabilities, 33
triatomic molecule, 92
twisted form, 50

uncertainty principle, 22, 133
uncertainty principle in classical mechanics, 112
unfolding theorem, 338, 340
unitary subgroup of Sp(n), 87
universal covering, 175, 221

van Vleck's density, 306
van Vleck's density (continuity equation), 310
van Vleck's determinant, 305
vector potential, 40
velocity boosts, 60
volume, 119
volume form (standard), 108
volume-preserving, 107

wave optics, 251
    ray optics, 100
wave-form, 196, 227
wave-function (optics), 251
Weyl, 8
Wigner quasi-distribution, 346